The Global Cable Industry

The Global Cable Industry

Materials, Markets, Products

Edited by

Günter Beyer

WILEY-VCH

Editor

Dr. Günter Beyer
Fire and Polymer
CEO
Auenweg 14
4700 Eupen
Belgium

Cover Design: Wiley
Cover Image: © Aeriform / Alamy Stock Photo

Library of Congress Card No.:
applied for

British Library Cataloguing-in-Publication Data
A catalogue record for this book is available from the British Library.

Bibliographic information published by the Deutsche Nationalbibliothek
The Deutsche Nationalbibliothek lists this publication in the Deutsche Nationalbibliografie; detailed bibliographic data are available on the Internet at <http://dnb.d-nb.de>.

Print ISBN: 978-3-527-34627-1
ePDF ISBN: 978-3-527-82228-7
ePub ISBN: 978-3-527-82227-0
oBook ISBN: 978-3-527-82226-3

Typesetting SPi Global, Chennai, India
Printing and Binding CPI Group (UK) Ltd, Croydon, CR0 4YY

Printed on acid-free paper

C094266_240321

Contents

About the Editor *xv*

1 **Overview of the Global Cable Industry – Markets and Materials** *1*
Astrid Aupetit
1.1 Demand for Polymeric Material *3*
1.1.1 Main Companies Profile *3*
1.1.1.1 Prysmian *4*
1.1.1.2 Nexans *4*
1.1.1.3 Southwire *5*
1.1.1.4 Sumitomo Electric Industries *5*
1.1.1.5 Furukawa Electric Co., Ltd. *5*
1.1.1.6 LS Cable & System *6*
1.1.1.7 Leoni AG *6*
1.1.1.8 Hengtong Group *6*
1.1.2 Global Demand *7*
1.2 Asia and Australasia *10*
1.2.1 Demand for Cable *10*
1.2.2 Demand for Polymeric Material *11*
1.3 Europe *12*
1.3.1 Demand for Cables *12*
1.3.2 Demand for Polymeric Materials *13*
1.4 The Middle East and Africa *14*
1.4.1 Demand for Cables *14*
1.4.2 Demand for Polymeric Materials *14*
1.5 North America *16*
1.5.1 Demand for Cables *16*
1.5.2 Demand for Polymeric Materials *17*
1.6 South and Central America *18*
1.6.1 Demand for Cables *18*
1.6.2 Demand for Polymeric Materials *19*

2 Thermoplastics for Cables *21*
Theo Geussens
2.1 Introduction *21*
2.2 Polyolefin Materials *21*
2.2.1 Polyethylene *22*
2.2.1.1 Manufacturing Processes *25*
2.2.1.2 Cable Applications *26*
2.2.2 Polypropylene (PP) *32*
2.2.2.1 Manufacturing Processes *33*
2.2.2.2 Applications *33*
2.3 Chlorinated Polymers *35*
2.3.1 Polyvinylchloride (PVC) *35*
2.3.2 Chlorinated Polyethylene (CPE) *36*
2.4 Fluoropolymers *37*
2.4.1 Fluoro-Ethylene Propylene polymer *37*
2.4.2 Polytetrafluoroethylene (PTFE), Ethylene Tetrafluoroethylene (ETFE), and Perfluoroalkoxy Polymer (PFA) *38*
2.4.3 Ethylene Chlorotrifluoroethylene (ECTFE) *40*
2.4.4 Polyvinyldifluoride (PVDF) *40*
2.5 Polyamide (PA) *40*
2.6 Polyesters *41*
2.6.1 Polybutylphtherephtalate (PBT) *41*
2.6.2 Polyester Elastomers *41*
2.7 Thermoplastic Polyurethane *42*
References *43*

3 Elastomers for Cables *57*
Burkhard Herpich
3.1 Introduction *57*
3.2 Rubber Compounds *63*
3.2.1 Rubber *63*
3.2.2 Fillers *66*
3.2.3 Plasticizer *68*
3.2.4 Stabilizers *70*
3.2.5 Classical Cross-linking Systems *71*
3.2.5.1 Cross-linking with Sulfur Systems *71*
3.2.5.2 Cross-linking with Peroxide Systems *71*
3.2.5.3 Moving Die Rheometers (MDRs) *72*
3.2.6 Other Cross-linking Systems *73*
3.2.6.1 Cross-linking by Irradiation *73*
3.2.6.2 Cross-linking with Silanes *73*
3.3 Compounding *73*
3.4 Extrusion *76*
3.5 Cross-linking/Vulcanization *78*
3.5.1 Vulcanization in Saturated Hot Water Steam *78*
3.5.2 Vulcanization in Liquid Salt Mixtures under Pressure *80*
3.5.3 Electron Beam Cross-linking *81*
References *82*

4 **Extrusion of Cables** *85*
Stéphan Puissant
4.1 Historical Introduction to Cable Extrusion *85*
4.2 Extruder in Cable Lines *87*
4.2.1 Description of the Single Screw Extruder *87*
4.2.1.1 Different Functional Screw Zones *87*
4.2.1.2 Description of a Screw Geometry *88*
4.2.2 Feeding Zone of the Extrusion Screw *89*
4.2.2.1 Friction-Based Feeding Mechanism *89*
4.2.2.2 Simple Modeling of the Feeding Zone *90*
4.2.2.3 Improvement of the Feeding Zone: Use of Helical Grooved Barrel *95*
4.2.3 Thin Film Plastification *97*
4.2.3.1 Melting on the Backside of Flight *97*
4.2.3.2 Initiation of Liquid Film *98*
4.2.3.3 Melt Flow Rate in the Liquid Film *99*
4.2.3.4 Influence of Different Parameters on Melting Length *99*
4.2.3.5 Thin Film Melting with the Help of a Barrier Zone *103*
4.2.3.6 Barrier Zone, Its Advantages, and Its Drawbacks *103*
4.2.4 Metering Zone *104*
4.2.4.1 Representation of the Metering Zone *104*
4.2.4.2 1D Analysis *105*
4.2.5 Example of Results for 1D Model Including the Three Zones of the Screw *105*
4.2.5.1 Influence of Friction Coefficients on the Screw Characteristics *105*
4.2.5.2 Interaction Between Compression and Friction in the Feeding Zone *106*
4.3 Accessories for Extruders *107*
4.3.1 Mixing Zones *107*
4.3.1.1 Observed Defaults *107*
4.3.1.2 Devices Selection Criteria *109*
4.3.1.3 Example of Results for Finathene HDPE *111*
4.3.2 Melt Filtration Systems *113*
4.3.3 Melt Gear Pumps *114*
4.4 Extrusion Heads or Dies *115*
4.4.1 Description of the Extrusion Head *115*
4.4.1.1 Extrusion Head Function *115*
4.4.2 Distributors *117*
4.4.2.1 Head with Coat Hanger Type Distribution Channels *117*
4.4.2.2 Distribution Function Through Flattened Distribution Channels *119*
4.4.2.3 Distributor with Helical Channels *119*
4.4.3 Diameter Adaptation Function (Tooling) *119*
4.4.3.1 Tube Tooling, DDR *120*
4.4.3.2 Tube Tooling, DRB *121*
4.4.4 Relation Between Pressure and Average Temperature Increase in an Extrusion Head *121*
4.5 Cooling *122*
4.5.1 Cooling Length Analytical Calculation for Fine Wires *122*

4.5.2 Cooling Length Finite Difference Calculation for a Wire of Radius >1 mm with Copper Core and PE Insulation *123*
4.6 Quality *124*
4.6.1 The Quality Parameter and Its Measurement *124*
4.6.1.1 Diameter and Product Circularity Measurement *125*
4.6.1.2 Concentricity Measurement *126*
4.6.1.3 Insulation Defects Measurements (Cable or Wire) *126*
4.6.1.4 Capacity Measurement (Telecommunication Wire) *127*
4.6.1.5 Sheathing Wall Thickness Measurement *128*
4.6.1.6 Periodicity of Measurement Analysis *128*
4.6.2 Common Production Defects, Causes, and Remedies *130*
References *130*

5 Foam Extrusion *133*
Horst A. Scheid
5.1 Motivation *133*
5.2 Physical Basics *133*
5.3 Selection of Polymer *136*
5.4 Selection of Blowing Agents *139*
5.5 Extrusion Equipment *141*
5.5.1 Extruder and Screw *143*
5.5.2 Dosing Station *145*
5.5.3 Gas Injection *145*
5.5.4 Melt Transport from Screw to Die *148*
5.5.5 Cooling Trough *149*
5.5.6 Measurement Devices *151*
5.6 Processing *151*
5.6.1 Extrusion and Die Setup *152*
5.6.2 Process Control Modes *153*
5.6.3 PBA Handling *154*
5.6.4 Maximum Void and Bubble Size *154*
5.6.5 Inline Analysis by FFT *156*
Glossary *157*
References *158*

6 Flame Retardancy of Cables *161*
Günter Beyer
6.1 Introduction *161*
6.2 Flame Propagation Tests for Wires and Cables *163*
6.3 Smoke, Corrosivity, and Toxicity Tests for Wires and Cables *164*
6.4 Circuit Integrity and Functional Integrity for Security Cables *165*
6.5 Laboratory Tests for the Flammability of Wire and Cable Materials *167*
6.6 Polymers for Flame-Retardant Wires and Cables *167*
6.7 Flame Retardants for Flame-Retardant Wires and Cables *168*
6.8 Flame-Retardant PVC *169*

6.8.1 Flame Retardants for PVC and Flame-Retardant PVC Cable Formulations *169*
6.8.1.1 Phthalate-Based Plasticizers and Other Plasticizers *169*
6.8.1.2 Antimony Trioxide *169*
6.8.1.3 Brominated Phthalate Plasticizers *170*
6.8.1.4 Chlorinated Paraffins *171*
6.8.1.5 Aluminum Hydroxide (ATH) and Magnesium Hydroxide (MDH) *171*
6.8.1.6 Zinc Borate *172*
6.8.1.7 Phosphate Plasticizers *172*
6.8.1.8 Smoke Suppressants *172*
6.8.1.9 Nanocomposites *173*
6.9 Flame-Retardant Polyolefins *174*
6.9.1 Flame Retardants for Polyolefins and HFFR Cable Formulations *175*
6.9.1.1 Aluminum Hydroxide (ATH) and Magnesium Hydroxide (MDH) *175*
6.9.1.2 Zinc Borate and Polysiloxanes *176*
6.9.1.3 Nanocomposites *176*
6.9.1.4 Ceramifiable Compounds *177*
6.10 CPR (Construction Products Regulation) *178*
References *179*

7 CPR Testing of Cables *181*
Franck Poutch
7.1 Introduction *181*
7.2 FIPEC Program *182*
7.2.1 FIPEC Approach *183*
7.2.1.1 Real-Scale Scenario *184*
7.2.1.2 Cable Selection *185*
7.2.1.3 Real-Scale Fire Tests *186*
7.2.1.4 Full-Scale Fire Test *186*
7.2.1.5 Capability Study *188*
7.2.1.6 Correlation Between Real- and Large-Scale Test *189*
7.3 Construction Product Regulation (CPR) Framework *190*
7.3.1 EN 13 501-6 *193*
7.3.2 EN ISO 1716 *196*
7.3.3 EN 60332-1-2 *196*
7.3.4 EN 50399 *198*
7.3.5 EN ISO 61034-1 (Apparatus) and -2 (Test Procedure) *199*
7.3.6 EN 60754-2 *201*
References *206*
7.A Measuring Heat Release Rate (HRR) by Oxygen Consumption Technique, and Smoke Density *208*
7.A.1 Measure of the Heat Release Rate (HRR) by Oxygen Consumption Technique *208*
7.A.1.1 Burning of Methane *208*
7.A.1.2 Determination of the Mass Flow Rate ($\dot{m}_a$) *210*
7.A.1.3 Smoke Opacity *211*
7.A.1.4 Calculation of FIGRA and SMOGRA Index *212*

8 **Crosslinking Technologies** *215*
Ron Goethals
8.1 Introduction *215*
8.2 Crosslinking, Curing, Vulcanizing *216*
8.3 Crosslinking Processes *217*
8.4 The Silane-Crosslinking Process *218*
8.4.1 The Sioplas® Process *218*
8.4.2 The Monosil® Process *219*
8.4.3 Silane Copolymers *220*
8.4.4 Reactivity of the Silane Crosslinking Process *220*
8.4.5 Advantages and Disadvantages of Silane Crosslinking *221*
8.5 The Peroxide Crosslinking Process (CV Curing) *222*
8.5.1 Polymer Selection for Peroxide Crosslinking *223*
8.5.2 Advantages and Disadvantages of Continuous Vulcanization *224*
8.6 e-Beam Crosslinking *225*
8.6.1 Awareness of e-Beams in Our Daily Life *225*
8.6.2 The Principle of an e-Beam *227*
8.6.3 Creating a High Voltage e-Beam *228*
8.6.4 Scan Horn and Beam Window *228*
8.6.5 Under-Beam Handling System (UBHS) *229*
8.6.6 Control System *231*
8.6.7 Safety *232*
8.6.7.1 Radiation and Radioactivity *232*
8.6.7.2 Bremsstrahlung or X-rays *232*
8.6.7.3 Shielding and Radioactivity *233*
8.6.7.4 Other Safety Systems *234*
8.6.7.5 Ozone *235*
8.6.7.6 IAEA *235*
8.6.8 Most Important Parameters During e-Beam Crosslinking *235*
8.6.8.1 Voltage *235*
8.6.8.2 Amperage *237*
8.6.8.3 Radiation Dose *237*
8.6.9 Capacity of an e-Beam *241*
8.6.9.1 Functional Capacity *241*
8.6.9.2 Efficiency *241*
8.6.9.3 Dose *241*
8.6.10 Temperature Rise *242*
8.6.11 Compound Design *242*
8.6.11.1 Polymers *242*
8.6.12 Costs Indications of e-Beam Crosslinking *243*
8.6.12.1 Annual Costs *243*
8.6.12.2 Annual Throughput and Costs per Kilograms *244*
8.6.13 Advantages and Disadvantages of e-Beam Crosslinking *244*
8.7 Conclusions *245*
Further Reading *246*

Silane Crosslinking 246
Peroxide Crosslinking 246
Electron-Beam Crosslinking 247
Hot-Set-Elongation 248

9 Nuclear Power Station Cables *249*
Frank Liu
9.1 Development of Nuclear Power in the World *249*
9.1.1 Development Stage of Nuclear Power in the World *249*
9.1.2 Current Status of Nuclear Power Development All Over the World *251*
9.1.3 Technical Characteristics and Differences of Nuclear Power Plants *256*
9.1.3.1 Technical Characteristics of Various Reactors of Nuclear Power Plants *256*
9.1.3.2 Technical Development of Nuclear Power Plant *257*
9.1.3.3 Technical Characteristics of Typical Nuclear Power Plants in the World *258*
9.1.3.4 Characteristics of CPR1000 Nuclear Power Plant *260*
9.2 Development of Cables for Nuclear Power Plants *261*
9.2.1 Development History of Cables for Nuclear Power Plants *261*
9.2.2 Classification of Cables for Nuclear Power Plants *263*
9.2.2.1 Classification by Use Function *263*
9.2.2.2 Classification by Safety Level *263*
9.2.2.3 Classification by Technical Types of Nuclear Power Plants *264*
9.2.2.4 Classification by Use Occasion *264*
9.2.3 Characteristics of Cables for Nuclear Power Plants *264*
9.2.4 Standards for Nuclear Cables *267*
9.2.5 Qualification and Test of Nuclear Cables *267*
9.2.5.1 Significance of Class 1E Equipment Qualification in Nuclear Power Plants *267*
9.2.5.2 Method of Class 1E Equipment Qualification in Nuclear Power Plants *269*
9.2.5.3 Content and Difference of Class 1E Cable Qualification Test for Nuclear Power Plants *270*
9.2.5.4 Key Points for Qualification of Class 1E Nuclear Power Cables *270*
9.2.6 Research on Nuclear Cable Compounds *277*
9.2.6.1 Application of Available Compounds *277*
9.2.6.2 Discussion on Future Materials *281*
9.3 Future Developments of Nuclear Cables *284*
9.3.1 Development Trend of Global Nuclear Power *284*
9.3.2 Development Trend of Nuclear Power in China *285*
9.3.3 Development Trend of Nuclear Cables *286*
References *289*

10 Submarine Cables *291*
George Georgallis
10.1 Introduction *291*

10.2 Submarine Power Cable Applications *291*
10.3 Submarine Power Cable Design Overview *292*
10.4 Power System Considerations *293*
10.5 Submarine Power Cable Elements *293*
10.5.1 Cable Conductor *293*
10.5.2 Cable Insulation *294*
10.5.2.1 Lapped Insulation *294*
10.5.2.2 Extruded Insulation *295*
10.5.2.3 HVDC Cables *296*
10.5.3 Metallic Sheaths *296*
10.5.4 Armor Layers *297*
10.5.5 Other Cable Elements *297*
10.6 Submarine Cable Manufacturing *298*
10.7 Submarine Power Cable Accessories *299*
10.8 Submarine Power Cable Testing *301*
10.8.1 Dynamic Submarine Cable Testing *303*
10.8.2 Submarine Power Cable Sizing and Other Design Considerations *305*
10.8.3 Submarine Power Cable Installation, Protection, and Commissioning *306*
10.9 Submarine Power Cables – What Next? *308*
References *308*

11 MV and HV Cables *311*
Detlef Wald
11.1 History of Cables *311*
11.2 Today's Cables *311*
11.2.1 Paper Cables *311*
11.2.2 Extruded Cables *312*
11.3 Conductor Design *312*
11.4 Conductor Screen *312*
11.5 Insulation *313*
11.5.1 Production of Polyethylene *315*
11.5.1.1 Influence of the Chain Transfer Agent *315*
11.5.1.2 Cross-linked Polyethylene *317*
11.5.1.3 Cleanliness *321*
11.5.2 Insulation Screen *321*
11.5.3 Metallic Screen *323*
11.5.3.1 Different Types of Metallic Protections *324*
11.5.3.2 Corrugated Screens *325*
11.5.4 Comparing Different Metallic Screens *327*
11.6 Jacketing/Sheathing *328*
References *329*

12 Coaxial Cables *331*
Timothy Cooke
12.1 History *331*

12.2 Design: Components and Principles *332*
12.2.1 Components *332*
12.2.2 Principles *336*
12.3 Characteristic Impedance (Z_o) *336*
12.4 Velocity of Propagation *338*
12.5 Capacitance *339*
12.6 Attenuation *339*
12.7 Impedance Mismatch – Reflection Coefficients and Return Loss *342*
12.8 Applications *343*
12.9 Video, TV, and Broadband Applications *344*
12.10 Automotive Coaxial Cable Applications *345*
12.11 Conclusions *348*
References *349*

13 Optical Fiber Cables *351*
Chen Baoping, Yao Di, and Qian Feng
13.1 Optical Communication System *351*
13.2 Fiber Main Features and Transmission Theory *353*
13.2.1 Principle of Total Reflection and Optical Transmission Waveguide *353*
13.2.2 The Classification and Main Features of Fiber *355*
13.2.2.1 Classification of Fiber *355*
13.2.2.2 Main Features of Fiber *359*
13.2.3 Development of Optical Fiber Technology *359*
13.2.3.1 Large Effective Area and Ultralow Loss Fiber *359*
13.2.3.2 Photonic Crystal Fiber *361*
13.2.3.3 Next Generation of Communication Fiber *361*
13.3 Cable Design and Manufacturing Technology *363*
13.3.1 Optical Cable Structural Design *363*
13.3.1.1 Design Principle *363*
13.3.1.2 Design Calculation of Optical Cable Structure *365*
13.3.2 Manufacturing Process Route for Optical Cable *370*
13.3.2.1 Technology of Fiber Coloring *371*
13.3.2.2 Technology of Fiber Ribbon *372*
13.3.2.3 Fiber Secondary Coating *372*
13.3.2.4 Stranding Cable Technology *374*
13.3.2.5 Sheath Extrusion of Optical Cable *374*
13.3.3 Optical Cable Standards and Key Test Items *375*
13.3.3.1 Cabled Fiber Attenuation Test *376*
13.3.3.2 Test Principle and Methods for Cabled Fiber PMD *376*
13.3.3.3 Cable Tensile Curve *377*
13.4 Classification of Optical Cables and Typical Application *378*
13.4.1 Classification of Optical Cables *378*
13.4.1.1 Structure Characteristics of Indoor Optical Cable *379*
13.4.1.2 Structure Characteristics of Outdoor Optical Cable *380*

13.4.1.3 Structure and Technical Characteristics of Electric Optical Cable *381*
13.4.2 Introduction to Typical Cable Structure and Application Scenarios *382*
13.4.2.1 Typical Cable Structure and Application Scenarios *382*
13.4.2.2 Introduction to New Special Optical Cable *385*
13.5 Scale and Market of Fiber and Cable Industry *386*
References *387*

Index *389*

About the Editor

Dr. rer. nat. Günter Beyer received his PhD in organic chemistry and photochemistry in 1984 by RWTH Aachen University (Germany) and started to work at Kabelwerk Eupen (Belgium) in the same year. He was responsible for the R&D activities of material developments and headed the chemical–physical laboratories. With more than 30 years of experience in polymer science and applications, Dr. Beyer is regularly acting as chairman and speaker at many international conferences, especially in the field of the cables, flame retardancy, nanocomposites, and polymer science. In 2003 and also in 2004 he received the Jack Spergel Memorial Award by IWCS for his fundamental work on nanocomposites by organoclays and carbon nanotubes as new classes of flame retardants for polymers.

1

Overview of the Global Cable Industry – Markets and Materials

Astrid Aupetit

Applied, Market Information, AMI Consulting, 1 Brunswick Square, Bristol BS2 8PE, UK

The cable extrusion industry makes up only a small part of the global plastics industry, accounting for just under 3% of plastic consumption. However, it is a high-value market, where producers commit to the industry for the long term; it is not a sector for opportunistic business.

Geographically, the market is dominated by Asia, and more specifically, China. In terms of market size by polymeric materials, Asia and Australasia accounted for 55% of the tonnage used in cable extrusion in 2018. There are around 8000 cable manufacturers in the world, with the number in China estimated between 4000 and 6000. China has not yet fully consolidated its market, but it has reduced its number of manufacturers from 7000 to around 5000. Most are small or medium-size companies.

The cable industry is a complex market segment, due to the number of applications under the main sub-segments: power and communications. Consequently, there are a large number of cable constructions, which can easily reach more than 2000 per manufacturing site. The cable business is strongly technology driven and therefore requires a good technical understanding by all the parties involved.

Electric cable constructions consist of a copper or aluminum conductor, often with steel armoring, and polymeric materials for insulation and jacketing.

Owing to the range of end-use applications, a variety of polymeric materials are used. As a result, any producers decide to specialize in only a few types of cables to reduce investment in machinery.

Further complexity arises from different national standards, which make it harder for foreign producers to enter markets if their cables do not already meet local requirements. Indeed, in Europe, standards are in place for specific cables, notably in construction (Construction Product Regulation [CPR]) and transportation, which producers have to take into account when manufacturing cables. India and China are also in the process of putting in place legislation for the construction and building sector.

The Global Cable Industry: Materials, Markets, Products,
First Edition. Edited by Günter Beyer.

Cable manufacturing involves various stages of production, including metal drawdown, compounding, insulation and sheathing material extrusion, armouring, and eventually cross-linking processes.

The three main sectors that influence the demand for cable are infrastructure, transportation, and construction. Investments in these sectors have a direct impact on a country's demand for cable. Regarding the construction sector, the current trend of urbanization is leading to larger cities, placing with greater demands on power transportation and distribution networks.

The trend toward more eco-friendly solutions for transportation such as e-mobility is also creating additional opportunities for the cable industry.

Renewable energy is also a sector with increasing demand for cable, not least as the electricity generated must feed into the power grid. Indeed, investment in renewable energies is driving particularly strong growth in the medium voltage (MV) and high voltage (HV) segments, related to energy transmission and distribution. In 2017, global cumulative installed renewable power capacity was 2179 gigawatts (GW), with wind energy representing over 23.5%.[1] The Paris Agreement (2016) gathered nations to meet a level of use of renewable energies and has heightened the demand for power cables across the board.

In the Americas and Europe, demand is being driven by the need for power connection between countries and transmission efficiency. In emerging markets such as Southeast Asia and the Indian Subcontinent, rapid economic development and urbanization are key growth drivers.

According to the United Nations Organization,[2] around 60% of the world population will be living in urban areas by 2030 and 68% by 2050, leading not only to new network needs but also to the renovation of existing infrastructure.

In China, the recent slowdown is not expected to last too long as long as government investment picks up.

Investment in construction is forecast to grow globally at an average rate of 3% per year until 2022 (Figure 1.1). This growth is spread quite evenly between civil engineering, and nonresidential and residential construction, with the latter seeing the highest investment level. Unsurprisingly, however, cable demand growth is not homogeneous across regions.

Rapid economic development and urbanization, especially in Southeast Asia, is driving strong construction investment growth. In more mature markets such as Europe and North America, growth is typically in the low single digits. Currently, the level of urbanization in Asia is 50%, while it is 82% in North America and only 43% in Africa.

Another important and bellwether sector is automotive. Global automotive investment growth is forecast to be marginally slow in 2019–2022 compared to the previous years (Figure 1.2). However, structural trends such as the development of electric vehicles and the trend toward e-mobility are having a positive effect on the cable sector.

---

1 https://irena.org/publications/2018/Mar/Renewable-Capacity-Statistics-2018.

2 https://www.un.org/development/desa/publications/2018-revision-of-world-urbanization-prospects.html.

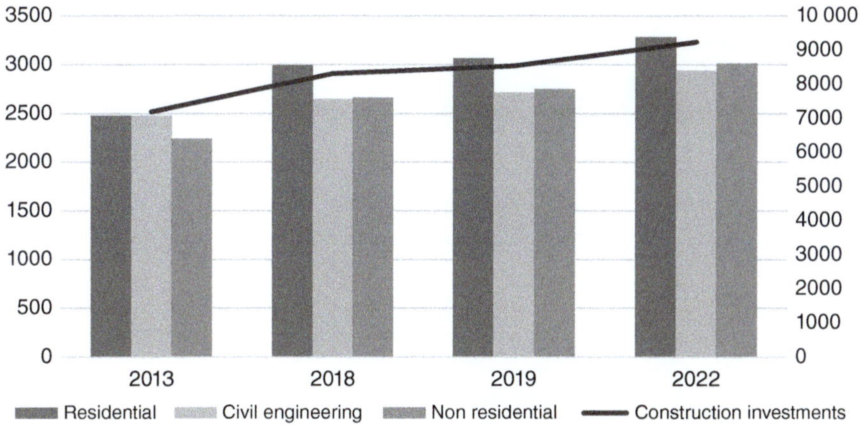

Figure 1.1 Investment in construction in 2013–2022 worldwide.

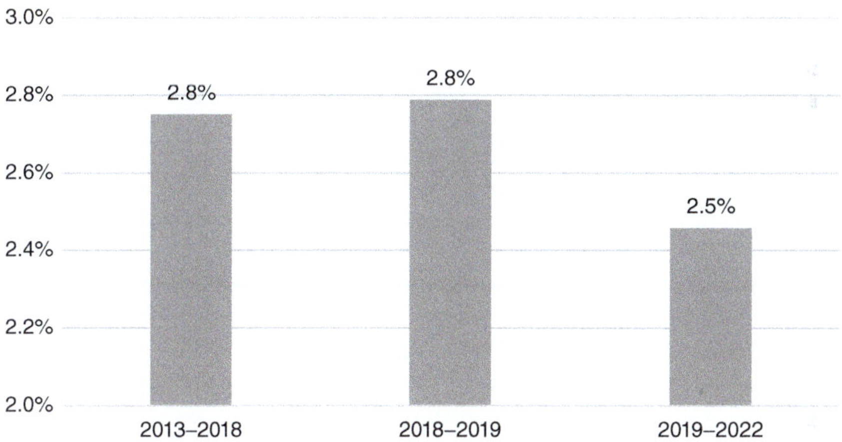

Figure 1.2 Global evolution of automotive production in 2013–2022.

Major concerns for the industry include a range of geopolitical uncertainties. Chief among them are US–China trade frictions and Brexit. Such events are, at best, causing investment to be merely delayed.

1.1 Demand for Polymeric Material

In 2018, Asia and Australasia accounted for almost 60% of global demand, with China alone representing just over 40%. Europe (17%), North America (12%), and Middle East and Africa (10%) all had double-digit shares. South America represented just 2% of the global market (Figure 1.3).

1.1.1 Main Companies Profile

The cable industry is spread worldwide, although it is dominated by a few companies that are present on most continents.

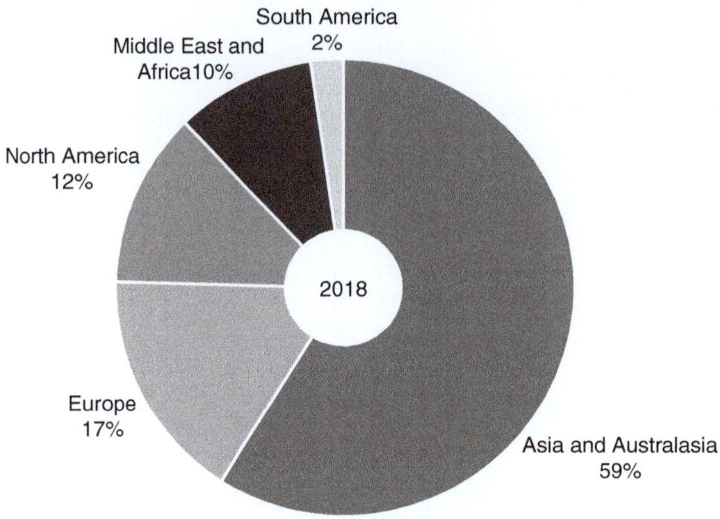

Figure 1.3 Global usage of polymeric materials by region. Source: AMI Consulting 2019.

1.1.1.1 Prysmian

Prysmian is the largest cable manufacturer in the world by revenue. Since the acquisition of General Cable in 2018, their combined sales amounted to €10.158 billion in the year. The company was created in 2005 in Milan, Italy, by Goldman Sachs from the cables and systems division of Pirelli & C. S.p.A. and the group was created in 2011 after Prysmian acquired Draka. Prysmian Group is a public company and offer products and solutions in power grids (including submarine and HV and extra high voltage [EHV] cable systems), oil and gas, telecoms (optical and copper cables), construction and infrastructure, transportation and mobility, and industrial applications. It serves the energy, military, mining, nuclear, and renewable energy industries. The group now has 112 plants in 50 countries. Overall, Prysmian has 29 000 employees not just in the manufacture of cables, but also in cable assembly and wiring and cable systems. Sales are spread as follows: 58% Europe, Middle East, and Africa; 24% North America; 8% Latin America; and 10% Asia Pacific.[3] They also operate a network of 25 R&D centers globally.

1.1.1.2 Nexans

Nexans is the second largest cable producer worldwide. Its headquarters are located in Paris, France, and the company has 51 plants in 34 countries, most of them in Europe. It employs 27 000 people.

Nexans manufactures power (low voltage [LV], MV, HV, EHV), telecom, industrial, and transportation cables as well as cable solutions (i.e. harnesses). It supplies the energy, transportation, telecom, industrial, construction, infrastructure, and aerospace industries. Nexans owns three plants in China and has managed to penetrate the Chinese market by winning projects working for the

3 The same included General Cable for the period 01 June to 31 December 2018.

State grid. In 2018, Nexans achieved €6.490 billion in sales (current metal price). The geographic split of revenue is as follows: 37% Europe, 14% North America, 12% Asia-Pacific, 6% Middle East, Russia, and Africa, 5% Central and Latin America (11% harness, 15% HV). The largest application segment is building and infrastructure, accounting for 43% of revenues.

1.1.1.3 Southwire

Southwire is based in Carrollton, Georgia, in the United States. Out of its 19 plants, only one is outside of North America. The company has 8000 employees and in 2017, their sales were US$ 5.5 billion. It is privately owned and is the largest North American manufacturer after Prysmian, which, thanks to the acquisition of General Cable, has increased its presence on the continent.

Southwire offers power cables (LV, MV, HV, and EHV), telecommunications cables, and industrial cables made of copper rod, aluminum, and magnetic materials.

The company also provides engineering, fabrication, and inventory management services. Southwire supplies to the building sector, utility companies, and original equipment manufacturers for the automotive, electrical, and industrial equipment industries.

1.1.1.4 Sumitomo Electric Industries

Sumitomo Electric Industries is part of the Sumitomo Corporation, a huge Sogo Shosha general trading company that comprises five business units: automotive, information and communications, electronics, environment and energy, and industrial materials. The company is present in 15 countries, with consolidated sales at US$ 9.8 billion in 2017[4] and employs over 255 000 people.

Its headquarters are in Chūō-ku, Osaka, Japan. The company was founded in 1897 to produce copper wire for electrical uses.

The automotive business unit accounts for 50% of Sumitomo Electrics' annual sales, the main part being wire harnesses. Info-communication provides products for optical communications, such as optical fibers, cables, and connectors. Sumitomo has invested heavily in R&D and was one of the first companies to manufacture optical fiber. The environment and energy business unit provides electric wire and cable products that are used for energy supply, including LV, MV, and HV power cables, as well as magnet wire.

1.1.1.5 Furukawa Electric Co., Ltd.

Furukawa Electric Co., Ltd. is a Japanese electric and electronic equipment manufacturer. Headquartered in Chiyodo, Tokyo, the company has 89 plants in seven countries and employs 51 925 people (9 consolidated).

Company sales amounted to US$ 8.8 billion in 2018. Its main activity is telecommunication cable production, with fiber-optic accounting for US$ 1.909 billion of sales.

Furukawa also produces underground submarine industrial and power cables and accessories (total sales: US$ 927.2 million). In addition to cables, it also manufactures batteries, wire harness, and components.

---

4 Consolidated: 3 082 247 million yen, non-consolidated: 1 084 165 million yen (as of March 2018)

1.1.1.6 LS Cable & System

LS Cable & System is by far the biggest cable manufacturer in South Korea (including all its subsidiaries). It also has a very strong presence in South Asia. Headquartered in Anyang, the company has 16 manufacturing plants, with eight in South Korea and the rest across Asia. Its 4203 employees generated revenues of US$ 3.56 billion in 2018.

Following a move into the frontier market of Myanmar in 2017, where no other cable extruder is present, it also plans to enter the African and South American markets. The company offers power cables including EHV with cross-linked polyethylene (XLPE) insulation, submarine and super-conductivity cables, industrial, telecommunication with an emphasis on optical cables (fiber to the antenna [FTTA]), automotive cables and harnesses, and military cables. In addition, LS Cable & System provides engineering services, installation and commissioning of HV and EHV landlines as well as turnkey submarine cabling project execution. LS Cable & System also owns the magnet wire and data cable manufacturer Superior Essex.

1.1.1.7 Leoni AG

Leoni AG, based in Nuremberg, Germany, is a cable and harness manufacturing company. In 2018, sales amounted to €5.1 billion. Leoni AG has 44 manufacturing plants in 31 countries, not just for cable manufacturing but also for automotive cable and wire solutions. It employs 92 549 people (2018). The company supplies all types of cables: power (energy, infrastructure, solar, and software), telecom (fiber optic cables), industrial (instrument cables, hybrid cables, thermocouple cables, LV/MV cables, bus cables, and flexible control electronic cables), automotive cables, rolling stock, maritime, bus, coaxial cables, and special cables (hybrid, sensor cables). As with most big players, Leoni AG also offers wire products and solutions, wiring systems, and related services.

1.1.1.8 Hengtong Group

Hengtong Group is a power and fiber-optic cable manufacturer and the first Chinese company entering the top 10 global cable manufacturers by revenue. It manufactures industrial wire and cable. Hengtong Group Co., Ltd. was founded in 1991 and is based in Jiangsu, China. The group possesses 50 wholly owned companies and holding companies. In 2017, sales amounted to US$ 3.86 billion.[5]

Hengtong offers LV, MV, HV, and EHV cables, railway contact line, and optical ground wire products.

Hengtong Optic-Electric Co., Ltd. specializes in optical fiber and has announced a joint venture with Leoni to make single-mode fibers for telecommunications and data networks in Jena, Germany.

The polymeric material supply chain of the cable industry is versatile and includes both direct sales, distribution, and sales through compounders. Borealis leads polyolefin sales together with Dow DuPont, Ineos, ExxonMobil, LyondellBasell, and Repsol, while LG, Hanwha, Anwil, and Sabic are among the top suppliers of polyvinylchloride (PVC).

---

5 25.95 billion yuan – http://www.hengtonggroup.com.

1.1.2 Global Demand

As mentioned earlier, cables are complex constructions that involve a conductor, typically copper or aluminum, and several layers of polymeric material for insulation and jacketing. They are produced by means of extrusion. More specifically, the conductor is covered by the insulation layer and then the individual coated wires are twisted together into a cable core. This is typically done using a twisting machine or by in-line SZ twisting. The resulting cable core and other construction elements are then jacketed together.

The process is similar for optical cables, but the core material is replaced by fiber. Two materials are used for optical fiber: plastic (mainly PMMA) and glass. They are utilized in different applications – plastic fibers are used for indoor, short-range consumer applications, while glass, which is more expensive, is used for longer range as well as medium range (multi-mode) telecommunications.

In 2017, the number one supplier of fiber-optic cable was Corning (USA); it led the industry and was followed by Yangtze Optical FC (China), and Furukawa (Japan). These three companies represented over 40% of market share.[6]

The main compounds used in cable insulation and jacketing are based on the resins below:

- *Polyvinylchloride*: PVC is a thermoplastic material that is made from two starting materials. Fifty-seven percent of its molecular weight is derived from salt and the rest derived from hydrocarbon feedstocks (ethylene from oil or natural gas). PVC can be combined with different kinds of additives such as plasticizers and fillers, making it a highly versatile polymer.
- *Polyethylene (PE)*:
  - *Low-density polyethylene (LDPE) (below 0.930 kg/m^3)*: Conventional LDPE manufactured by a high pressure, high-temperature process. The material has a highly branched structure, the short branches disrupting the crystalline structure and giving a low-density material.
  - *Linear low-density polyethylene (LLDPE) (between 0.890 and 0.960 kg/m^3)*: This is produced by a low-pressure process. The addition of various comonomers produces a material with short side branches.
  - *Medium-density polyethylene (MDPE) (between 0.930 and 0.940 kg/m^3)*: Comprises polyethylene manufactured by either high- or low-pressure processes, which has a density between 0.930 and 0.940 kg/m^3.
  - *HDPE (above 0.940 kg/m^3)*: High-density polyethylene is manufactured by a low-pressure process and has few short branches and no long chain branches.

    Conventional LDPE has seen its market share decrease, largely because it has been replaced by linear low density (LLDPE) grades in LV power insulation and jacketing, as they can offer improved mechanical properties, increased melting temperature, far better stress crack resistance, and faster cross-linking, and are generally cheaper to produce.
- *Cross-linked polyethylene*: XLPE is produced by forming links between single PE molecules in both the crystalline and the amorphous phases of the polymer

6 Source: Network Telecom Research Institute (http://www.nti.news).

matrix. A three-dimensional network is formed as a result, improving a range of physical properties.

- *Ethylene vinyl acetate* (*EVA*): EVA, the copolymer of ethylene and vinyl acetate. The weight percent of vinyl acetate usually varies from 10% to 40% with the remainder being ethylene. EVA is a polymer that approaches elastomeric materials in softness and flexibility, yet can be processed like other thermoplastics.
- *Polypropylene* (*PP*): PP includes all homopolymers and PP-based copolymers.
- *Thermoplastic elastomer* (*TPE*): TPEs are mainly products with elastomeric properties, but capable of being processed like thermoplastic materials. TPEs included in this category include cable grade TPEs based on blends of different polymers such as EDPM/PP.
- *Rubber*: Includes all cable grade synthetic and natural vulcanized rubbers. In the cable industry, the two main rubbers (elastomers) used are cross-linked EPDM for insulation and cross-linked CPE for jacketing.
- Other polymers such as polyamide (PA) and polybutylene terephthalate (PBT) and thermoplastics polyurethanes (TPUs) and fluoropolymers, such as FEP, PTFE, PFA, PVDF, MFA, and ETFE, are used in high-performance cables, mainly in automotive, aerospace, and military applications.

HFFR/LS0H: The products included in the low smoke zero halogen (LS0H) and halogen-free flame-retardant (HFFR) material sector are polymers based on polyolefins, often combined with EVAs, which have been compounded to give low levels of smoke in the event of a fire.

The total global tonnage of polymeric materials used in cable extrusion is expected to see an annual growth rate of 4.5% until 2023.

Low smoke zero halogen/halogen-free flame-retardant compound is expected to grow the fastest in the next four years. LS0H/HFFR consumption is being driven by increasing concerns over the effects that halogen released from cable materials may have in case of fire. Current technology of thermoplastic LS0H/HFFR compounds is mostly based on PE copolymers or a blend of LLDPE/EVA and filled with flame-retardant fillers such as alumina trihydrate (ATH), magnesium hydroxide (MDH), and also synergists such as zinc borate, silica, and also organoclays. With the organoclays, nanocomposites are formed with a strong reduction in heat release and enhanced char formation.

Prime areas of application for cables with halogen-free compounds are public buildings and transportation infrastructure where a high degree of protection against fire and fire damage must be provided with the demand of low smoke generation. The trend started in Europe in the early 2000s but it has gained global momentum and is spreading across the rest of the world although at different rates. North America, for example, is lagging behind with HFFR accounting for less than 5% of its total demand while in Europe it represents over 15%. In addition, the demand for more material to be recycled is inciting producers to use more halogen-free additives that can be recycled. The Fraunhofer Institute for structural durability and system reliability LBF in Germany conducted a project between 2015 and 2018 to boost mechanical recycling of plastic

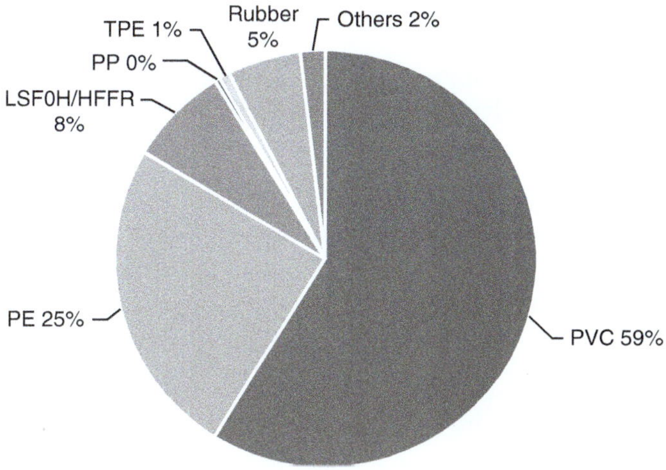

Figure 1.4 Polymeric material consumption in the Global Cable Industry, 2013.

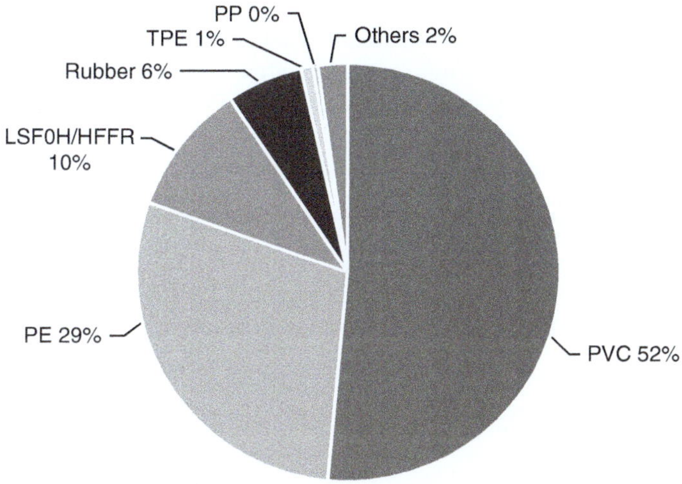

Figure 1.5 Polymeric material consumption in the Global Cable Industry, 2019.

containing halogen-free retardants in partnership with Forschungsgesellschaft Kunststoffe e.V. and PINFA (Figure 1.4–1.6).[7]

Polyvinylchloride compounds used in the cable industry are by far the slowest growing polymeric materials globally. Its 2019–2023 CAGR is just 1.1%. Despite being gradually replaced by other materials, PVC compounds remain the most used polymeric material in the global cable industry and in 2018 it accounted for 53% of overall tonnage. Legislation such as European CPR, fully implemented in July 2017, is encouraging other materials such as XLPE, PE, and HFFR compounds to replace PVC compounds.

7 http://fgkunststoffe.de/?q=content/recycling-von-halogenfrei-flammgesch%C3%BCtzten-kunststoffen.

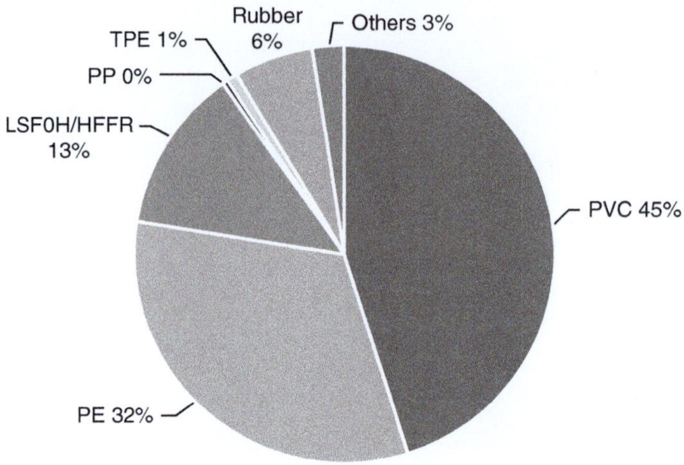

Figure 1.6 Polymeric material consumption in the Global Cable Industry, 2023.

In 2018, PP compounds accounted for less than 1% of global polymeric material tonnage consumed for cable extrusion. While it has a solid, if not spectacular, 2019–2023 CAGR of 5.4%, this figure is somewhat more uncertain than for other materials. Cable industry stakeholders note that PP compounds, over the long run, could increasingly replace XLPE compounds for insulation, with the former being cheaper, easier to work with, and recyclable. The issue is that leading extruders have invested heavily in machines and an overall production process that makes use of XLPE compounds, and would have to invest heavily once more to shift to PP. Unless Prysmian (which drives the market of PP for insulation) introduces a new formulation to the market, XLPE will remain the most used insulation material for MV and HV cables in particular, at least over the medium term.

1.2 Asia and Australasia

1.2.1 Demand for Cable

The construction market in Asia and Australasia is predicted to grow at around 2.5% in 2019 (Figure 1.7).

The pace of growth is slowing down in China while the reverse is true for India. Several big transportation projects are in the pipeline, for example, the PAN Asia railway network linking China, Singapore, and all mainland Southeast Asian countries.

Global Construction 2030 (a global study of the construction and engineering industry published by Global Construction Perspectives and Oxford Economics) predicts that Southeast Asia's construction market will exceed US$ 1.0 trillion by 2030.

Despite a slowdown in China, the 13th Five-Year Plan and Made in China 2025 are expected to boost cable demand growth going forward.

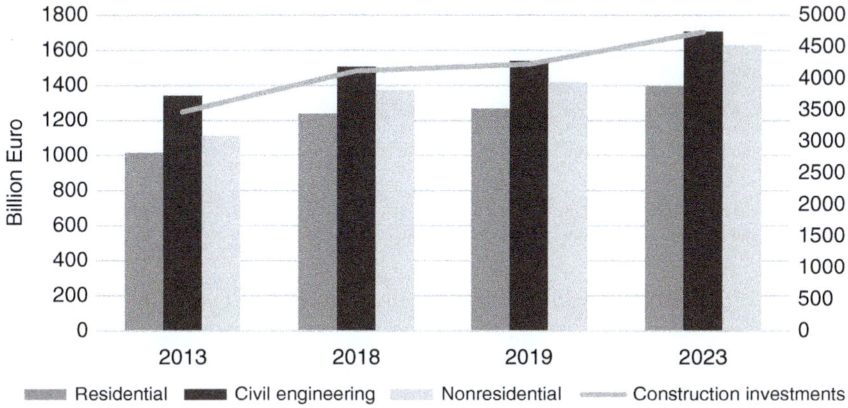

Figure 1.7 Construction investment in Asia and Australasia 2013–2022.

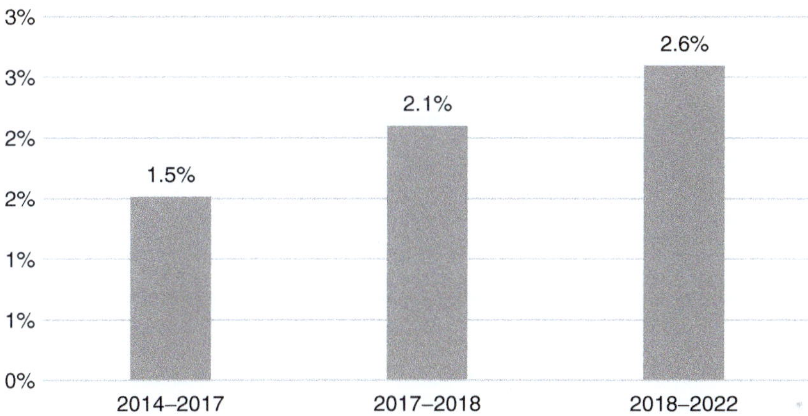

Figure 1.8 Evolution of automotive production in Asia and Australasia in 2014–2022.

According to voices in the industry, India's construction sector will achieve a CAGR of 7–8% from 2019 to 2023, with 10–12% growth in infrastructure. In terms of power grid development, government targets state that India's inter-regional transmission and distribution network capacity should increase from 86 GW in fiscal 2018 to 130 GW in fiscal 2023.

As for the automotive sector, declining car production in Japan will be offset by growth in automotive production in India. For the region, the annual growth rate for automotive production is expected to slow down to 2.6% between 2018 and 2022 (Figure 1.8). New car sales in Southeast Asian countries rose by 5% in 2017 but were predicted to experience a slowdown due to a looming tax hike in the Philippines (Tax Reform for Acceleration and Inclusion Act) and currency fluctuation.

1.2.2 Demand for Polymeric Material

Asia and Australasia accounted for almost 60% of global polymeric material demand related to cables production in 2018. Demand for polymeric material in the region is being driven by strong economic growth leading to robust construction and infrastructure investment. Southeast Asian nations have pledged US$ 323 billion on infrastructure developments over the next few years, and will be a particularly strong sub-region for growth. The ASEAN Smart City Network will support urbanization and will serve as a platform to share best practice and link cities with private investment and secure funding.

Asia and Australasia are the biggest consumers of polymeric materials for cable extrusion, and despite a slowdown in the last few years, they clearly outperform other regions in terms of growth.

China dominates global and regional demand for polymeric materials, representing 70% of the regional demand. While China's Hengtong is one of the world's 10 largest cable extruders by revenue, the country is nevertheless a highly fragmented market, with more than 6000 cable manufacturing sites. The top five companies are estimated to represent less than 15% of the total market.

Asia and Australasia's top 10 cable producers by material demand are dominated by local players. The top five are Chinese-based companies, while South Korea and Japan only have one producer each in the ranking. Prysmian is the only European company in the top 10, ranking 9th.

Power cable production in Asia and Australasia will be primarily driven by major new urban and infrastructure developments across the region, necessitating power grid expansion.

Japan's demand for polymeric material slowed down in 2018 due to decreased demand for power, telecommunication, and automotive cables. Cable demand is expected to barely increase over the forecast period.

However, in India, demand for polymeric material is expected to grow faster than any other country in Asia and Australasia until 2023.

Halogen-free flame-retardant compound is the fastest growing polymeric material in the region and is expected to carry on being so for the foreseeable future. XLPE is replacing PVC in the insulation of power cables and its use has been extended where there is a need for more heat resistance. It represents 14% of the tonnage, far behind PVC's leading share of 54%.

1.3 Europe

1.3.1 Demand for Cables

In Europe, overall construction investment is predicted to grow in the low single digits over the next four years (Figure 1.7).

Superior investment forecasts related to civil engineering projects (road, rail, airport, and maritime) coupled with favorable structural trends are driving higher growth rates for transportation-related cables. While annual residential construction investment growth of just over 1% is behind overall growth expectations for power cables, the 2019–2023 CAGR for MV and HV cables of above 5% is being driven by renewables and power grid investment (Figure 1.9).

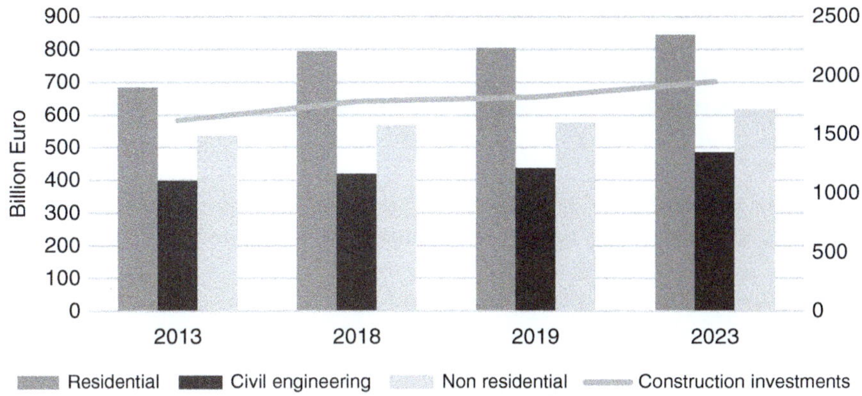

Figure 1.9 Construction investment in Europe in 2013–2023.

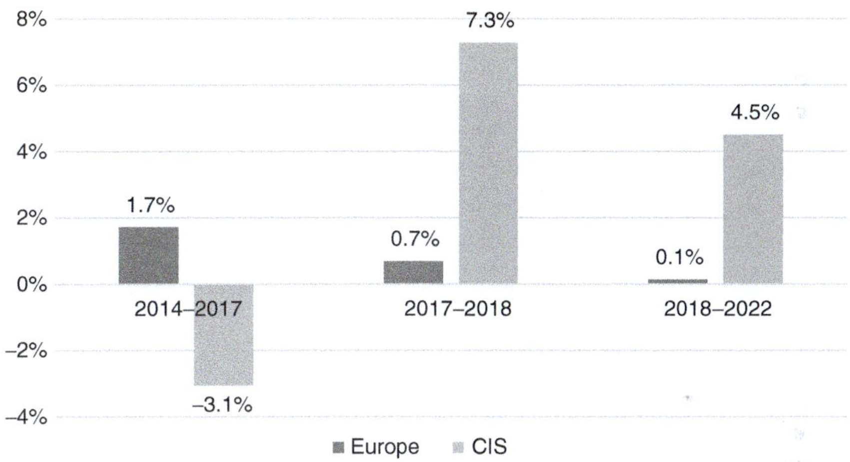

Figure 1.10 Evolution of automotive production in Europe in 2014–2022.

Automotive production growth is set to slow down in Europe over the next three years (Figure 1.8). Germany has been particularly hit by the threat of US tariffs on its cars as well as new EU rules on car emissions, particularly diesel. Production growth in the CIS region is predicted to be in the mid-single digits by 2022, a far stronger performance compared to 2014–2017, a period when Russia was experiencing significant economic difficulties (Figure 1.10).

1.3.2 Demand for Polymeric Materials

European cable production accounted for 16.3% of global polymeric material demand related to cables production in 2019.

Low smoke zero halogen/halogen-free flame-retardant compound is forecast to grow the fastest in Europe. Its consumption is planned to reach over 20% of the total polymeric material used in cable production in Europe by 2023.

Conversely, the share of PVC compounds is decreasing. It represents 39.5% of the market in 2019 down from over 48% in 2013 and it is expected to decrease to under 35% by 2023.

Ongoing concern over the effect that halogens released from cable materials have in the event of fire and approval of the CPR are driving the dynamics of LS0H/HFFR and PVC compounds in particular.

Countries such as Spain and Italy were early adopters of HFFR compounds in the early 2000s, followed by Scandinavia. Germany, France, and the United Kingdom were behind but are increasing their consumption.

Four countries (Russia, Italy, France, and Germany) have regional material demand market shares of above 10%. Italy is Prysmian territory, France is Nexans', and Germany is the home of Leoni and NKT, which have a significant amount of their manufacturing sites there.

Power cable applications account for around two-thirds of European material demand. Telecom is around one-fifth, while transportation and appliances and industry are both just under 10%.

Prysmian, with its major acquisitions of Draka and more recently General Cable, is by far the largest source of material demand in Europe, accounting for almost 25% of the market.

France-based Nexans is almost three times smaller. Some of the big players in the region have their plants located in one country (Tele-Fonika, Romcab, etc.). Ongoing consolidation is helping to mitigate overcapacity and profitability issues; market concentration is increasing.

1.4 The Middle East and Africa

1.4.1 Demand for Cables

Civil engineering accounted for almost half of Middle East and Africa's construction investment in 2018. Nonresidential and residential investments were of approximately equal size, accounting for around a quarter of investment each. Construction investment growth will be highly contingent on the price of oil for the next four years (Figure 1.11). Therefore it is difficult to predict growth.

As for the automotive sector, Africa manufactures around 1 million vehicles/yr. South Africa accounts for over half of this, with Morocco representing most of the remainder.

Except for Turkey and Iran, both of which manufacture over 1 million vehicles/yr, the Middle East has very little automotive production. Iran's growth prospects have been severely curtailed by the reintroduction of American sanctions.

Growth prospects for 2018–2022 in both the Middle East and Africa are considerably weaker compared to 2014–2017 growth (Figure 1.12).

1.4.2 Demand for Polymeric Materials

Middle East and Africa's cable production accounted for 9.8% of global polymeric material demand related to cables production in 2019.

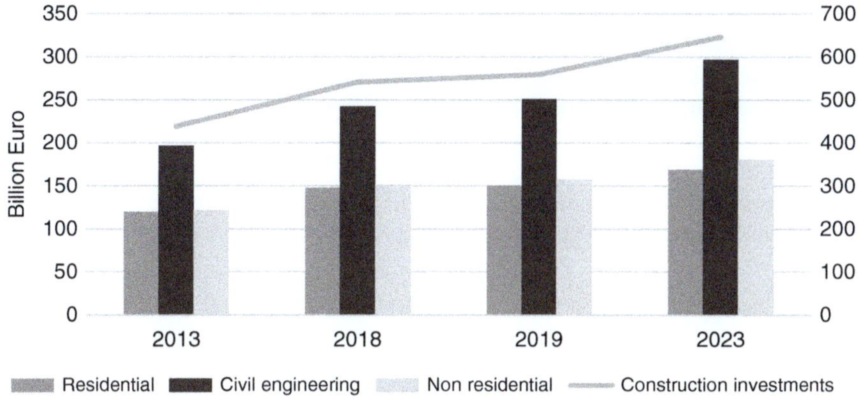

Figure 1.11 Investment in construction in the Middle East and Africa in 2013–2023.

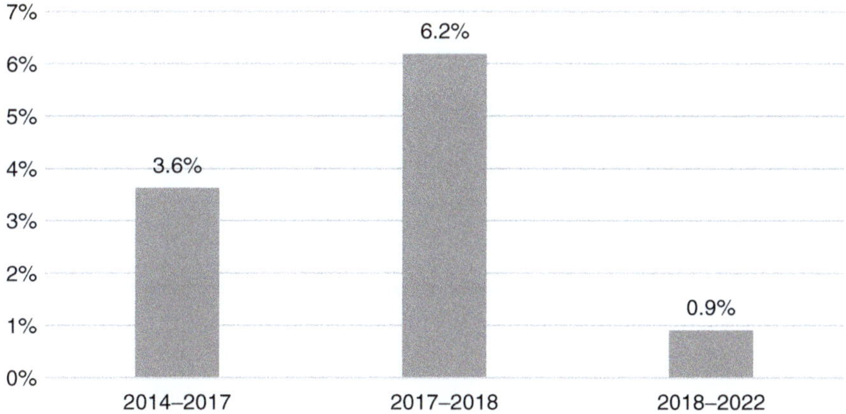

Figure 1.12 Evolution of automotive production in the Middle East and Africa in 2014–2022.

By polymeric material, for 2019–2023, the annual growth rates forecast for LS0H/HFFR and XLPE are highest. Market dynamics in North Africa are significantly influenced by Europe. It is becoming an ever more popular "near-sourcing" location for Europe, as is the case with Turkey, where demand growth for HFFR compound is forecast to be particularly strong.

Indeed, of the largest cable manufacturing countries in the Middle East, Turkey (8.3%) has by far the strongest growth rate forecast for polymeric material tonnage overall, followed by the UAE, while Saudi Arabia's growth is notably lower (2.5%).

North African countries dominate Africa's cable production, with South Africa and Nigeria being the main markets elsewhere. While cable plants themselves are ramping up production in North Africa to supply Europe directly, they are also feeding nearby manufacturers. For example, the region is home to a substantial automotive industry, which also exports heavily to Europe.

Middle East and Africa's top 10 cable producers by material demand are dominated by local players. Two of the top three are Egypt-based companies, El

Sewedy Cables and Energya Cables, while Saudi Arabia is the home of Riyadh Cables and Saudi Cables. The United Arab Emirates is the home of just one of the top 10, while Turkey has three manufacturers in the top 10.

Power cable production in the Middle East is primarily being driven by major new urban and commercial developments across the region. Numerous Gulf states are launching various long-term "Visions" to develop their economies, with the focus on diversifying away from hydrocarbons. The success of these plans remains highly contingent on the price of oil going forward.

1.5 North America

1.5.1 Demand for Cables

Residential investment accounted for almost 45% of North American construction investment in 2018. Nonresidential and civil engineering (road, rail, airport, and maritime) investments accounted for 30% and 25%, respectively. The ratio of investments between these three sub-segments does not change much over time (Figure 1.13).

As for the automotive industry, the United States is the largest vehicle manufacturer in the region, manufacturing over 11 million units in 2018. Mexico's production is almost three times smaller with over 4 million vehicles in 2018. Canada accounts for just 12% of the regional market share, with the production of 2.1 million units in 2018, bringing regional production overall to 18 million units in 2018.

North America is forecast a compound annual growth rate of 0.22% between 2018 and 2022 (Figure 1.14). President Trump's administration is threatening taxes on General Motors for importing vehicles to the United States from Mexico. Ford announced it would not open its planned facility in San Luis Potosi, Mexico; instead, it will invest US$ 700 million in a plant in Michigan, USA.

The unpredictable government policy environment is making investment decisions for regional automotive manufacturers difficult, making North American vehicle production difficult to forecast.

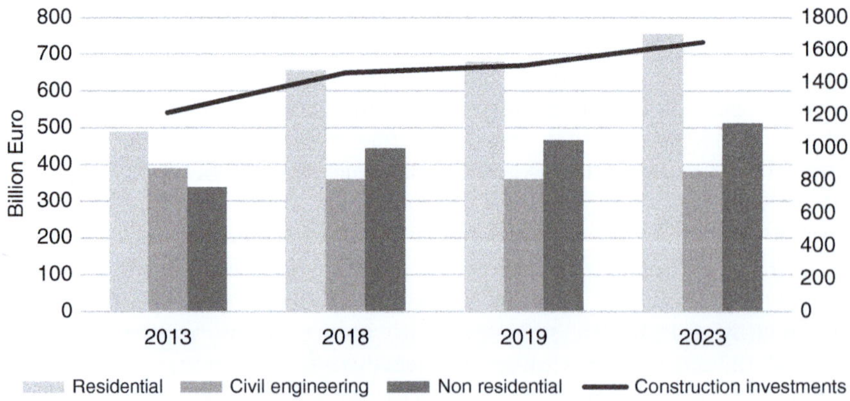

Figure 1.13 Investment in construction in North America in 2013–2023.

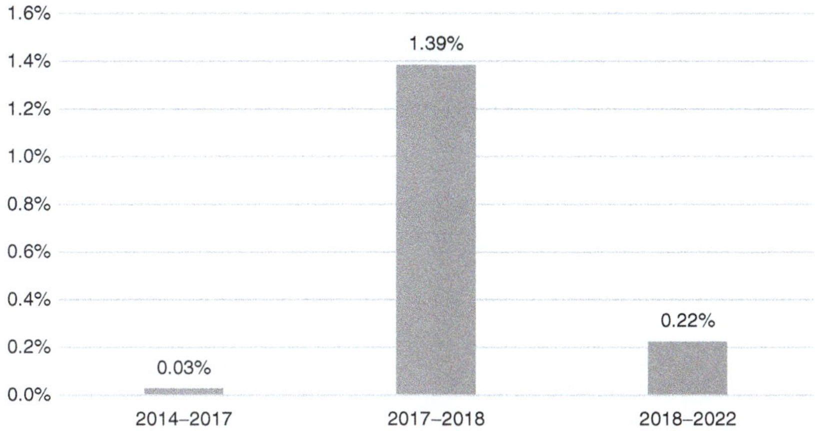

Figure 1.14 Evolution of automotive production in North America in 2014–2022.

1.5.2 Demand for Polymeric Materials

North America's polymeric material consumption accounted for 12.3% of global polymeric material demand related to cables production in 2019.

Marginally negative growth forecasts for PVC compound are largely explained by other materials such as XLPE, PE, and HFFR compounds being used as substitutes. HFFR compounds have the highest regional growth rate primarily due to the substitution of PVC compounds in cables for public buildings and public transport. XLPE and PE compounds will be in high demand from the ongoing need to maintain and replace parts of the power grid affected by natural disasters.

Rapid growth in renewable energies is also expected to stoke a strong demand for MV/HV cables, also explaining the relatively strong growth expectations for XLPE, PE, and rubber compounds. According to the 2019 Sustainable Energy in America Factbook, in 2018, 18% of all electricity in the United States was produced from renewable sources, up from 11% in 2009. Renewables generation grew by 5.1% in 2018, with a 13% increase in wind and solar offsetting a small decline (−2.7%) in hydropower output. A study from the Department of Energy's National Renewable Energy Laboratory showed that the United States can generate most of its electricity from renewable energy by 2050. This will obviously necessitate a full commitment from the government along with the right policies and measures in place.

The United States accounted for approximately 80% of North American polymeric material demand in 2018. Indeed over 85% of the total number of cable extrusion facilities of the region are in the largest economy of the region. For 2018–2023, it is forecast that the United States and Canada will suffer a deceleration in their GDP growth, relative to 2013–2018, primarily due to uncertainty from trade frictions and oil price fluctuations. The reverse is true for Mexico where the economy continues to grow despite the uncertainty linked to recent elections and the country's trading relationship with the United States.

By country, Mexico has the strongest growth rate forecast for polymeric material demand between 2019 and 2023 due to an acceleration of the economy and a continuous focus on construction projects.

North America's top 10 cable extruders by material demand are led by Italy's Prysmian, thanks to its acquisition of General Cable, followed by the American producer Southwire.

By application, power cable production in North America represents over three-quarters of the market. It is forecast to be the fastest growing end-use application, with an annual rate of 2.7% from 2019 to 2023 with the HV and EHV sectors growing the fastest.

1.6 South and Central America

1.6.1 Demand for Cables

Residential investment accounted for almost 40% of South America's construction investment in 2018. Nonresidential and civil engineering investments were of approximately equal size, accounting for around 30% of investment each. Construction investment growth across South America is expected to improve over the next five years compared to 2013–2018 (Figure 1.15). This is the case for all three sub-segments. Brazil's economic difficulties primarily account for the poor performance of the region for the last five years. Investments in the sector will increase demand for power, communications, and industrial cables.

Brazil is the main regional vehicle manufacturer, producing over 2.3 million cars in 2018, which represents a 6.7% uplift from 2017.[8] General Motors is expected to invest US$ 2.7 billion in Brazil in two of its plants over the next five years. Argentina is far smaller, producing around 210 000 cars in 2018 and 260 000 commercial vehicles. Other countries have very little production.

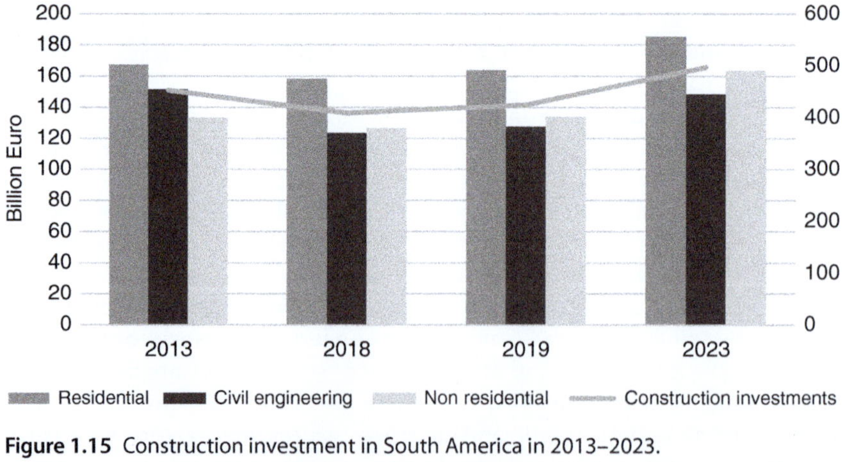

Figure 1.15 Construction investment in South America in 2013–2023.

8 Total automotive production 2 880 724 units including cars, light commercial vehicles, trucks, and buses.

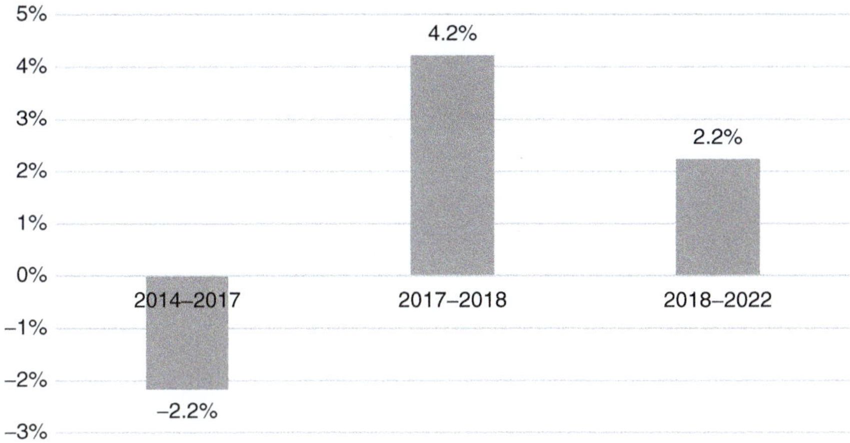

Figure 1.16 Evolution of automotive production in South America in 2014–2022.

South America is recovering from a big fall in production. The region produced over 4 million vehicles in 2012, whereas the output was below 3 million in 2017 and 2018. Growth prospects for 2018 to 2022 in South America are considerably stronger, primarily due to Brazil's expected recovery (Figure 1.16).

1.6.2 Demand for Polymeric Materials

South American cable production accounted for 2.2% of global polymeric material demand related to cables production in 2019.

Low smoke zero halogen/halogen-free flame-retardant and XLPE compounds are expected to grow fastest between 2019 and 2023, both at an annual average rate of 8.1%. HFFR is not forecast to grow as fast as in other regions, but in Chile, the need for special cables for the mining sector is driving an especially strong demand for non-halogenated cables.

Among the largest cable manufacturing countries, Chile has by far the strongest growth forecast due to new projects in mining and renewables, followed by Peru, while Brazil's growth is notably lower. Brazil and Argentina dominate cable production in the region, accounting for 43% and 30% respectively of overall material consumption.

South America's largest markets have struggled with economic and political instability in recent years, which could resurface going forward. Hence, forecasts for this region have a high degree of uncertainty.

South America's top 10 cable producers by material demand are dominated by foreign players. The two largest are Prysmian and Nexans, while most of the local producers in the top 10 are Brazilian companies, i.e. Nambei, Corfio, and Megacabos.

Power cable production in South America is still the largest end-use market, but the fastest growth is forecast for the transportation sector, with an annual growth rate of 4.8% expected between 2019 and 2023. However, transportation is still a small market compared to power and telecom meaning that this growth will only contribute an additional 3000 tons to material demand in absolute terms.

The biggest economy in the region, Brazil, has a 2019–2023 CAGR of over 3%. After going through a deep recession and a slow recovery, some have hope in the new president's policies.

Renewable energies will be an important sector driving cable demand going forward. In 2018, 79% of the electricity generated was from renewable sources. In Brazil's Energy Master plan 2016–2026, it is expected to install 18.5 GW of additional wind power generation. Currently, wind energy is growing at an average of 2 GW/yr. As for oil and gas, Petrobras has unveiled multi-billion plans in Brazil, with oil production growth forecast at 10% in 2019 due to new operations.

Numerous projects in the mining, renewable energy, and automotive sectors will also be important for stoking material demand.

2

Thermoplastics for Cables

Theo Geussens

Dow Europe GmbH, Wire & Cable Technical Service & Development, Bachtobelstrasse 3, Horgen 8810, Switzerland

2.1 Introduction

A thermoplastic material is a polymer that becomes moldable or extrudable at elevated temperatures and becomes a solid upon cooling to room temperature. They can be re-melted and reshaped, which is an important factor in the ability to recycle these materials.

Thermoplastic polymers are mainly used for telecom and specialty cable insulation, some power cable insulations, components in fiber optic cables, and in general cable jackets.

The five main types of thermoplastic materials that commonly used in cable applications are polyolefins, chlorinated polymers, fluoropolymers, polyamides, and polyesters.

2.2 Polyolefin Materials

Polyolefins are polymers made of olefinic monomers; the official IUPAC nomenclature for these polymers is poly(alkene).

The two most common polyolefins, which are used as thermoplastic cable materials, are polyethylene and polypropylene; other polyolefins such as polybutylene and polyisobutene are not used in cables.

Polyolefin elastomers and ethylene copolymers such as ethylene vinyl acetate (EVA) and ethylene butyl acrylate (EBA) are often used as base resins in halogen-free flame retardant (HFFR) compositions or in crosslinkable compounds for power cable; they will not be discussed in this section; see the Chapter 4 by Beyer in this book. Other polyolefins like ethylene propylene rubber (EPR) and ethylene–propylene diene monomer (EPDM) are always cured for use in cable applications.

The Global Cable Industry: Materials, Markets, Products,
First Edition. Edited by Günter Beyer.

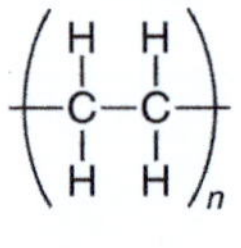

Figure 2.1 Repetitive units in polyethylene.

2.2.1 Polyethylene

In its simple form polyethylene consists of a chain of carbon atoms, covalently bond to each other and to two hydrogen atoms (Figure 2.1) (except for the terminal carbon atoms that are bonded to three hydrogen atoms). Polyethylenes only consisting of ethylene monomers are called homopolymers. In the low pressure polymerization process, the density and stiffness of the polymer can be lowered by incorporating comonomers. Comonomers having a double bond, such as propylene, butene, hexene, or octene, can be added in the polymerization reactor [1]; the resulting polymer is called an ethylene copolymer. When such comonomers are polymerized into the polymer backbone, a small side chain will stick out of the backbone, for butene this will be an ethyl side-group, for octene a hexyl side-group. Such side groups are also called short chain branches, to distinguish them from long chain branches created for instance in a high pressure polymerization process or when certain constrained-geometry catalysts are being used in a low pressure polymerization process (Figure 2.2).

The amount of comonomer built into the polymer backbone determines the level of crystallinity of the polyethylene. As the crystalline phase in polyethylene has a density of 1.0 g/cm^3 and amorphous polyethylene has a density of 0.85 g/cm^3, as a rough approximation the percentage of crystallinity will determine the polymer density. In this way linear-low density, medium or high density polyethylene (HDPE) can be created [2]. Metallocene catalysts can incorporate higher amounts of comonomer (up to 35%) so that densities in the range of 0.85–0.915 can be obtained; polymers with such low densities are called Plastomers and Elastomers [3]. Metallocene catalyst distributes the comonomer more homogeneously over the polymer chains than, for instance, a

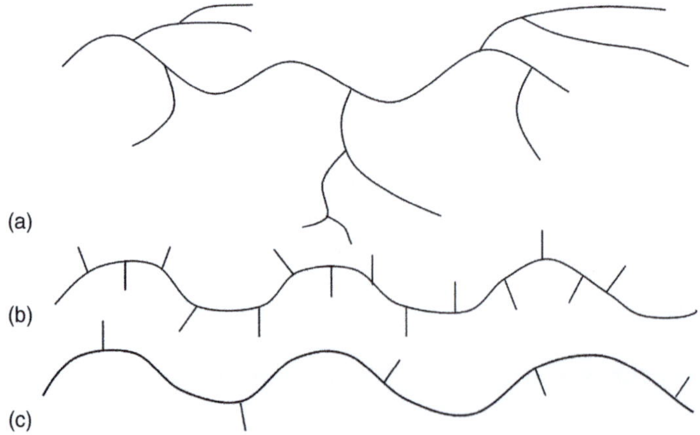

Figure 2.2 The structure of three polyethylene types; (a) LDPE. (b) LLDPE, and (c) HDPE.

Table 2.1 Classes of polyethylene.

Polyethylene type	Density range (g/cm³)	Type (ASTM D1248)	Data block 4 (ISO 17855)
Elastomer	0.855–0.885	0	00
Plastomer	0.885–0.915	0	00 and 01
VLDPE or ULDPE	0.890–0.915	0 or I	00, 01, or 08
LDPE	0.910–0.925	I	13, 18, 23
LLDPE	0.915–0.925	I	13, 18, 23
MDPE	0.926–0.940	II	27, 33, 38
HDPE	0.941–0.966	III and IV	44, 50, 57, 62

Ziegler–Natta catalyst, so that a single melting peak can be observed in a DSC trace.

Based on the density range, polyethylene is usually divided into HDPE, Medium Density Polyethylene (MDPE), Linear-Low Density Polyethylene (LLDPE), Very-Low Density Polyethylene (VLDPE) or Ultra-Low Density Polyethylene (ULDPE), Plastomers, and Elastomers (see Table 2.1). To classify a polyethylene in Wire and Cable applications ISO 17855-1 [4] (which replaced the old ISO 1872-1) or ASTM D1248 [5] is used; the classification into types (ASTM) or the data block 4 in the ISO designation corresponds to the classical terminology used, but splits the density range of VLDPE, low density polyethylene (LDPE), MDPE, and HDPE into three or four sections.

Polyethylene crystallizes in chain folded lamella crystals. When an alkene comonomer is incorporated into the polymer backbone, the backbone will not nicely fit into the folded lamella structure of polyethylene crystals. The larger the short chain, the lower is the likelihood that that piece of the polymer chain fits into a polyethylene crystal and will reside in the amorphous phase of the polymer. If the polymer chain is long enough, it can be introduced into another growing polyethylene crystal and form a tie molecule. These tie molecules are important for the environmental stress crack resistance (ESCR) and strain hardening behavior in the tensile curve, leading to higher tensile strengths at break.

The properties of polyethylene are controlled by three basic parameters: density, melt index, or average molecular weight (MW) and the molecular weight distribution (MWD).

With density the stiffness or flexibility of polyethylene is controlled. A higher density polyethylene is stiffer, has a higher temperature resistance (Vicat softening point, pressure test at high temperature), a lower coefficient of friction and is harder, but has a lower ESCR, see Figure 2.3. As a higher density indicates a more crystalline structure and faster crystallization rates, higher density grades, having similar rheological characteristics, will show a higher shrinkage.

The melt index or melt flow rate of a polymer like polyethylene is a measure for the average molecular weight; a high melt index represents a low average

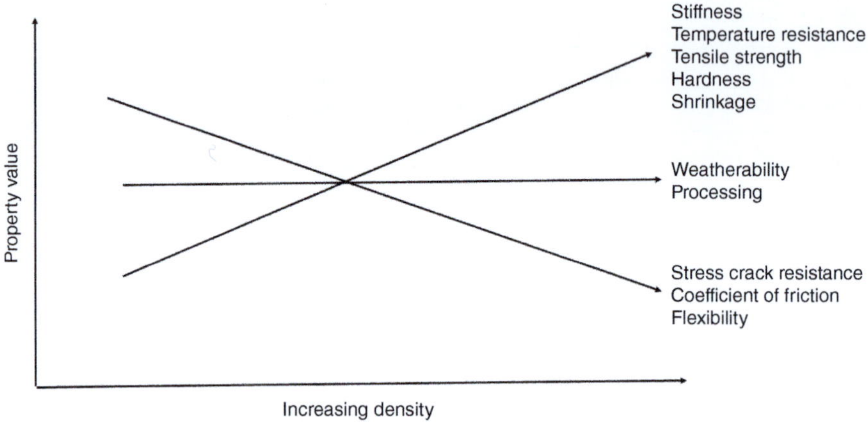

Figure 2.3 The effect of increasing density on the properties of polyethylene.

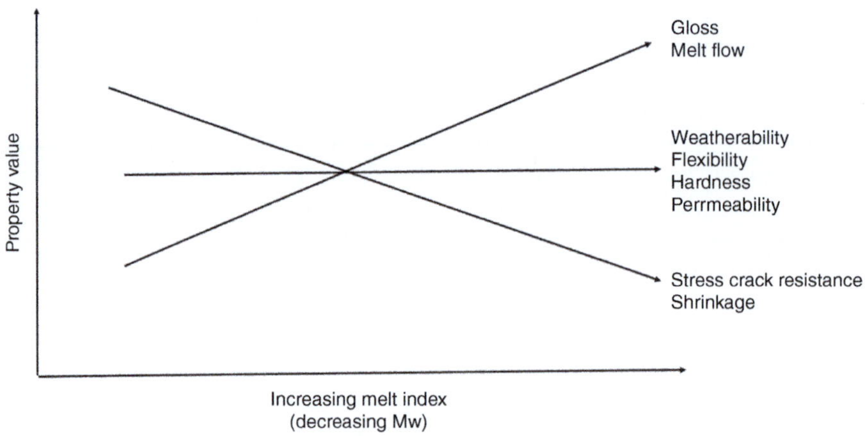

Figure 2.4 Effect of increasing the melt index on polymer properties.

molecular weight, a low melt index indicates a high molecular weight. The melt index of polyethylene is measured at 190 °C and typically (especially for cable materials) using a load of 2.16 or 5 kg as defined in ISO 1133 or ASTM 1238. With the melt index or molecular weight of a polymer both the physicomechanical properties and the rheology or processability of a polyethylene are controlled. A higher melt index grade has an easier flow, resulting in a glossier surface and a lower post-extrusion shrinkage (see Figure 2.4). Longer molecules (lower melt index) will build more tie molecules, improving the ESCR performance. However a lower melt index means higher viscosities and longer relaxation times so shrinkage will increase.

Another important parameter is the MWD of a polymer; an easy way to get an idea of the MWD is to measure the melt flow rate with different weights and calculate the ratio, for instance (melt flow rate [190 °C/21.6 kg])/melt index [190 °C/2.16 kg]). A higher melt flow ratio indicates a broader MWD.

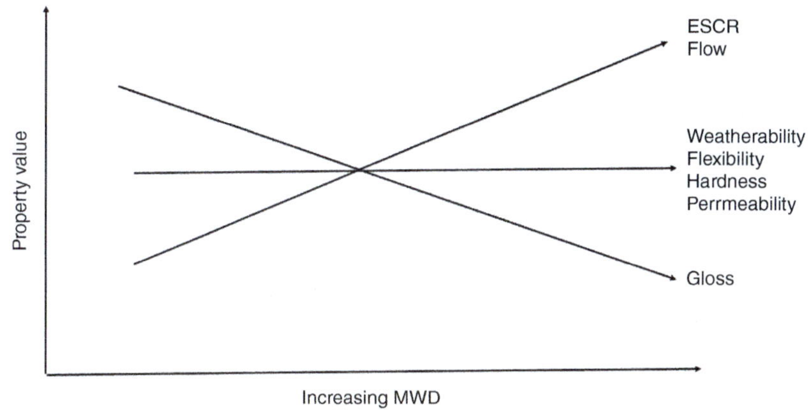

Figure 2.5 The effect of increasing molecular weight distribution (MWD) on product properties.

The MWD has a primary effect on the processability of a polyethylene; the viscosity curve of a broad MWD material will give more shear thinning than a polyethylene having the same molecular weight but a narrower MWD. Increasing the MWD has some effect on gloss, on ESCR (the longer chains in a broad MWD polyethylene provide tie molecules to connect the crystals) (see Figure 2.5) but could potentially lead to higher post extrusion shrinkage because the longer molecules have longer relaxation times.

In a high pressure polymerization process, other types of comonomer can be incorporated, which in a low pressure polymerization process would deactivate the catalyst, such as vinyl acetate (VA), ethyl acetate (EA), butyl acetate (BA), acrylic acid (AA), and methacrylic acid (MA) but also vinylsilane can be incorporated into the polymer backbone to give a moisture curable material.

2.2.1.1 Manufacturing Processes

Polyethylene was discovered by R. Gibson and E. Fawcett at the Imperial Chemical Industries (ICI) when they found deposits in an autoclave in which they heated ethylene under high pressure [6].

When ethylene is heated under very high pressures in the presence of atmospheric oxygen or peroxide initiators, the double bonds in an ethylene molecule will open, giving radicals that react with each other to form new covalent C—C bonds. In such conditions many radicals are formed, not only of ethylene monomers but also of growing polymer chains so the polymer made in this way contains many long chain branches. The density of such a polyethylene is low (typically around $0.92\,g/cm^3$) and is called Low Density Polyethylene (LDPE). In the high pressure process the density of the polyethylene could go as high as 0.930; such polymer would be called a Medium Density Polyethylene.

In the late 1950s Ziegler and Natta developed independently from each other a catalyst system, based on titanium to produce isotactic polypropylene [7]. The same catalyst can also be used to make linear polyethylene at low pressure and temperatures.

Ziegler–Natta catalysts can be used in a variety of manufacturing processes, such as slurry autoclave, slurry loop reactors, the solution process, the gas-phase process, or combinations thereof such as the Borstar process of Borealis, which consists of a slurry loop reactor in series with a gas-phase reactor [8]. An overview of polyethylene suppliers and manufacturing processes can be found in [3, 9].

Besides catalyst systems based on titanium, also catalyst systems based on chromium as the coordination catalyst or various metallocene catalysts can be used in low pressure processes. The chromium oxide catalysts were developed by Phillips Petroleum [10].

Polyethylene can be modified by grafting other molecules on the polymer backbone, such as vinyl silane (Sioplas® process) or maleic acid to create polymers, which are curable or could be used as compatibilizers, for instance, in HFFR formulations.

2.2.1.2 Cable Applications

Polyethylene is a polymer that is widely used in various applications such as flexible and rigid packaging, pressure pipes, and cables; the application with the highest volume being flexible packaging. It is chemically relatively inert and has good low temperature properties.

2.2.1.2.1 Cable Insulation The electrical properties of polyethylene are excellent [11], as mentioned before. The dissipation or loss factor of a virgin polyethylene resin depends upon residues from the manufacturing process such as catalysts or chemical impurities.

Soon after its discovery LDPE was used for cable insulation and cable jackets. Before the start of World War II, polyethylene was used for submarine cable insulation; with the start of the hostilities nearly all polyethylene was used as an electrical insulator in radar applications [7]. It became the insulation material of choice after World War II [12, 13]. It replaced gutta-percha and paper insulated cables; at the initial stage a small amount of polyisobutylene was added to polyethylene to improve processing and low temperature properties. With improvements in the manufacturing process, higher molecular weight grades could be polymerized and an additional compounding step could be avoided [9].

One of the first commercial applications as an insulation material was in a submarine coaxial cable across the English Channel in 1947. In this cable the conductor was kept in place by a solid spiral tube, so it actually was a partially air insulated cable. In the 1950s polymer insulated cables (PICs) appeared and soon became the dominant telecom cables, once the price of polyethylene decreased rapidly.

Thermoplastic polyethylene, for most applications this means HDPE, is used for solid and cellular insulation of copper telecom cables. HDPE has the highest stiffness and the lowest absorption of filling compounds, but also has the lowest adhesion to the conductor (copper). This low adhesion is caused by the high crystallization rate of HDPE; in extreme causes the insulation can even shrink away from the conductor. The adhesion to the conductor can be improved by using conductor pre-heating; conductor pre-heating also avoids a shock cooling of the molten insulation, which results in very low elongations at break.

The adhesion to copper can be improved by adding special additives or polar copolymers (EEA, EAA, terpolymers); the addition of polar copolymers might have some minor effect on the dissipation factor. A compromise between stiffness, gel absorption, and a certain level of adhesion is offered by MDPE insulation compounds.

Important for an insulation grade is the processability of the material and to avoid melt fracture, causing a rough surface which might create dust during the twisting operation. The processability is related to the rheology of the insulation material. Unimodal Cr-catalyzed HDPEs have a broad MWD [3] and are being used in this application for many years with success. In the past decade bimodal HDPE grades have been offered as telecom insulation material [14, 15].

Other requirements for telecom insulation materials are the absence of particles, which could create small holes in the insulation, leading to spark failures [16] and chemical cleanliness to maintain a low dissipation factor [17]. The same is true for power cable insulation where particles can act as stress concentrators and cause electrical breakdowns and a high dissipation factor will lead to warming up of the cable, which can result in thermal runaway and electrical breakdowns.

One of the first applications was underground telephone cables. To avoid water ingress the interstices of the cable are filled with a filling compound, usually a petroleum jelly [18, 19].

Low density polyethylene and LLDPE insulation materials can be used as flexible telecom insulation. Sometimes these are used for solid insulation but more often these materials are used in the inner skin layer of skin/foam/skin structures, providing adhesion to the conductor.

Cables have life times, which often span more than one decade. In order to guarantee the functionality of the insulation (and jacket) over the lifetime of a telecom cable, the right type and amount of antioxidants need to be added [20]. Several aging tests can be performed. In Europe telecom insulations must pass the requirements of EN 50290-2-23 [21]. Standard aging requirements for telecom insulation are 10 days at 100 °C. However there is also a long term heat aging test, which lasts for 42 days at 100 °C. As it is often impossible to separate the insulation from the conductor after the oven aging test, a wrapping test can be performed instead of a tensile test, especially after 42 days of oven aging. Another option is to perform aging tests at various temperatures in the aging oven and measuring at each temperature the time to failure. With the help of the Arrhenius equation a lifetime can be predicted.

Polyethylene requires the addition of antioxidants to be resistant against oxidative degradation. Several antioxidant systems are offered by suppliers like BASF, Clariant, Addivant, Songwon, and Adeka [22–26].

For telecom insulation applications, the addition of a metal deactivator [27] is needed. When polyethylene is in contact with copper, copper induces degradation of the polymer [28, 29]. Metal deactivators will form complexes with copper ions and prohibit the auto-degradation of the polymer. Probably the best known metal deactivator, originally developed by Ciba-Geigy, is BASF Irganox™ MD 1024 [30].

Often the Oxidation Induction Time (OIT) measurement is used as an indication of the long-term performance of polyolefins [31, 32]; OIT data should however be regarded with care: they relate well with the antioxidant loading for a given stabilization system, but they should not be used to compare different stabilization systems [33].

Copper telephone cables in North America and also in certain other geographies need to pass more stringent heat aging requirements than in other geographies, as defined in cable standards such as Telcordia GR 421 CORE. The aging test in this specification was set up after the experience that in pedestals the temperature can raise to high values and antioxidants could be extracted from the insulation by the filling compound. This results in premature degradation and cracking of the insulation [32, 34]. To avoid such accelerated degradation minimum requirements for the stabilization system of telephone wires were developed [35–42]. Insulation grades with stabilizer systems designed to pass these stringent requirements are available from various suppliers of the cable industry [43–50].

Solid HDPE insulation is predominantly used in copper telecom cables and lower category twisted pair data cables.

Although polyethylene in principle has excellent electrical properties such as a low dissipation factor, residues from the manufacturing process, mainly from the polymerization catalyst, as well as the type and amount of stabilizers added to protect the polymer will negatively affect the dissipation factor. The best insulator is air, so it is no surprise that if the polymer in an insulation is (partly) replaced by air, the electrical properties improve; both the dielectric constant and the dissipation factor decrease [51, 52].

One way of producing a foamed insulation is to incorporate a chemical foaming agent into the polymer. The most commonly used chemical foaming agent in cables is azodicarbonamide (ADCA) [53–55]. This chemical substance gives a high gas yield and has an ideal decomposition temperature of 200 °C; the decomposition temperature can be lowered by adding certain coagents. The decomposition temperature of ADCA falls in a temperature range that allows the incorporation into HDPE or MDPE during the compounding process without premature foaming but at the same time allows the processing of the insulation at low enough melt temperatures. LDPE cellular insulation, used for flexible coaxial cables, can use other foaming agents, as LDPE is normally processed at lower melt temperatures than HDPE.

The major challenge for the use of ADCA as a foaming agent was the incorporation of the substance on the List of Substances of Very High Concern, under the REACH legislation in the European Union [56], because the substance was identified as a respiratory sensitizer, although some sources dispute the decision of the European Chemicals Agency to classify ADCA as such [57]. The incorporation in the List of Substances of Very High Concern is a first step in a process that can lead to the incorporation of this substance in the List for Authorization, which means that a dossier with relevant data needs to be submitted to the European Chemicals Agency to authorize its further use or which will stop its use after a sunset date if no authorization is obtained. Potential alternatives could be endothermic foaming systems [58].

With chemical foaming degrees, which is the ratio of the foam density divided by the density of the solid polymer × 100, of 35% up to maximum 50% in most ideal conditions, can be achieved. The rheology of the resin is important to avoid cell coalescence [59].

The insulation structure is either a foam-only layer or a skin/foam coextruded structure. The outer solid skin layer provides a smooth surface, mechanical properties and can be colored for wire identification, without having a significant effect on insulation properties [59, 60]. Chemically foamed HDPE insulations are used in classical twisted pair telephone cables, railway telecom cables as well as in data cables (mostly category 6 cables). The thermal stability of HDPE cellular insulation decreases by 35% compared with solid insulation; this is probably due to the larger surface to volume ratio of the polymer and the fast diffusion of oxygen into the cells of cellular insulation. In contact with filling compounds, used in twisted pair copper telephone cables, the thermal stability decreases further [61]. Similar to solid insulation, the higher oven aging requirements imposed by Telcordia lead to special versions of the HDPE insulation grades offered by the major material suppliers such as The Dow Chemical Company [62], LyondellBasell, and Borealis.

The chemical foaming process has two main limitations: solid decomposition products of ADCA can be formed that negatively impact the dissipation factor of the insulation and the maximum foaming rate is limited to 50% in the ideal case, as already mentioned. The use of a hydrocarbon as a liquid foaming agent was investigated in the past, but did not become a current practice [63].

The solution for these two limitations is to inject an inert gas into the polymer melt [64, 65]; either nitrogen or carbon dioxide is used. Foaming rates of up to 84% can be achieved with this technology. In order to get a fine, uniform cell structure, a three-component system is being used [66–68]:

- An HDPE base resin which provides foamability and foam stiffness. Foamability means that the zero-shear rate viscosity should be kept low [69].
- An LDPE component that provides the necessary melt strength for the foaming process and prevents the coalescence of growing cells [70–74].
- A nucleator masterbatch. The nucleator system used can be exothermic, endothermic, or passive. The type of nucleator used will strongly determine the cell structure of the foam [75–77].

In order to get a proper foam structure, the elongational viscosity should show an increase (also called: strain hardening) [78], as shown in Figure 2.6.

The proper composition should be chosen for an application to provide the optimal balance between cell size and electrical properties [79–81].

Physically foamed insulation is either an inner skin/foam or a skin/foam/skin structure. Such structures are used for coaxial cable as well as for category 6A to category 8 data cables [82–84].

2.2.1.2.2 Cable Jackets Polyethylene is also widely used for cable jackets. For a large part black grades are used for cable jackets. Carbon blacks with a well-defined particle size below 20 μm [85] offer a superior UV resistance at

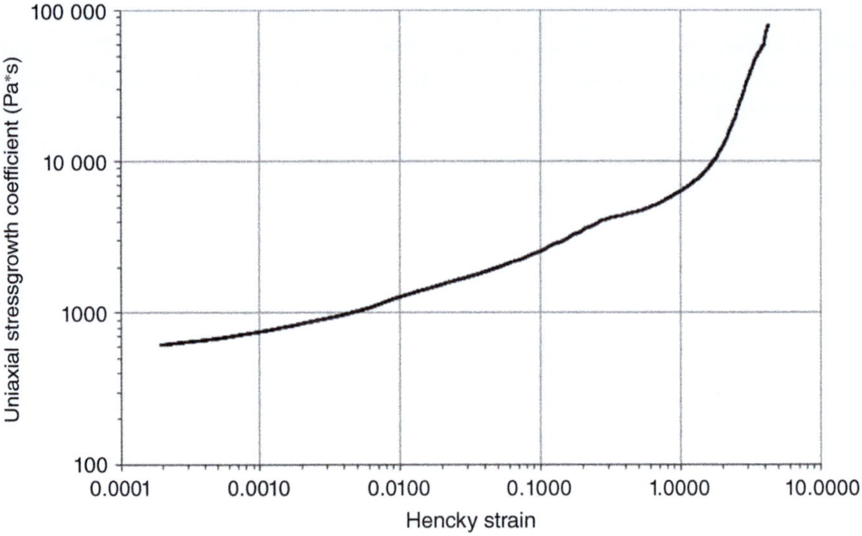

Figure 2.6 Example of the elongational viscosity curve of a typical HDPE resin for physical foaming.

concentrations above 2.0 wt%. Typically 2.5–2.6 wt% of carbon black is added to protect against the effect of sunlight.

One of the work horses for cable jackets is black LLDPE [86, 87]. This material is used for power cable jackets, especially for MV cable jackets in Nordic countries (because of the flexibility, which is appreciated at low temperatures) and in various telecom cable applications such as fiber optic cables for direct buried applications, radiofrequency coaxial cables, and copper telephone cables. Especially in the latter application LLDPE jackets replaced LDPE jackets; they offer more flexibility than MDPE or HDPE, a good processability, and low shrinkage, but also lower temperature resistance and lower mechanical properties. LLDPE jacket compounds exhibit high ESCR failure times but LDPE itself has a poor ESCR behaviour. High ESCR performance is achieved by modifying the LDPE with, for instance, polar copolymers (high vinyl acetate EVA), which form a rubbery phase that stops the growing cracks or by blending with LLDPE [88–90]. Very old jacket extruders were designed to run LDPE jackets; they can have problems when trying to extrude LLDPE at high line speeds.

If more stiffness and temperature resistance is needed, MDPE jackets were traditionally favored. MDPE jackets are used for medium voltage cables and fiber optic cables. For power cables, depending on the geography and specifications, either LLDPE, MDPE, or HDPE jackets are used [91].

Traditional unimodal HDPE have a limited share in jacket applications. Because of the high density, less comonomer is incorporated and less tie molecules are formed, which results in a moderate ESCR performance. A potential solution to achieve higher ESCR values is to use a higher molecular weight grade (lower melt index) [92]; that would however have an impact on processability and jacket shrinkage [93, 94]. This dilemma was resolved by using bimodal HDPE grades [95–98]. These grades offer more design flexibilities. The

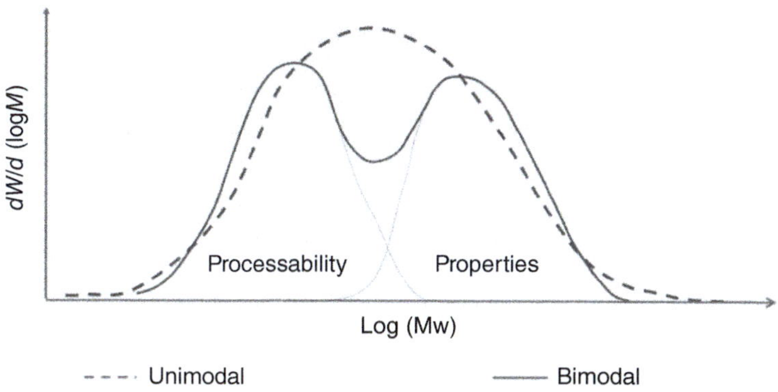

Figure 2.7 Comparison of typical MWD curves for uni- and bimodal polyethylene.

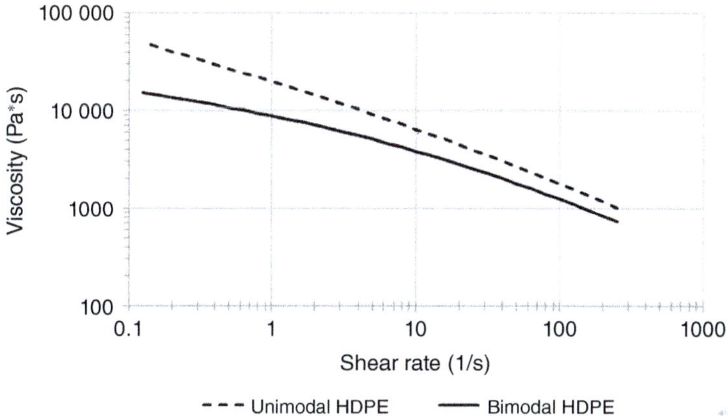

Figure 2.8 Viscosity curves at 190 °C of a typical unimodal and bimodal jacket grade.

comonomer can be incorporated in the high MW fraction (Figure 2.7), thus maximizing the formation of tie molecules [99, 100]. The low MW fraction, which principally acts as a lubricant, can be made with a high density, which allows to lower the density of the high MW fraction for a given overall resin density. This bimodal design allows to use a lower molecular weight to be used in comparison with unimodal grades, giving a lower viscosity (Figure 2.8).

Another advantage of a bimodal HDPE is that the high molecular weight tail can be reduced, which results in a faster relaxation and lower shrinkage [101]. That is especially important for fiber optic mini- and micro-cables [102–104], which are used for access and metropolitan networks [105, 106]. The jacket of these cables, with a small diameter and thin jackets, is extruded at high shear rates, which results in higher shear stresses in the die. For this particular application, a special HDPE jacket grade is used, reducing the shear stresses in the die and having shorter relaxation times [107]. HDPE doesn't only give these cables the needed stiffness but also a lower coefficient of friction for optimal jetting performance.

High density polyethylene jackets are also used for power cables, for medium voltage cables in Europe and high voltage (HV)/extra high voltage (EHV) cables in general. These cables develop heat when in operation (the maximum conductor temperature is 90 °C in normal operation), and to make sure that power cable jackets do not deform, the pressure test at high temperature is one of the properties specified for these applications. This test is specific for cable applications and measures the temperature resistance of a jacket or insulation material. Whereas other tests like the Vicat softening point (test method ASTM D1525 or ISO 306) or the Heat Distortion Test (HDT, test method ASTM D648, or ISO 75) measure the temperature at which a certain deformation is reached, the pressure test at high temperature, as specified in the international standard IEC 60811-508, measures the deformation at a set temperature after a specified time and with a specified force that depends on the thickness and outer diameter of the jacket or insulation [108]. LLDPE jackets can pass the pressure test at 110 °C, MDPE and HDPE jackets are recommended if the pressure test at 115 °C needs to be passed [109].

Metallocene technology can also be used to manufacture polyethylene jacket grades with improved performance [110].

Another advantage of HDPE jackets is the higher hardness of the material. This does not only offer more abrasion resistance but will also offer protection against less aggressive termites, such as the European species [111].

2.2.2 Polypropylene (PP)

The development of Ziegler–Natta catalysts made the production of isotactic polypropylene possible.

Polypropylene homopolymer is stiffer than HDPE and has a higher melting point (160 °C), which makes it better suited for higher temperature applications. Polypropylene becomes less suited when used at low temperatures, below −20 °C. Homopolymer polypropylene becomes brittle at temperatures around 0 °C (Figure 2.9 and Table 2.2).

In order to improve the impact and low temperature performance impact, heterophasic copolymers are used [112].

Either an elastomeric phase is introduced during the polymerization process or an elastomer can be incorporated in a post-extrusion compounding step. The temperature resistance is not significantly affected by the incorporation of an elastomeric phase. A third type of polypropylene is a random copolymer PP, in which ethylene comonomer is incorporated into the PP backbone. Random copolymer PP will have a lower stiffness and a lower melting point as more ethylene is incorporated.

Polypropylene can be foamed with chemical foaming agents such as ADCA [113]. Standard PP has a low melt strength; for excellent foaming performance a high melt strength PP is preferred, either a modified PP or a bimodal PP.

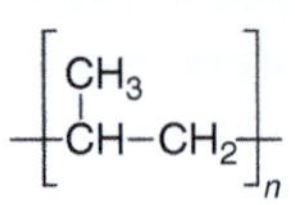

Figure 2.9 Monomer units in polypropylene.

Table 2.2 Some characteristic properties of polypropylene [223, 224].

Property	Homopolymer	Impact copolymer	Random copolymer
Density	0.904–0.908	0.904–0.908	0.898–0.900
Melting point (°C)	160–165	160–165	135–159
Flexural modulus (MPa)	1400–1800	900–1200	900–1300
Brittleness temperature (°C)	5–15	40–10	−10 to 15
Dielectric constant	2.1–2.3		2.1–2.3
Dissipation factor at 1 MHz	<0.0005	<0.0005	<0.0005

Sources: Refs. [223, 224].

2.2.2.1 Manufacturing Processes

Polypropylene can be produced in a gas-phase reactor (for instance, the UNIPOL process) [114], a dual reactor process (Borstar process), or the Montell Spheripol process [115].

Standard polypropylene grades have a low melt strength, making them less suited for applications requiring high melt strengths like cellular insulation. The melt strength of polypropylene can be improved by using bimodal PP or modifying polypropylene by coupling reactions.

When polypropylene degrades, it undergoes chain scission, which decreases the average molecular weight. This characteristic property of polypropylene is used to do visc-breaking; in this way the melt index of the polymer is increased by peroxide-induced chain scission reactions.

Polypropylene has the tendency to degrade more readily than polyethylene when exposed to high temperatures; this means that the stabilization needs for PP are higher compared with typical stabilization packages for polyethylene.

2.2.2.2 Applications

In cable industry polypropylene is used in four applications: as insulation of twisted copper wires, in particular data cables, in buffer tubes in fiber optic cables, as a HFFR automotive cable with a rating of 125 °C [116, 117], and as thermoplastic insulation of power cables.

The low dielectric constant of polypropylene, the higher stiffness and improved temperature resistance versus HDPE offers certain advantages. Polypropylene is mainly used for solid insulation of telecom wires and twisted pair data cables [118–120]. Cellular insulation made of polypropylene is less common, first of all because of the lower melt strength of standard PP [121], which makes getting a fine cell size more challenging. The melt strength of PP can be increased by using bimodal PP or by introducing long chain branches [122, 123]. An additional challenge for chemical foaming of PP is that the common foaming agents used in the cable industry like ADCA or endothermic foaming systems have a decomposition temperature of 200 °C or lower. Since polypropylene melts at 160 °C, the melt temperatures in the compounding extruder are typically as high if not higher than the decomposition temperature of the foaming agent.

Despite these challenges, occasionally cellular PP-based insulation is used [124]. The main advantage is the higher crush resistance of the PP cellular insulation compared with cellular HDPE insulation.

Impact copolymer PP is also used in fiber optic cables [125–131], especially in North America and Asia; in other geographies PP has been used in central tube cable designs. PP offers a few advantages over PBT such as the lower price and more flexible buffer tubes, which gives improved ease of installation. One challenge that PP has is its compatibility with filling compounds in loose tube and central tube applications. Standard gels, which are based on mineral oils, perform well in terms of compatibility with PBT, thanks to the polar nature. In combination with PP however a higher absorption occurs and the buffer tubes swell and become softer. In order to avoid this higher absorption more expensive special filling compounds have to be used. Flooding compounds are not generally used in fiber optic cables anymore (their role has been taken over by the use of water swellable tapes) but especially European utilities still prefer filled buffer tubes. The evolution of fiber optic cables in the metropolitan and subscriber access network to lightweight, downgauged cables, which can be jetted into microducts, requires the use of stiffer materials for buffer tubes such as PBT or PBT/polycarbonate (PC) coex structures. This explains why PP based buffer tubes are not common in Europe. In North America fully dry design fiber optic cables are widely accepted and PP has a larger share in buffer tube applications. The role of the filling compound as water blocking element is taken over by incorporating water swellable yarns in the buffer tube.

During extrusion of PP buffer tubes, controlling the excess fiber length (EFL) or shrinkage of the tube is more difficult than, for instance, for PBT [132]. Extrusion lines for PP buffer tubes must therefore have a clinching or clenching unit to reduce EFL and shrinkage.

A new application, which has been developed in the last decade, is the use of polypropylene-based thermoplastic insulation for medium voltage cables. The first company to introduce a thermoplastic, polypropylene-based thermoplastic insulation, originally targeted for medium-voltage applications, was Prysmian with their P-laser cables [133]. One advantage of thermoplastic insulation is its recyclability. Polypropylene-based insulation can be used for conductor temperatures of 90 °C, just like XLPE. Another advantage of thermoplastic insulation is the absence of peroxide decomposition products. This means that no degassing step is required, which means that a complete medium voltage cable can be produced in a single process, the one-step process, in which the insulation system is applied over the conductor, neutral wires can be applied and the jacket extruded. This could give cost advantages as the cable manufacturing process is significantly reduced and less energy is consumed, for instance, for heating the degassing room, which is required for HV and EHV AC cables as well as for high voltage direct current (HVDC) cables [134, 135]. Degassing is required to remove methane, which is one of the decomposition products coming from the peroxides like dicumyl-peroxide used to crosslink polyethylene; its presence could create a safety hazard because methane migrates through the cable and can accumulates in manholes. The absence of other peroxide decomposition products like acetophenone and cumylalcohol, which increase the electrical losses in AC insulation and which act a space charge carriers in DC cables, could be an advantage in such applications.

Although initially commercialized by Prysmian for medium voltage cables, the P-laser technology is now being pushed into HV applications. Nexans recently commercialized a PP-based thermoplastic insulation for their POWERBOOST cable and more cable makers have been working on a thermoplastic MV insulation material based on PP.

In the Suscable project, a cooperation between National Grid, University of Southampton, Gnosys, and Dow Europe, a compound composition and manufacturing process was developed to produce a PE-based [136] and a PP-based thermoplastic insulation with improved electrical breakdown strength compared with standard XLPE insulation, while still maintaining a reasonable flexibility. Critical is the choice of a combination of PP homopolymer and elastomer, which is then cooled at a rate within a certain range to create a microstructure consisting of fine spherulites [137–141].

2.3 Chlorinated Polymers

2.3.1 Polyvinylchloride (PVC)

PVC is predominantly manufactured by the suspension polymerization process, in which the vinylchloride monomer (VCM) is introduced under pressure into a sealed reactor where it is finely dispersed in water; an initiator is added and the mixture heated to 60–70 °C for the polymerization to start (Figure 2.10). When the conversion reaches 90%, the reaction is stopped by discharging the slurry into a degasser. The mass polymerization process does not use water. VCM is pre-polymerized in a first reactor giving 10% conversion, the solution is then transferred into an autoclave where more VCM and initiator are added and the mixture heated [142]).

The PVC resin is inherent stiff and brittle. A plasticizer is added to make PVC more flexible and the addition of impact modifiers results in a tough and impact resistant PVC. Traditionally phthalate plasticizers are used, but, for instance, trimellitate (for high-temperature applications), phosphates, and sebacate (for low temperature applications) plasticizers [143], and various bio-based plasticizers [144–148] are offered as alternatives. These bio-based plasticizers are typically based on organic oils, which are sometimes epoxidized like epoxidized soybean oil (ESO), epoxidized linseed oil. The use of bioplasticizers might require an optimized thermal stabilization package [149].

Lead-based stabilizers were used for flexible PVC in the past but due to concerns about environmental impact and toxicity of these stabilizers, alternatives are used for quite some years now [150]

Flexible PVC compounds are used both for cable insulation [151] and cable jackets; its first use in cable insulation took place during the World War II, especially in the US Navy [152] (9-2); at the end of the war the first LV underground cable with PVC insulation was put into service in Germany [153]. PVC cable

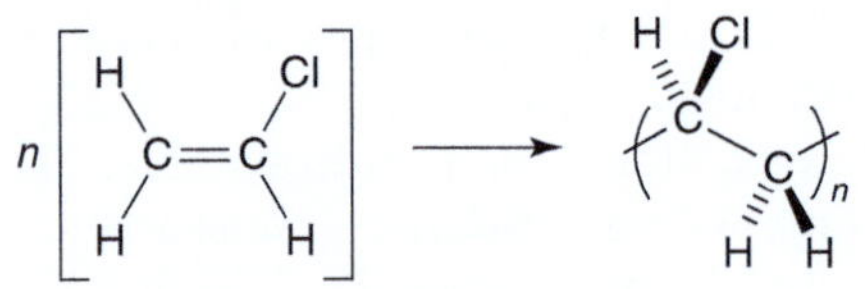

Figure 2.10 Polymerization from vinylchloride to PVC. Sources: Refs. [223, 224].

compounds generally have a low cost and process easily; PVC also offers a certain level of UV resistance.

PVC by itself burns less easily than polyethylene, but to make it flame retardant (FR), FR additives, typically antimony oxide or metal hydroxides, need to be added [154]. When PVC burns, quite some smoke is generated; as smoke suppressant molybdates can be used [155, 156]. The flame retardant and smoke suppressant can show synergistic effects [157]. The addition of high amount of these stabilizers can affect the heat stability [158]. The main disadvantage of PVC is the formation of HCl when the polymer burns; HCl will react with metals in electronic equipment and affects the health of people. That is the reason why in particular in public buildings PVC is largely replaced by Low Smoke Zero Halogen compounds, also called Halogen Free Flame Retardant (HFFR) compounds, in Europe and Asia.

Currently PVC is used as building wire (low voltage) insulation and in general as jacket material for indoor telecom cables and low voltage cables. As an insulation material PVC can be used for applications with a maximum conductor temperature of 70 °C [159]; it can withstand a short circuit temperature of 160 °C for one second. This is a disadvantage over XLPE, which can be used at a maximum conductor rating of 90 °C, sometimes 105 °C. Since plasticized PVC is used for insulation and jackets, one important factor in the aging performance of PVC insulation is the loss of plasticizers [160]; other research suggest that the aging performance of PVC insulation is depending on the way it was processed, the type of colorants used, etc. [161].

The use of PVC for the insulation of MV cables has been abandoned; PVC is not such a good insulator [162, 163]. PVC can be foamed by adding a blowing agent; in this way a lightweight jacket can be produced [164]. PVC jackets are in general not well suited at low temperatures, although special versions do perform better than standard grades [165]. The fire performance can be improved by adding FR additives [166, 167]. Various mechanisms can play a role in the ignition of PVC [168].

As a jacket material for medium voltage cables it is largely replaced by polyethylene [169]. As a jacket for indoor data cables, PVC is replaced by HFFR formulations, especially in Europe. PVC is not used as an insulation material in copper telecom cables anymore, due to its poor electrical properties.

PVC is also used as tight buffering for indoor fiber optic cables, although it is also here being replaced by HFFR compounds. The implementation of the Construction Products Regulations in Europe certainly pushed the substitution of PVC by HFFR compounds.

2.3.2 Chlorinated Polyethylene (CPE)

Chlorinated polyethylene is manufactured by the reaction of chlorine with HDPE. Hydrogen atoms are replaced by chlorine atoms.

As the structure becomes more polar, CPE based jackets have improved oil resistance. It is high-temperature, flame, and chemical resistant like PVC [170]. High and low temperature performances of CPE compounds are better than those of PVC. CPE is not used in its pure form, but as part of a compound composition that includes filler, stabilizers. CPE based compounds can be used for thermoplastic jackets or as a plasticizing component in PVC formulations; in many applications CPE based jackets are cured.

Thanks to its excellent oil, chemical, UV, and heat resistance CPE-based jackets are used in cable jackets for household appliances, mine, marine, and wind-power cables [171–173].

2.4 Fluoropolymers

The typical fluoropolymers used in cable applications are fluorinated ethylene propylene (FEP) polymer, perfluoroalkoxy (PFA), ethylene tetrafluoroethylene (ETFE) polymer, ethylene chlorotrifluoroethylene (ECTFE) polymer, and in some applications polytetrafluoroethylene (PTFE) [174].

The properties of some typical fluoropolymers used in the cable industry are given in Table 2.3.

Fluoropolymers are extensively used in cable applications, especially in North America [175, 176].

2.4.1 Fluoro-Ethylene Propylene polymer

Fluoro-ethylene propylene polymer is a halogenated polymer that consists of tetrafluoroethylene and hexafluoropropylene (Figure 2.11) and is used in those applications where improved flame retardant properties are required, combined with excellent dielectric properties, heat resistance, toughness, and flexibility. The most common tradenames are Teflon® FEP from DuPont, who invented the polymer, Neoflon® FEP (Daikin), and Dyneon® FEP (Dyneon). From an electrical property point of view, PTFE, FEP, and PFA have similar dielectric constants, the dielectric strength of PFA is higher than that of FEP (see Table 2.3). FEP has a higher dissipation factor than both PTFE and PFA.

Fluorinated ethylene propylene can be used in insulation and cable jackets requiring high stress crack resistance. FEP jackets have a slightly higher coefficient of friction and lower tensile strength than PTFE or PFA. It has a high temperature resistance and is rated for service up to 200 °C, so it can be used in automotive applications [177].

Especially in North America indoor copper data cables are insulated with FEP to meet UL plenum cable requirements (CMP rating as defined in the UL 910 standard). FEP can be used as solid insulation, chemically or physically foamed insulation for premise and data cables [178–181]. These applications in which foamed FEP insulation is used don't only cover twisted pair data cables but also direct attach (twinax) data cables [182] and coaxial cables, for instance, jumper cables [182]. Foamed FEP or MFA insulated twisted pair copper cables offer a good balance of electrical, burn, and heat stability characteristics [183].

Table 2.3 Some characteristic properties of fluoropolymers [225].

Property	FEP	PTFE	ETFE	ECTFE	PFA	PVDF
Melting point (°C)	265	327	250–280	220	306	180–210
Processing temperature (°C)	360–395	370–400	300–350	240	370–395	220–225
Density	2.12–2.17	2.17	1.7	1.7	2.12–2.17	1.7–1.8
Flexural modulus (MPa)	586	496	1172	1240	586	
Dynamic coefficient of friction	0.08–0.3	0.05–0.1	0.3–0.4	0.19	0.21–0.29	0.2–0.4
Dielectric constant	2.0	2.1	2.5–2.6	2.5	2.06	6.0
Dissipation factor at 1 MHz	0.0006	<0.0001	0.007	0.03–0.002	<–0.0001	0.18
Dielectric strength (100 μm film) (V/μm)	53	18	79		80	120–140
Limiting oxygen index	95%	95%	30%	52%	95%	44%

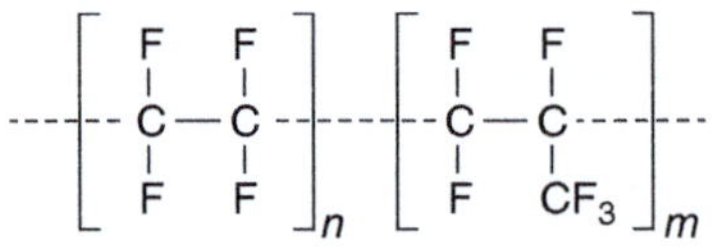

Figure 2.11 The repetitive unit in FEP.

Other applications include insulation for aerospace, signal, control, and instrument cables, exhibiting excellent dielectric properties and a low dissipation factor.

Used in jacket applications FEP exhibits a low coefficient of friction and a high level of flexibility, which results in easier installations, certainly when cables need to be pulled around corners. In addition FEP jackets have a good heat, weather, and UV resistance. A typical application is space cable buffer tube and jacket [184, 185].

2.4.2 Polytetrafluoroethylene (PTFE), Ethylene Tetrafluoroethylene (ETFE), and Perfluoroalkoxy Polymer (PFA).

Other fluoropolymers that are used in the cable industry are PTFE, ETFE, and PFA, a copolymer of tetrafluoroethylene and perfluorinated vinyl ether [186]. PFA has a lower melting point than PTFE. Both polymers are chemically inert and have a low coefficient of friction (Figures 2.12–2.14).

Ethylene tetrafluoroethylene is a copolymer of ethylene and tetrafluoroethylene and is melt processable. It might be better known under the most common tradenames Tefzel® (former DuPont, now Chemours), Fluon® (Asahi Glass Company), and Neoflon ETFE (Daikin). It is nearly as inert as PTFE and has

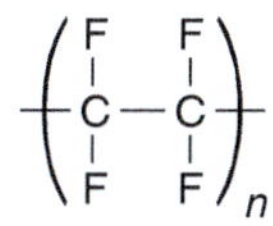

Figure 2.12 Repetitive unit in PTFE.

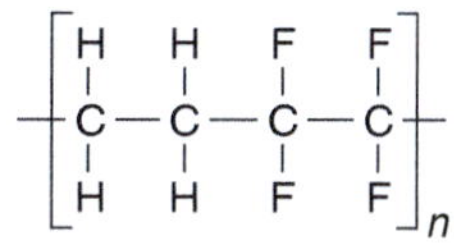

Figure 2.13 Building blocks of ETFE.

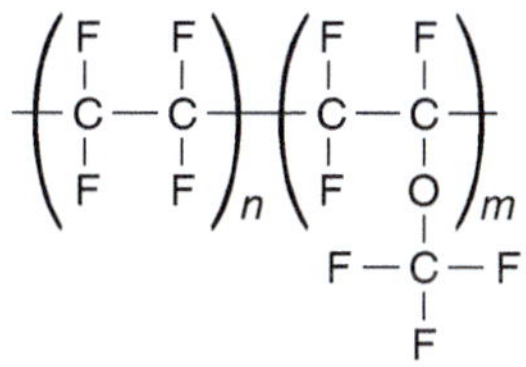

Figure 2.14 Building blocks of PFA.

high mechanical, thermal, chemical, and electrical properties. It is used as jacketing and insulating material in heater cables and automotive wiring and for other applications where short-term temperatures of 200 °C are experienced or where repeated mechanical stresses at the rated temperature of 150 °C are anticipated. Because of its high-energy radiation resistance, ETFE can be used as insulation of critical nuclear power plant applications. Because of the easier flow ETFE resins provide higher rates with less draw-down compared WITH other fluoropolymers as FEP and PFA.

Perfluoroalkoxy is better known under its various tradenames: Dyneon PFA and Hostaflon® (Dyneon), Hyflon® PFA (Solvay), Neoflon PFA (Daikin), and Teflon PFA (DuPont); DuPont first introduced PFA into the market in 1972. Like FEP, PFA can be used for high temperature applications such as thermocouple cable; both PFA and PTFE can be used for continuous service temperatures up to 260 °C; PFA also offers excellent mechanical and electrical properties. As insulation it offers a low dissipation factor and excellent dielectric properties. In cable jackets PFA offers flexibility, high abrasion and tear resistance so that cables rather slide than tear when being pulled as well as low smoke and low flame-spread properties. PFA is for instance used as a jacket material for space cables and as insulation and jacket for thermocouple cables.

To extrude PFA, the same melt processing technique as other thermoplastics can be used; the extruder must be made of corrosion resistant material, such as chrome and nickel plated metals.

Fluorinated ethylene propylene and PFA fluorocarbon resins can be used to produce cellular insulation by gas injection (physical foaming) [187]. Applications cover both coaxial cables and twisted par cables (higher category data cables). Expanded PTFE is used as insulation material for aircraft cables (ethernet cables, leaky feeder cables) [188]. Other applications for PTFE include high performing RF [189] and microwave coaxial cable insulation complying with military specs and downhole cable.

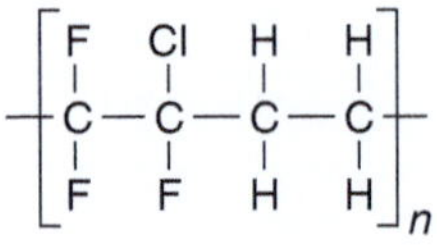

Figure 2.15 Repetitive units in ECTFE.

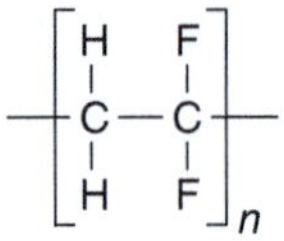

Figure 2.16 Vinylidene monomer building block for PVDF.

2.4.3 Ethylene Chlorotrifluoroethylene (ECTFE)

Ethylene chlorotrifluoroethylene (see Figure 2.15), such as Halar® from Solvay is both used in insulation and jacket applications. It is melt processable, offers flame retardancy, but the electrical and physical properties are less than those of FEP. It is predominantly used in high temperature cable [190].

Ethylene chlorotrifluoroethylene offers good resistance to UV exposure and resistance to aromatic hydrocarbons. It also has a low temperature brittleness. Typical applications include cables in fuel tanks, hook-up wires and heating cables for car seats [191], and special twisted pair data cables [192].

2.4.4 Polyvinyldifluoride (PVDF)

PVDF (see Figure 2.16) is synthesized in an emulsion polymerization process. PVDF is known under different tradenames such as Kynar® (Arkema) and Hylar® (Solvay).

PVDF is used in similar insulation and jacket applications as other fluoropolymers but offers improved processability [193]. Often a copolymer is used, which is based on HFP as the comonomer.

Its brittleness temperature is higher than that of ECTFE. Its chemical resistance and burning characteristics are excellent so the product can be used for cables intended to be installed in corrosive and extreme temperature environments. Other characteristics are a high abrasion resistance, UV resistance, inherent flame resistance, and low smoke generation. PVDF can meet the requirements of NFPA 262 (UL910), the test procedure for evaluating the flame travel and smoke of cables and wires housed in plenum and other air-handling spaces.

With this balance of properties PVDF is used in plenum cables, cathodic protection cable, and insulated down well cables. PVDF can also be foamed to produce lightweight jacket [194].

PVDF can be e-beam cured to improve cut-through resistance [195] and to comply with military specifications such as MIL 81044. Crosslinked PVDF can be used for military, aerospace, automotive, locomotive, shipboard, and industrial cables.

2.5 Polyamide (PA)

Polyamide can be manufactured via two routes: the synthesis via ring opening of cyclic monomers such as caprolactam followed by polycondensation or by a

Figure 2.17 Polyamide 12 building blocks.

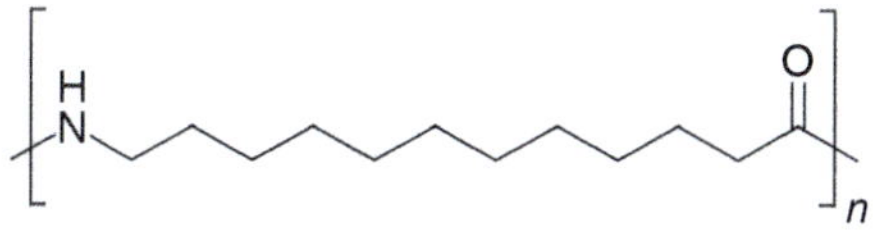

condensation reaction of one or two different monomers (diamines and diacids) such as polyhexamethylene diapide and dicarboxylic acid [196–198]. Aliphatic polyamide or Nylon, the name given by DuPont, exists in various versions such as PA6, PA66, and PA12 depending on the number of CH_2 groups in the repeating unit: polyamide XY, in which X refers to the number of carbon atoms in the diamine monomer and Y refers to the number of carbon atoms in the diacid monomer and polyamide Z, in which Z refers to the number of carbon atoms in the monomer [199] (Figure 2.17).

Yarns made of aromatic polyamides, also called aramids, such as Kevlar or Twaron, are used as strength members in fiber optic cables.

For cable applications polyamide 12 is mostly used. Polyamide 12 has the highest hardness (Shore D of 70), which makes the material very suited for termite-resistant jackets [200], and stiffness. Other typical applications for which polyamide jackets are used are THHN and THWN type cables in the North-American market [201, 202]. Polyamide jackets can also be used for cables within nuclear power plants [203].

Another advantage of polyamide is the low coefficient of friction, which can further be improved by adding a silicone containing masterbatch [202]; for this reason it is sometimes used for fiber optic cables, which are jetted through microducts [204].

Polyamide 12 can also be used to produce central tubes [204] and tight and loose buffer tubes in fiber optic cables [205, 206].

2.6 Polyesters

2.6.1 Polybutylphtherephtalate (PBT)

Polybutylphtherephtalate (Figure 2.18) is a stiff polymer which is predominantly used for loose buffer tubes in fiber optic cables [207]. PBT combines a high stiffness, needed to give the buffer tube the crush and impact resistance needed to protect the optical fibers inside, with a low thermal expansion and a compatibility with a variety of filling compounds. The shrinkage of the buffer tube or excess fiber length can easier be controlled for PBT than for PP. Because of these properties PBT is the preferred material for buffer tubes in Europe, where the buffer tubes still contain a filling compound as a water and gas barrier.

Coextruded structures of PBT and PC can offer a larger stiffness than a monolayer of PBT [208].

2.6.2 Polyester Elastomers

Thermoplastic polyester elastomers (TPEEs) like HYTREL from DuPont, are used in cable applications.

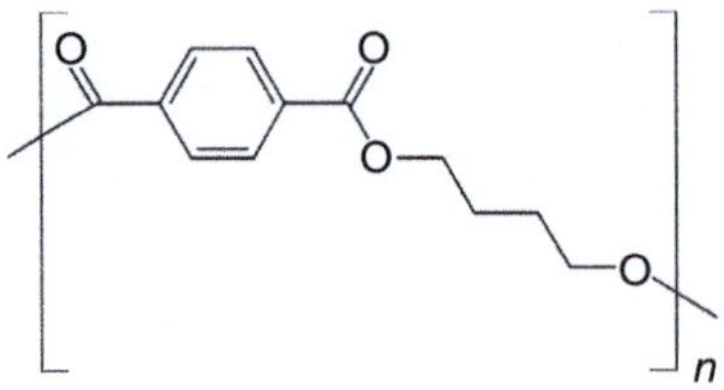

Figure 2.18 Building blocks of PBT.

They can be used in those applications requiring high modulus and high strengths, in addition to heat, abrasion, and cut growth resistance. Because of the balance in properties such as creep resistance, impact and fatigue resistance, high impact strength, and good low temperature flexibility, with a continuous service temperature range from −50 to 125 °C, resistance to chemicals, oil, solvents and weathering, high tensile strength, and high abrasion resistance TPEE can be used for specialty cable jackets such as ship and boat, mining, military, and aerospace applications. TPEE are not flame retardant by themselves, but can be made flame retardant by adding the right FR additives such as a combination of organo-phosphinate and melamine-polyphosphate containing flame retardants [209].

Another important application is the insulation of tinsel telephone wire, flexible cables, and coil cords, offering excellent flex-fatigue, and creep resistance, coupled with superior recovery from mechanical deformation and stability to copper.

Other applications include tight and loose buffer tubes in fiber optic cables. One typical application is the use in micromodule fiber optic cables. Very flexible, finger strippable buffer tubes or micromodules provide superior bending characteristics in combination with bend-resistant fibers [210–214], without the kinking seen with PBT buffer tubes.

2.7 Thermoplastic Polyurethane

TPU is predominantly used in flexible, abrasion, and scratch-resistant cable jackets [215–218]. TPU also has outstanding mechanical properties and good chemical and oil resistance and excellent low temperature properties. Because of these properties TPU jacket are, for instance, used for mining and oil ridge cable jackets, automotive, and industrial cables. For industrial applications halogenated or non-halogenated FR additives are used; the trend to e-mobility pushed the developments of HFFR TPU for automotive cables [219].

Polyester based TPU is sensitive to hydrolysis; polyether based TPU's are resistant to hydrolysis and can be used for automotive cables such as ABS, sensor, and control and battery cables [220]. In this sector there is a need for TPU jackets giving improved long-term heat aging and hydrolytic stability, which can be used close to gearboxes and the car's engine [221].

Because of the excellent wear resistance TPU jacketed data cables are used for harsh environments [222].

References

1 Mark, H. and Krooschwitz, J. (1985). *Encyclopedia of Polymer Science and Engineering*, vol. 6. Wiley.
2 Murphy, E. and Morgan, S. (1937). The dielectric properties of insulating materials. *Bell System Technical Journal* 16 (4): 493–512.
3 Ronca, S. (2017). *Polyethylene. Brydson's Plastic Materials*, 8e. Elsevier.
4 ASTM D1248-16. (2016). *Plastics – designation system and basis for specifications*. American Society for Testing and Materials.
5 ASTM D1248-16. (2016). *Standard specification for polyethylene plastics extrusion materials for wire and cable*. American Society for Testing and Materials.
6 Whiteley, K. (2012). Polyethylene. In: *Ulman Encyclopedia of Industrial Chemistry*. Wiley-VCH.
7 Peacock, A. (2000). *Handbook of Polyethylene. Structure, Properties and Applications*. Marcel Dekker.
8 Albertsson, A.-C. (2004). *Long Term Properties of Polyolefins*, Advances in Polymer Science, vol. 169. Springer-Verlag.
9 Vasile, C. and Pascu, M. (2005). *Practical Guide to Polyethylene*. RAPRA.
10 Malpass, D. (2010). *Introduction to Industrial Polyethylene: Properties, Catalysts, and Processes*. Shrivener Publishing LLC, Wiley.
11 Underwood, J. (1948). *Television Networks*, 11–32. The Michigan Technological (April 1948).
12 Windeler, A.S. (1959). Polyethylene insulated telephone cable. *Bell System Technical Journal* 32 (5): 1245–1256.
13 Bartnikas, R. and Srivastava, K. (2000). *Power and Communication Cables*. IEEE Press, Wiley Interscience.
14 Rogestedt, L., Martinsson H.-B., Thorn, L., Dammert, R. (1999). Insulating composition for communication cables. US Patent Office 6 180 721 B1, filed 3 March 1999 and issued 30 January 2001.
15 Buryal, A., Kumar, A., and Ali Ahmed, I. (2017). Multimodal polyethylene. US Patent Office US 2017/0349685 A1, filed 29 December 2015 and issued 7 December 2017.
16 Silverman, R. and Roberts, D. (1994). *Troubleshooting the Foam/Skin Insulation Extrusion Process*. LyondellBasell.
17 Schmidt, G. and Bopp, L. (1980). Identification of particles in high density polyethylene that cause spark failures during high speed extrusion. In: *Proceedings 21th IWCS*, 331–349.
18 Jachimowicz, L., Olszewski, J., and Kolodny, I. (1973). Transmission properties of filled thermoplastic insulated and jacketed telephone cables at voice and carrier frequencies. *IEEE Transactions on Communications* 21 (3): 203–209.
19 Mitchell, D. and Sabia, R. (1980). Development, characterization and performance of an improved cable filling compound. In: *Proceedings 29th IWCS*, 15–25.

20 Chanda, M. (2018). *Plastics Technology Handbook*. CRC Press, Taylor & Francis Group.

21 European Committe for Electrotechnical Standardization (2013). *EN 50290-2-23 Communication cables, Part 2-23: common design rules and construction – PE insulation*. CENELEC.

22 Ciba-Geigy (1988). Processing and long-term heat stabilization of polyolefins. Publication No. 28622/e.

23 Al-Malaika, S. (2004). Perspectives in stabilization od polyolefins. *Advances in Polymer Science* 169: 121–150.

24 Bowmer, T., Miner, R., Plitz, I. et al. (1990). Thermal stability tests for polyolefin insulations. In: *Proceedings 39th IWCS*, 316–325.

25 Tolinsky, M. (2015). *Additives for Polyolefins*, 2e. Elsevier.

26 Pearson, K. and Lee, R. (2003). Polypropylene cable stabilization. In: *Proceedings 52nd IWCS*, 555–561.

27 Hansen, R., Russell, C., De Benedictis, T. et al. (1964). Inhibition of the copper-catalysed oxidation of polypropylene. *Journal of Polymer Science Part A: General Papers* 2: 587–609.

28 Allasta, D.L. and Chan, M.G. (1976). Mechanisms of inhibition against the copper-catalyzed oxidation of polyethylene: structures and catalytic reactivities of copper-inhibitor complexes. *Journal of Polymer Science: Polymer Chemistry Edition* 14: 1857–1876.

29 Bowmer, T. (1988). Cracking of foam-skin polyethylene insulations in pedestals. In: *Proceedings 37th IWCS*, 475–483.

30 Zweifel, H. (1998). *Stabilization of Polymeric Materials*. Berlin, Heidelberg: Springer-Verlag. ISBN: 978-3-642-80307-9.

31 Howard, J. (1973). DTA for control of stability in polyolefin wire and cable compounds. *Polymer Engineering and Science* 13 (6): 429–434.

32 Howard, J. and Gilroy, H. (1975). Some observations on the long term behavior of stabilized polyethylene. *Polymer Engineering and Science* 15 (4): 268–271.

33 Pauquet, J., Todesco, R., and Drake, W. (1993). Limitations and applications of oxidative induction time (OIT) to quality control of polyolefins. In: *Proceedings 42nd IWCS*, 1/9–9/9.

34 Bowmer, T., Hjorth, T., Miner, R., and Gebizlioglu, O. (1988). Stability of polyethylene insulations in the field and laboratory. In: *Proceedings 37th IWCS*, 490–498.

35 Hawkins, W. (1969). Multiconductor communications cable. US Patent Office 3 668 298, filed 10 December 1969 and issued 6 June 1972.

36 Hendrickson, L. and Connole, K. (1995). Review of stabilization of polyolefin insulated conductors. Part I: Theory and factors affecting stability. *Polymer Engineering and Science* 35 (2): 211–217.

37 Hendrickson, L. and Connole, K. (1995). Review of stabilization of polyolefin insulated conductors. Part II: Characteristics of good stabilizers and methods of evaluation. *Polymer Engineering and Science* 35 (2): 218–221.

38 Lee, R., Doumen, C., Keck-Antoine, K., and Johnson, B. (2002). New concepts in the stabilization of polyolefin compounds for wire and cable applications. *Plastics, Additives and Compounding (Dec. 2002)*: 30–32.

39 Neri, C., Riva, R., Lee, R. et al. (2001). Stabilization of polyolefin compounds for wire & cable applications. In: *American Chemical Society – Rubber Division, Wire and Cable Conference*, Paper 41. Providence, RI.

40 Fech, J. and De Witt, A. (1980). Copper inhibitor and antioxidant levels in high density polyethylene: their measurement by high performance liquid chromatography and relation to insulation life. In: *Proceedings 29th IWCS*, 327–330.

41 Brown, G. (1987). Performance of HDPE insulation antioxidants in filled telephone cable applications. In: *Proceedings 36th IWCS*, 337–343.

42 Brown, G., Cogen, J., and Keogh, M. (1996). Evaluation of the enhanced thermo-oxidative aging performance of an improved telephone wire insulation compound. In: *Proceedings 45th IWCS*, 167–176.

43 Keogh, M., Brown, G., and Cogen, J. (1995). Telephone cable. US Patent 4 453 322, filed 3 June 1994 and issued 26 Sep 1995.

44 Cogen, J., Keogh, M., and Brown, G. (1998). Telephone cables. US Patent 5 807 635, filed 24 January 1997 and issued 15 September 1998.

45 Schmutz, T., Krumer, E., Zweifel, H., Rotzinger, B. (2003). Process for the stabilization of polyolefins in permanent contact with extracting media. US Patent 6 541 547, filed 9 September 1996 and issued 1 April 2003.

46 Lee, C. and Schloemer, T. (2007). High density polyethylene and insulation compositions for wire and cable. US Patent 7 238 765 B2, filed 5 August 2004 and issued 3 July 2007.

47 Bilda, P., Hammel, R., and Warschewski, W. (1999). Stabilized polyolefin, World Intellectual Property Organization. WO 99/19397, filed 2 October 1998 and issued 22 April 1999.

48 Lupia, J. and Fay, J.; World Intellectual Property Organization (2000). Stabilized telecommunications cable insulation composition. WO 00/58975, filed 21 Feburary 2000 and issued 5 October 2000.

49 Reid, D. (1975). Stabilized polyolefin useful as wire insulation in petroleum-jelly filled cables. US Patent Office 3 864 431, filed 12 October 1973 and issued 4 February 1975.

50 Lee, C. and World Intellectual Property Organization (2002). Polyolefin insulation compositions having improved oxidative stability. WO 02/36673 A1, filed 28 September 2001 and issued 10 May 2002.

51 Windeler, A. (1953). Polyethylene insulated telephone cable. *The Bell System Technical Journal* 32: 1245–1256.

52 Marvian, N., Behravesh, A., and Golzar, M. (2008). Extrusion foam coating of coaxial cables using chemical blowing agent. *Cellular Polymers* 27 (1): 27–41.

53 Wasserman, S., Chaudhary, B., Cogen, J. et al. (2018). Wire and cable applications of polyethylene. In: *Handbook of Industrial Polyethylene and Technology* (eds. M. Spalding and A. Chatterjee), 1125–1154. Scrivener Publishing LLC.

54 Normaton, J. (1972). Expandable insulation compositions. US Patent 3 645 929, filed 22 July 1969 and issued 29 Feburuary 1972.

55 Normanton, J. (1978). Extrusion of telephone cable insulation using expandable medium density polyethylene compounds. In: *Proceedings 27th IWCS*, 182–187.

56 Suolalehto, H., Malo, J.-L., and Cullinan, P. (2018). The classification of vazodicarbonamide (ADCA) as a respiratory sensitiser; adding to the weight of evidence. *Regulatory Toxicology and Pharmacology* 94: 330–331.

57 Arts, J. and Kimber, I. (2017). Azodicarbonamide (ADCA): a reconsideration of classification as a respiratory sensitizer. *Regulatory Toxicology and Pharmacology* 89: 268–278.

58 Sadik, T., Pillon, C., Carrot, C., and Ruiz, J. (2018). Dsc studies on the decomposition of chemical blowing agents based on citric acid and sodium bicarbonate. *Thermochimica Acta* 659: 74–81.

59 Tsujikawa, A., Yashuhara, H., Suzuki, T. et al. (1980). Comparison of PEF and foam-skin cores. In: *Proceedings 29th IWCS*, 9–14.

60 Gouldson, E., Farago, M., and Baxter, G. (1978). Foam-skin, a composite expanded insulation for use in telephone cable. In: *Proceedings 27th IWCS*, 158–172.

61 Chanda, M. and Roy, S. (1989). *Plastics Technology Handbook*, 3e. New York, NY: Marcel Dekker.

62 Brown, G., Gau, Y., Leech, R., Maki, S. (1998). HDPE insulation developments for advanced copper cable systems. *Plastics in Telecommunications VIII*, London (14–16 September 1998).

63 Sato, Y. and Yoshida, Z. (1978). A process of manufacturing foamed polyethylene insulation by using low hydrocarbon as blowing agent. In: *Proceedings 27th IWCS*, 173–181.

64 Scarola, L. (1976). Reduction of coaxial cable attenuation through the use of gas injected dielectric. *Kabelitems* 149: 1–7.

65 Maki, S. (2000). Improved low loss materials for high performance coaxial cable. *Wire and Cable Technology International, March 2000*: 73–75.

66 Ishihara, K., Sakamoto, T., and Maki, S. (1998). Novel material and extrusion technology for highly expanded insulation of coaxial cables. In: *Proceedings 47th IWCS*, 307–316.

67 Abe, M., Yamazaki, T., Yokoyama, K. et al. (2003). Development of low-loss highly-foamed polyethylene coaxial cable (LHPX) for base stations for cellular phones. *Hitachi Cable Review* 22: 32–37.

68 Maki, S. (1997). High expansion insulation compounds for coaxial cables. *Kabelitems* 166: 1–9.

69 Watanabe, R., Kondo, T., and Watanabe, T. (2008). Low-loss coaxial cable with highly foamed insulator. *Fujikura Technical Review 37*: 43–49.

70 Watanabe, R., Kondo, T., and Watanabe, T. (2006). Development of low-loss foamed coaxial cable. In: *Proceedings 55th IWCS*, 172–177.

71 Kang, D. (2006). Melt strength of cable compounds and its effect on extrusion process. In: *Proceedings 55th IWCS*, 164–171.

72 Eklind, H., Martinsson, H.-B., Westling, L. et al. (1998). Gas injection compound for high frequency insulation. In: *Proceedings 47th IWCS*, 317–324.

73 Flory, A., Kmiec, C., and Zinkweg, D. (2013). A low density polyethylene gas injection compound for low signal loss radio frequency cable. In: *Proceedings 62th IWCS*, 192–206.

74 Ruinaard, H. (2005). How to choose a polyolefin grade for physical foaming. *Blowing Agents and Foaming Processes*, Stuttgart (10–11 May 2005), Rapra Technology.

75 Chen, T., Ginger, R., Leech, J., and Maki, S. (1998). Analysis of the processing of foamed polyethylene for high performance coaxial cable applications. In: *Proceedings 47th IWCS*, 325–334.

76 Leung, S., Wong, A., and Park, C. (2008). Ideal surface geometries of nucleating agents to enhance cell nucleation in polymeric foaming processes. *Journal of Applied Polymer Science* 108: 3997–4003.

77 Yang, H.-H. and Han, C. (1984). The effect of nucleating agents on the foam extrusion characteristics. *Journal of Applied Polymer Science* 29: 4465–4470.

78 Ruinaard, H. (2006). Elongational viscosity as a tool to predict the foamability of polyolefins. *Journal of Cellular Plastics* 42: 207–220.

79 Flory, A., Sun, G., Esseghir, M., and Kmiec, C. (2017). Nucleated LDPE/HDPE blends for foamed insulation. In: *Proceedings 63th IWCS Conference*, 620–624.

80 Eaton, R. and Kmiec, C. (2008). Electrical losses in coaxial cable. In: *Proceedings 57th IWCS*, 515–520.

81 Flory, A. and Kmiec, C. (2010). Electrical loss in coaxial cable. In: *Proceedings 59th IWCS*, 353–356.

82 Sun, G. (2017). Enhanced foaming polyolefin insulation for high performance telecommunication cables. In: *Proceedings 66th IWCS*, 137–143.

83 Flory, A., Sun, G., Esseghir, M., and Kmiec, C. (2016). Evaluation of foaming technology for data cable insulations. In: *Proceedings 65th IWCS*, 188–192.

84 Miller, K., Watson, A., and Kontro, A. (2016). Alternative physically foamed insulation concepts for higher category data cable. In: *Proceedings 65th IWCS*, 193–200.

85 Williams, F., Jordan, M., and Dannenberg, E. (1965). Effect of the chemical and physical properties of carbon black on its performance in polyethylene. *Journal of Applier Polymer Science* 9: 861–886.

86 Kmiec, C. and Leech, J. (1996). Advanced LLDPE compounds in cable jacketing. In: *Proceedings 24th IWCS*, 177–182.

87 Graham, G. and Szaniszlo, S. (1995). Insulating and semiconductive jackets for medium and high voltage underground power cable applications. *IEEE Electrical Insulation Magazine* 11 (5): 5–12.

88 Zhang, Y. and Tianhua, L. (1988). Improvement of environmental stress-crack resistance of LDPE cable jacket materials. *China Plastic Industry* 05: 42–46.

89 Tonglin, X. and Zhang, L. (1995). Development of LDPE compound for cable sheathing. *Plastics* 1: 48–50.

90 Borisova, B. and Kressler, J. (2003). Environmental stress-cracking resistance of LDPE/EVA blends. *Macromolecular Materials and Engineering* 288 (6): 509–515.

91 Gau, Y., Kmiec, C., and Person, T. (1999). Advanced polyethylene jacketing compounds for medium and high voltage cable applications. In: *Jicable*, Paper B10.4, 638.

92 Lee, C. and Schloemer, T. (2000). Polyethylene based black jacketing compounds: structure-compounding-performance relations. In: *Proceedings 49th IWCS*, 243–247.

93 Chen, T. and Leech, J. (1999). Design of polyethylene cable jacket compounds of superior jacketing performance. In: *Proceedings 48th IWCS*, 807–814.

94 Brunner, E. and Westling, L. (1999). Development of a new uni-modal fibre optic MDPE jacketing compound. In: *Proceedings 48th IWCS*, 824–830.

95 Dammert, R., Rogestedt, L., and Martinnsson, H.-B. (1998). New Polyethylene manufacturing process for improved cable products. *Plastics in Telecommunications VIII* (14–16 September 1998).

96 Rogestedt, L. and Martinsson, H. (1997). A new generation of jacketing compounds for wire and cable applications. In: *Proceedings 46th IWCS*, 126–131.

97 Dammert, R., Kirchner, M., Phillipczyk, C., and Wald, D. (2003). Application of fifth generation jacketing technology for improved performance. In: *Jicable*, 789–795.

98 Fajardo, W. and Watson, A. (2012). Borstar enhanced polyethylene for communications cables. In: *Proceedings 61st IWCS*, 181–203.

99 Böhm, L., Enderle, F., and Fleisner, M. (1992). High-density polyethylene pipe resins. *Advanced Materials* 4 (3): 234–238.

100 Böhm, L. (2003). The ethylene polymerization with ziegler catalysts: fifty years after the discovery. *Angewandte Chemie International Edition* 42: 5010–5030.

101 Sandelin, M., Edde, U., Martinsson, H.-B., and Dammert, R. (2001). Shrinkage of Bi- and unimodal polyethylene cable jacketing materials – influence of molecular structure and processing conditions. In: *Proceedings 50th IWCS*, 311–315.

102 Lutomski, M. and Gawdzik, B. (2013). Polymer dimensional changes in optical cables. *Annales Universitatis Mariae Curie-Sklodowska* LXVIII (1–2): 33–48.

103 de Boer, H., Grosskurth, O., Watson, A., and Robinson, J. (2010). Low shrink HDPE for the sheathing of fibre optic minicable. In: *Proceedings 59th IWCS*, 176–182.

104 Wasserman, S. (1999). Small-scale rheological tests to describe wire and cable extrusion. In: *Proceedings 48th IWCS*, 190–198.

105 LAN-2093-AEN. (2016). Smaller Cables, Bigger Possibilities. Corning.

106 AFL. (2014). MicroCore blown fiber optic cable. AFL.

107 Geussens, T., Lee, D.-C., Yu, J., Esseghir, M. (2018). Development of a new jacket material for fiber optic mini and microcables. *Proceedings 67th IWCS*.

108 International Electrotechnical Commission. (2012). Electric and optical fibre cables – test methods for non-metallic materials – Part 508: Mechanical tests – pressure test at high temperature for insulation and sheaths. Geneva.

109 Sutton, S., Geussens, T., and Bolz, K. (2010). Cable oversheath selection: the right material for the right application. In: *China International Conference on Electricity Distribution*, 1–6.

110 Kale, L., Iacciono, T., and Bow, K. (1996). *Enhanced Polyethylene Resins in Cable Jacketing Applications*, 1647–1649. ANTEC.

111 21-07, Working Group (1994). Prevention of termite attack on HV power cables. *Electra* 157: 69–79.

112 Posch, W. (2017). Polyolefins. In: *Brydson's Plastic Materials* (ed. M. Gilbert). Elsevier.

113 Maier, C. and Calafut, T. (1998). *Polypropylene, the Definite User's Guide and Databook: Plastics Design Library*, PDL Handbook Series. Elsevier.

114 Sato, H. and Ogawa, H. (2009). Review on development of polypropylene manufacturing process. *Sumitomo Kagaku* II: 1–11.

115 . *Licensed Polyolefin Technologies and Services – Spheripol*. LyondellBasell.

116 Chang, D., Kim, S., Kim, I. et al. (2007). Investigation of polypropylene composite for automotive cable insulating applications. In: *Proceedings 56th IWCS*, 323–330.

117 Beisert, C., Ruder, J., Watson, A. et al. (2017). The missing link for 125 °C automotive wires-halogen free flame retardant polypropylene solution. *Proceedings 66th IWCS*.

118 Tran, C. (1999). Polypropylene in cable applications. In: *Polypropylene: An A-Z Reference* (ed. J. Karger-Kocsis). Dordrecht: Kluwer Publishers.

119 Lee, C. and Schloemer, T. (2002). Alternative polyolefin insulation materials for communication cable applications. In: *Proceedings IWCS*, 124–128.

120 Watson, A. and Fajardo, W. (2013). Tailored insulations for high transmission rate data cable. In: *Proceedings 62th IWCS*, 530–534.

121 Mohebbi, A., Mighri, F., Ajji, A., and Rodrigue, D. (2015). Current issues and challenges in polypropylene foaming: a review. *Cellular Polymers* 34 (6): 299–338.

122 Klier, J. Wright, D., and Chaudhari, B.; Worl Intellectual Property Organization. (2006). Polypropylene-based wire and cable insulation or jacket. WO 2007/019088 Al, filed 27 July 2006 and issued 15 February 2007.

123 Robinson, J., Loyens, W., and Verheule, B. (2007). New polyolefin insulation materials for high performance data cables. *AMI Cable Conference* (5–7 March 2007), Cologne.

124 Schroeck, P. and Minton, R. *Particle Nucleation in Chemical Foam Extrusion*. Reedy International Corporation www.readychemicalfoam.com/white-papers (accessed 6 July 2020).

125 Malm, B., Heikillä, P., Carlsson, C. et al. (2004). Olypropylene heterophasic copolymer for loose tube buffers – material choice, processing conditions and economy. In: *Proceedings 53rd IWCS*, 428–432.

126 Zhang, H., Liu, W., Li, L. et al. (2017). Design and performance study of PP all dry ADSS cable. In: *Proceedings 66th IWCS*, 440–444.

127 Dallas, G., Witt, G., Risch, B. et al. (2000). Thermal and mechanical optimization of easy-access flexible buffering materials. In: *Proceedings 49th IWCS*, 357–361.

128 Brown, G. and Bunker, S. (2003). Optimization of higher modulus polyolefin compounds for fiber optic buffer tube. In: *Proceedings 52nd IWCS*, 347–355.

129 Malm, B., Heikillä, P., Lagti, M., and Dammert, R. (2002). Polypropylene heterophasic copolymer for loose tube buffers. In: *Proceedings 51st IWCS*, 348–351.

130 Risch, B., Auton, J., and Tatat, O. (1998). Advanced impact polypropylene buffering material for superior long term hydrolitic and thermo-oxidative stability. In: *Proceedings 47th IWCS*, 698–704.

131 Holder, J. and McNutt, C. (2000). Polypropylene–polyethylene copolymer buffer tubes for optical fiber cables and method for making the same. European Patent Office EP 1 063 550, filed 20 August 2000 and issued 27 December 2000.

132 Kotala, J. (2002). Shrinkage of PBT, PP and PE in central tube buffering process. In: *Proceedings 51st IWCS*, 352–356.

133 Belli, S., Perego, G., Bareggi, A., Caimi, L. (1988). P-Laser: breakthrough in power cable systems. *Conference Record of the 1988 IEEE International Symposium on Electrical Insulation* (5-8 June 1988), Cambridge MA.

134 Zhou, Y., Yuan, C., Li, Q., and Jinliang, H. (2018). Recyclable insulation material for HVDC cables in global energy interconnection. *Global Energy Interconnection* 1 (4): 520–526.

135 Du, B., Hou, Z., and Li, J. (2018). A review of polypropylene and polypropylene/inorganic nanocomposites for HVDC cable insulation. In: *New Trends in High Voltage Engineering* (ed. R. Shariatinasab). IntechOpen.

136 Green, C., Vaughan, A., Stevens, G. et al. (2013). Recyclable power cable comprising a blend of slow-crystallized polyethylenes. *IEEE Transactions on Dielectrics and Electrical Insulation* 20: 1–9.

137 Green, C., Vaughan, A., Stevens, G. et al. (2015). Thermoplastic cable insulation comprising a blend of isotactic polypropylene and a propylene-ethylene copolymer. *IEEE Transactions on Dielectrics and Electrical Insulation* 22: 2, 639–648.

138 Yoshino, K., Demura, T., Kawahigashi, M. et al. (2004). Application of a novel polypropylene to the insulation of an electric power cable. *Electrical Engineering in Japan* 146: 8721–8879.

139 Huang, X., Fan, Y., Zhang, J., and Jiang, P. (2017). Polypropylene based thermoplastic polymers for potential recyclable HVDC cable insulation applications. *IEEE Transactions on Dielectrics and Electrical Insulation* 24 (3): 1446–1456.

140 Cao, X. and Andritsch, T. (2018). Modification of polypropylene-based cable insulation material. In: *3rd Asia-Pacific Electronics and Electrical Engineering Conference (17–18 November 2018), China*, 167–174.

141 Vaughan, A., Green, C., Hosier, I. et al. (2015). Thermoplastic high performance cable insulation systems for flexible system operation. In: *2015 Electrical Insulation Conference*, 543–546. Seattle, WA.

142 Polyvinylchloride production and manufacturing process. ICIS Chemical report. www.icis.com/explore/resources/news/2007/06/9076438, 6 January 2011.

143 Kiddoo, D. (2014). Case study of a phthalates alternative assessment for use in PVC-jacketed network cable. In: *Proceedings 63th IWCS*, 143–164.

144 Laufer, C., Chaudhary, B., Eaton, R., and Nguyen, B. (2010). Performance analysis of bio-based vs. non-bio-based plasticizers for PVC. In: *Proceedings 59th IWCS*, 152–156.

145 Mundra, M., Ghosh-Dastidar, A., and Eaton, R. (2012). Performance comparison of bio-based Plasticizers with selected phthalate and non-phthalate alternatives in Wire & Cable Jacket application. In: *Proceedings 61st IWCS*, 381–385.

146 Jia, P., Xia, H., Tang, K., and Zhou, Y. (2018). Plasticizers derived from biomass resources. *Polymers* 10: 1–27.

147 Feng, G., Hu, L., Ma, Y. et al. (2018). An efficient bio-based plasticizer for poly (vinyl chloride) from waste cooking oil and citric acid: synthesis and evaluation in PVC films. *Journal of Cleaner Production* 189: 334–343.

148 Brostow, W., Lu, X., and Osmanson, A. (2018). Nontoxic bio-plasticizers for PVC as replacements for conventional toxic plasticizers. *Polymer Testing* 69: 63–70.

149 Karmalm, P., Hjertberg, T., Jannson, A., and Dahl, R. (2009). Thermal stability of poly(vinyl chloride) with epoxidised soybean oil as primary plasticizer. *Polymer Degradation and Stability* 94: 2275–2281.

150 Schelling, T., Hoover, C., and Cox, M. (2007). Comparison of lead and non-lead stabilized flexible PVC compound technology for insulation and sheathing. In: *Proceedings 56th IWCS*, 331–339.

151 Barber, K. and Alexander, G. (2013). Insulation of electrical cables over the past 50 years. *IEEE Electrical Insulation Magazine* 29 (3): 27–32.

152 Nass, L. and Heiberger, C. (1986). *Encyclopedia of PVC*. CRC Press.

153 Arrighi, R. (1986). From impregnated paper to polymeric insulating materials in power cables. *IEEE Transactions on Electrical Insulation* EI-21 (1): 7–18.

154 Park, D., Ah, M., and Lee, G. Development of FR-PVC for compact and high performance UTP cable. In: *Proceedings 51st IWCS*, 391–396.

155 Walker, K. and Ho, W.-K. (2006). New molybdate smoke suppressant for PVC. In: *Proceedings 55th IWCS*, 404–409.

156 Walker, L., Ho, W.-K., and Mittleman, M. (2005). Smoke suppression in PVC plenum cable jackets. In: *Proceedings 54th IWCS*, 208–213.

157 Coakere, W. (2003). Fire and flame retardants for PVC. *Journal of Vinyl and Additive Technology* 93: 108–115.

158 Nam, G., Yoon, S., and Lim, H. (2003). Improvement of heat stability of flame retardant PVC compound for plenum rated UTP cable sheath by an experimental design method. In: *Proceedings 52th IWCS*, 357–361.

159 Metwally, I. (2012). The evolution of medium voltage power cables. *IEEE Potentials*: 20–25.

160 Brebu, M., Vasile, C., Antonie, S., and Chiriac, M. (2000). Study of the natural ageing of PVC insulation for electrical cables. *Polymer Degradation and Stability* 67: 209–221.

161 Shwehdi, M., Morsy, M., and Abugurain, A. (2003). Thermal aging tests on XLPE and PVC cable insulation materials of Saudi Arabia. In: *2003 Annual Report Conference on Electrical Insulation and Dielectric Phenomena*, 176–180.

162 Moore, G.F. (1999). *Electric Cables Handbook*, BICC Cables, 3e. Blackwell Science.

163 Cardello, P., Wade, R., and Solomon, R. (1983). Wire and cable products with improved jacketing. World Intellectual Property Organization, 1984000245 A1, filed 29 June 1983 and issued 19 January 1984.

164 Garbacz, T. (2012). Structure and properties of cellular thin-walled cable coatings. *Polimers* 57 (11–12): 865–868.

165 Josey, L. and Torrey, S. (2003). Improving cold temperature impact performance with PVC jackets. In: *Proceedings 52nd IWCS*, 562–567.

166 Park, D., Ahm, M.J., and Lee, G. (2002). Development of FR-PVC for compact and high performance UTP cable. In: *Proceedings 51st IWCS*, 391–396.

167 Coaker, W. (2003). Fire and flame retardants for PVC. *Journal of Vinyl and Additive Technology* 9 (3): 108–115.

168 Babrauskas, V. (2005). *Mechanisms and Modes for Ignition of Low-Voltage PVC Wires, Cables, and Cords*, 291–309. Interscience Communications Ltd.

169 Hampton, N., Hartlein, R., Lennartsson, H. et al. (2007). Long-life XLPE insulated power cable. Jicable'07, Versailles (24–28 June 2007).

170 Denko, S. Chlorinated polyethylene ELASLEN – application for Wire and Cable. www.sdk.co.jp (accessed 6 July 2020).

171 The Dow Chemical Company (2005). Tyrin Chlorinated Polyethylene – Tyrin in Thermoplastic Jackets.

172 Rich, B., Lovie, R., and Baird, P. (2015). Enhanced solvent resistance and flammability requirements for harsh environment and industrial optical fiber cables. In: *Proceedings 64th IWCS*, 755–759.

173 Wilkens, D. (1990). Telephone cable: overview and dielectric challenges. *IEEE Electrical Insulation Magazine* 6 (2): 23–28.

174 Ebnesajjad, E. (2015). *Fluoroplastics: Melt Processable Fluoropolymers*, vol. 2. Amsterdam, Boston, MA: Elsevier.

175 Chapman, G. (2004). The past and future of fluoropolymers in wire and cable. In: *Proceedings 53th IWCS*, 172–179.

176 Henry, J., Seiler, D., and Perrinaud, R. (2008). The "goods" on fluoropolymers in wire and cable. In: *Proceedings 57th IWCS*, 347–352.

177 Kenneflick, D. (2018). FEP as a dielectric material for multi-gigabit, single pair ethernet cable for automotive. *Proceedings 67th IWCS*, Paper 2–3.

178 Young, R., Johnston, F., Netta, J., and Walck, K. (2017). The effects of fluoropolymer selection on electrical performance in twisted pair cabling. In: *Proceedings 66th IWCS*, 115–122.

179 Lunardi, G., Bassi, M., Colaianna, P., and Hrivnak, J. (2007). Physical properties of perfluorinated polymers used in telecommunication copper cables. In: *Proceedings 56th IWCS*, 316–320.

180 Glew, C. (2012). Chemically foamable fluorinated ethylene propylene (FEP) – the beneficial technical and economic aspects of foamed FEP insulation for plenum cable applications. In: *Proceedings 61st IWCS*, 621–665.

181 Thuot, G. and Young, T. (2011). Foam fluoropolymer solutions and processing for insulating high performance cables. In: *Proceedings 60th IWCS*, 233–236.

182 Young, R. (2018). Fluoroplastics solutions in high-speed direct attach cabling. *Proceedings 67th IWCS*, Paper 2-1.

183 Glew, C., Glew, C.A., Speer, R. et al. (2018). Limited combustible foamable fluoropolymers for LAN cabling enabling high operating temperatures for power over ethernet. In: *Proceedings 67th IWCS*, 1–8.

184 Leidecker, H. and Plante, J. (2000). *ISS Fiber Optic Failure Investigation Report*. NASA.

185 Brochure W.L. Gore & Associates. (2016). GORE® aerospace cables and materials.

186 Ebnesajjad, S. (2015). *Fluoroplastics 2:4. Properties and Structure*, 22–38. Elsevier, Plastics Design library.

187 Young, R. (2007) Overview of Fluoropolymer Foam Technology, Wire New England '07, WAI New England Chapter.

188 Ebnesajjad, S. (2015). *Fluoropolymer Foams. Fluoroplastics*, vol. 2. Elsevier.

189 Kang, D. (2017). Outlook for the coaxial cables in the forthcoming 5G mobile networks. In: *Proceedings 66th IWCS*, 367–377.

190 Khanna, J., Sinilia, J., and Chandrasekaran, S. (1986). Molecular relaxations in Halar, an alternating copolymer of ethylene and chlorotrifluoroethylene. *Macromolecules* 19 (9): 2426–2431.

191 Drobny, J. (2001). *Technology of Fluoropolymers*. CRC Press.

192 Ringler, R. and Butkevich, P. (1986). Fluoropolymer foam for twisted pair cables. In: *Aerospace Avionics Equipment and Integration Conference Proceedings*, 179–187.

193 Goldbach, J., Amin-Sanayei, R., He, W. et al. (2017). Commercial synthesis and applications of poly(vinylidene fluoride). In: *Fluorinated Polymers: Applications*, vol. 2 (eds. B. Ameduri and H. Sawada). The Royal Society of Chemistry.

194 Pecsok, R. (1987). Foamable polyvinylidene fluoride and methods. US Patent 4 692 381, filed 9 March 1987 and issued 8 September 1987.

195 Henry, J., Kreh, D., Puglia, L., O'Brien, G. (2018). Crosslinking of PVDF resins for high temperature wire and heat shrink tubing applications. *Proceedings 67th IWCS*.

196 Thieblemont, V. and Thierry, J.-F. (2010). Polyamide manufacturing process. US Patent 8 735 530 B2, filed date 9 March 2010 and issued 27 May 2014.

197 Tobita, H. and Hamielec, A. (2016). Polymerization processes, 2. Modeling of processes and reactors. In: *Ullmann's Encyclopedia of Industrial Chemistry: Products and Processes* (ed. B. Elvers). Wiley-VCH.

198 Thue, W. (2012). *Electrical Power Cable Engineering*. CRC Press, Taylor Francis Group.

199 Deopura, B.L., Alagirusamy, R., Joshi, M., and Gupta, B. (2008). *Polyesters and Polyamides*, Woodhead Publishing Series in Textiles. CRC Press LLC.

200 Lenz, M., Creffield, J., Evans, T. et al. (2012). Resistance of polyamide and polyethylene cable sheathings to termites in Australia, Thailand, USA,

Malaysia and Japan: a comparison of four field assessment methods. *International Biodeterioration and Biodegradation* 66: 53–62.
201 Montes, S., Lopez, H., Murga, P., and Rodriguez, V. (2010). Easy-to-instal electrical cable . US Patent 8 912 253 B3, filed 25 March 2010 and issued 16 December 2014.
202 Peraka, S.K., Shukla, V., Sahoo, K. et al. (2016). Air blown cables for enhanced blowing performance. In: *Proceedings 65th IWCS*, 439–442.
203 Risch, B. and Babe, B. (2014). Impact of high radiation environments on polyamide cable materials. In: *Proceedings 63th IWCS*, 930–937.
204 Weiss, A., Berg, D., Lausch, P., and Nothofer, K. (2003). Design and performance testing of a new central tube cable for 7/5.5 mm microducts. In: *Proceedings 52th IWCS*, 34–37.
205 Siddhamalli, S. (2000) Polyamide/polyolefin fiber optic buffer tube material. US Patent Office 6 411 403 B1, filed 4 January 2000 and issued 25 June 2002.
206 Cruz, R., Caetano, M., da Silva, L. et al. (2008). Tight-buffer drop cable for FTTH deployments. In: *Proceedings 57th IWCS*, 445–448.
207 Bassaco, R., Scheiner, D., Massao, R. et al. (2018). Totally gel-free fiber optical cables manufactured with PBT. *Proceedings 67th IWCS*, Paper 2-7.
208 Kumar, S., Sahoo, K., Mishra, A. et al. (2018). Next-gen reduced diameter microcables. *Proceedings 67th IWCS*, Paper 22.
209 Gu, J. and Venkataswamy, K.; World Intellectual Property Organization. (2010). Flame retardant thermoplastic elastomers. WO 2010/039616 A2, filed 9 July 9 2009 and issued 14 January 2010.
210 Draka (Prysmian Group) (2009). *Draka FTTH FlexTube Solution.*
211 Consonni, E. et al.; World Intellectual Property Organization. (2011). Optical communication cable and manufacturing process. WO2011/035814 A1, filed 28 September 2009 and issued 11 September 2016.
212 Harbort, H., Widler, J., and Grögl, F. (1990). Optical cable having at least two separate multiple-fiber units each having its own soft plastic envelope. US Patent Office 4 909 593, filed 20 December 1988 and issued 20 March 1990.
213 Overton, B. (2010). Reduced-diameter, easy-access loose tube cable. US Patent Office US 2010/0092139 A1, filed 10 November 2009 and issued 15 April 2010.
214 Zhu, J., Lu, X., Xiong, Z. et al. (2019). Application of thermoplastic polyester elastomer in the preparation of optical cable loose tube. In: *Proceedings IWCS China*, Paper 10-3.
215 Elleithy, R. and Sardanopoli, A. (2005). Thermoplastic polyurethane applications in the wire and cable industry. In: *Proceedings 54th IWCS*, 524–527.
216 DeGross, J. (1999). Thermoplastic polyurethane materials for cable jacketing. In: *Proceedings 48th IWCS*, 831–845.
217 Makadia, C., Vontorcik, J., and Wilson, V. (2013). High temperature thermoplastics polyurethanes for cable jacketing applications. In: *Proceedings 62nd IWCS*, 718–721.
218 Grote, J. and Watson, R. (2006). Standard and innovative thermoplastic polyether polyurethanes for wire and cable jacketing. In: *Proceedings 55th IWCS*, 66–68.

219 Muehren, O. and Henze, O. (2011). Thermoplastic polyurethane solutions for low smoke zero halogen (LSZH) flame retardant cable applications. In: *Proceedings 60th IWCS*, 257–263.

220 Mühren, O. (2007). Thermoplastic polyurethane (TPU) for cable applications. *AMI Cables 2007 Conference* (5–7 March 2007), Cologne.

221 Henze, O. and Muehren, O. (2018). Thermoplastic polyurethane (TPU) solutions providing outstanding longterm aging properties for cable applications. *Proceedings 67th IWCS.*

222 Beninca, M., Nardi, M., Scheiner, D. et al. (2012). Electrical performance of category 6 U/UTP and F/UTP cable in harsh applications. In: *Proceedings 61th IWCS*, 573–579.

223 Tripathi, D. (2002). *Practical Guide to Polypropylene.* RAPRA.

224 Ineos Olefins & Polymers USA. (2014). Typical engineering properties of polypropylene.

225 Chemours. Fluoroplastic comparison – typical properties. www.chemours.com (accessed 17 June 2020).

3

Elastomers for Cables

Burkhard Herpich

Germany

3.1 Introduction

Elastomers are chemically cross-linked rubbers.

Like thermoplastics, rubbers are polymeric materials, generated by polymerization. Polymers are large filamentary molecules – macromolecules – composed of many repeated monomer units (Figure 3.1).

Since the early beginnings the cable industry is closely connected with polymeric materials. With the industrial revolution in the nineteenth century, electrical telegraphy, electricity, and their transportation became increasingly important for everyday life. It was the time of founding cables. Generally speaking, cables consist of a conductor and an insulating covering material. In the nineteenth century Gutta-percha was used as covering material for cables. Gutta-percha is a rubber-like natural material from the coagulated latex from the Gutta-percha tree (Palaquium gutta). In the earlier days Gutta-percha was roll-milled around the conductor. A clear disadvantage was that it always had easy-to-break lap joints. With the invention of the so-called Gutta-percha press by Werner von Siemens in 1847, Gutta-percha was pressed around the conductor without any lap joints [1, 2] (Figure 3.2).

From a chemical point of view Gutta-percha is very close to natural rubber, but not 100% identical. In both cases we speak about polyisoprene. While natural rubber is 1,4-*cis*-polyisoprene, Gutta-percha is 1,4-*trans*-polyisoprene. In addition, the molecular weight of Gutta-percha is much lower than that of natural rubber.

Natural rubber, a real biopolymer, is obtained from a runny, milky white liquid, called latex, from the *Hevea brasiliensis* tree (Figure 3.3). Approximately 99% of the natural rubber comes from these trees. There are various other plants producing latex, e.g. dandelion, but for different reasons they have no relevancy. The growing need for natural rubber aroused the interest in synthesizing rubber early [1, 2].

The Global Cable Industry: Materials, Markets, Products,
First Edition. Edited by Günter Beyer.

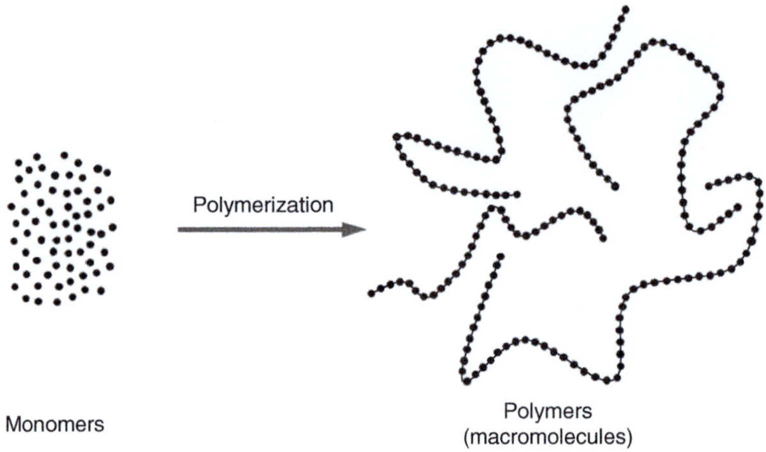

Figure 3.1 From monomers to polymers.

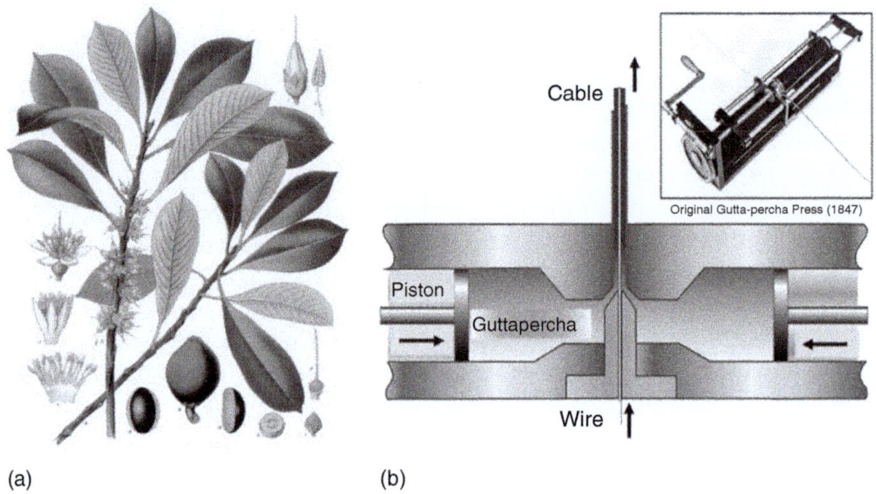

Figure 3.2 (a) Palaquium gutta. (b) Gutta-percha press. Source: (a) Köhler [3], (b) https://www.higgs.ch/eine-erfindung-veraendert-die-kabelwelt/23153/, © 2019, Scitec-Media GmbH.

(a) (b) (c) (d) (e) (f) (g) (h)

Figure 3.3 The way of processing natural rubber [4]. (a) *Hevea brasiliensis* tree. (b) Tapping. (c) Single latex collection. (d) Latex unification. (e) Coagulation. (f) Pre-drying. (g) Sheet shaping by milling. (h) Final drying and storing.

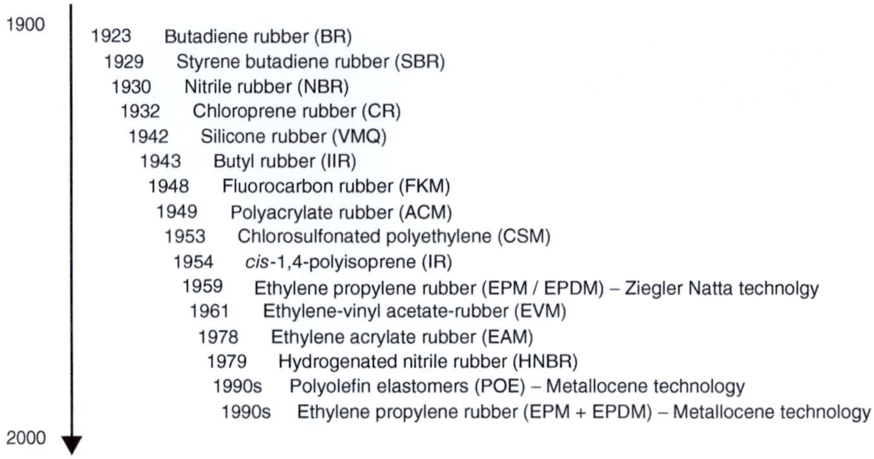

Figure 3.4 Commercialization of synthetic rubbers. Source: Röthemeyer [1].

The fundamental milestones in this path are as follows [1]:[1,2]

1860 – Isoprene by dry distillation (pyrolysis) of natural rubber by Charles Hanson Greville Williams.
1878 – First polymerization of isoprene by Gustave Bourchardat.
1900 – First fully synthetic rubber with di-methyl-butadiene in the laboratory by Iwan Kondadow.
1909 – First patent for manufacturing of synthetic rubber in a technical way by Fritz Hofmann.
1915 – First technical plant at BAYER in Leverkusen (methyl rubber). 2.500 tons of methyl rubber were produced until plant closure in 1918.

On the base of the rapid growing technical petrochemistry and acetylene chemistry in the first half of the twenties century and the upcoming macromolecular chemistry, established by Hermann Staudinger in the 1920s [5], numerous synthetic rubbers, used in the cable industry today, were commercialized during the last 100 years (Figure 3.4).

Elastomers are the result of chemical cross-linking processes of rubber. Those chemical cross-linking processes, classically called vulcanization, are non-reversible. Vulcanization, entitled after the Roman god of fire and forge, was discovered by Charles Goodyear in 1839 during the booming phase of natural rubber in the United States (Figure 3.5) [2, 6].

Owing to the cross-linking of the separate rubber chains viscous flow is prevented. The wide-meshed three-dimensional covalent network structure gives the elastomers their exceptional elastic properties. The unique so-called rubber elasticity is based on changes in the entropy of the rubber chains and chain parts during extending and relaxing [1] (Figure 3.6).

Beside thermoplastics and elastomers there is another group of polymeric materials, used for cables, the so-called thermoplastic elastomers (TPEs).

---

1 https://www.ir-dichtungstechnik.de/wissenswertes/zeittafel-der-kautschukgeschichte.
2 https://de.wikipedia.org/wiki/Synthesekautschuk.

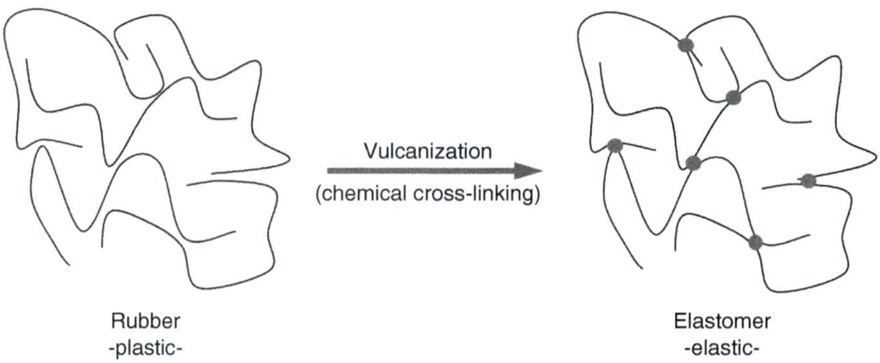

Figure 3.5 From rubber to elastomer.

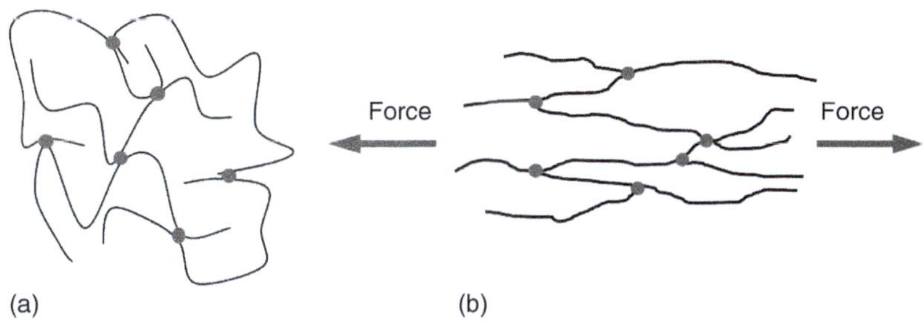

Figure 3.6 Rubber elasticity (a) entangled, (b) extended.

The TPEs seem to be the connecting link between thermoplastics and elastomers. The TPEs combine properties of thermoplastics and elastomers as well. At higher temperatures the TPEs can be processed like thermoplastics; at room temperature they show a rubber-like elastic behavior. Ideally, melting and cooling can be done as often as needed [7].

There are two main groups of TPEs:

- Block copolymers (e.g. thermoplastic polyurethanes [TPUs])
- Polymer blends with separated phases (e.g. polyvinylchloride [PVC]/nitrile rubber [NBR] blends)

At room temperature the TPEs have a multiphase morphology. Micro-phase separation is the result of the incompatibility of the different polymer blocks with respect to the used polymers. The domains were held together purely physically, for example, by van der Waals forces or hydrogen bridge bonds. Especially in the case of block copolymers, micro-phase separation leads to a three-dimensional physical network structure (Figure 3.7), similar to the three-dimensional network structure of the elastomers, formed by covalent chemical cross-linking processes (Figure 3.5). However, the chemical nature of the three-dimensional network structure determines the significant difference between both – non-thermo-reversibility in the case of elastomers and thermo-reversibility in the case of TPEs.

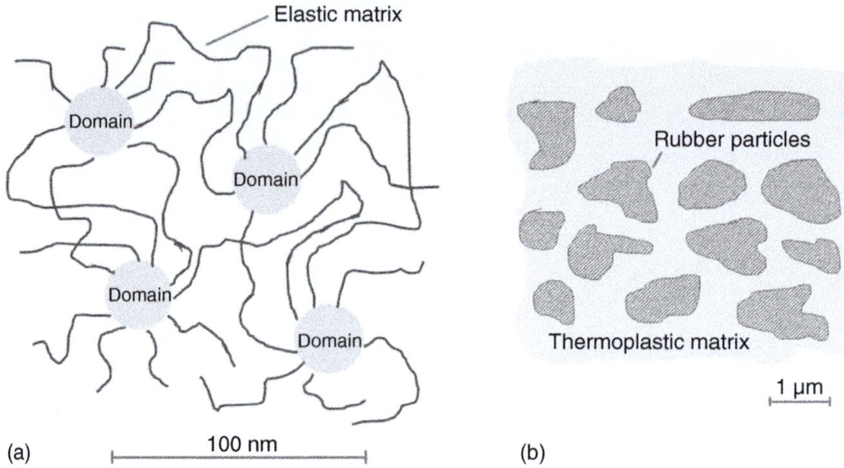

Figure 3.7 Structural models of thermoplastic elastomers. (a) Block copolymer. (b) Polymer blend. Source: Morton, Wegner, Rader and Stemper [8–10].

The block copolymers are synthesized directly by polymerization. Polymer blends are generated by blending a thermoplastic with rubber in a certain ratio. The thermoplastic must be the coherent phase, wherein the rubber is dispersed discretely (Figure 3.7). Thermoplastic vulcanizates (TPVs) are the result of reactively blending thermoplastics with rubber during in-situ-vulcanization.

It should be noted that not all block copolymers or polymer blends show thermo-reversible elastic behavior [7].

Although polyurethane was already commercialized in the 1940s, the structure–properties relationship of the TPEs was well investigated first in the 1960s on styrene–butadiene–styrene (SBS) triblock copolymers [11, 12].

Elastomers are dedicated mainly for flexible energy cables. During application these cables are dynamically stressed and normally not rigidly fixed avoiding any motion. So, a certain flexibility is required. In everyday language, cables based on elastomers are called rubber cables.

In contrast to thermoplastics, elastomers are used in cable industry in much lower quantities. Roughly one-quarter of the overall polymer consumption is based on rubber. No significant increase is expected in the future. Among others the complexity and needs of combined extrusion/vulcanization lines and the costlier recycling as well are influencing this trend significantly.

Rubber cables are used at rated temperatures between 90 and 150 °C normally. For higher rated temperatures various silicone rubber grades (180 °C) and fluoro-elastomers (>200 °C) are the choice.

Among the rubber materials silicone rubber has a special status. The backbone chain contains no carbon atoms.

Since many years natural rubber, the rubber used for cables first, especially in the first half of the twentieth century, has not been playing a significant role in the cable industry. Cable specifications and standards reflect this fact.

3.2 Rubber Compounds

As rubber and thermoplastic compounds are polymeric compounds as well, they are almost identical in design; just the cross-linking system in the rubber compounds makes the difference (Figure 3.8).

The basic requirements for a rubber compound:

- Compoundable
- Extrudable
- Cross-linkable
- Fulfilling specifications and requirements

Although this sounds very simple and logical, in everyday practice they are really challenging.

The centerpiece of a rubber compound is the rubber itself. In rubber industry it is common practice that in rubber compound recipes every ingredient is referred to rubber, to 100 weight parts of the used rubber(s), named parts per hundred rubber (phr). Table 3.1 shows a very abstract rubber compound recipe in order to just get a feel for the different ingredient quantities. More detailed recipe recommendations are available from the raw materials suppliers.

3.2.1 Rubber

Of course, the most important part of a rubber compound is the rubber itself.

Rubbers can be systematized and coded according to ASTM D 1418 [14]. The rubber codes consist of two to four letters. The last letter indicates the chemical characteristics of the polymer chain. The first letters describe the polymers (Table 3.2).

Inside each rubber family a lot of different grades are available and making the selection not easy indeed. Also, some chemical aspects such as microstructure

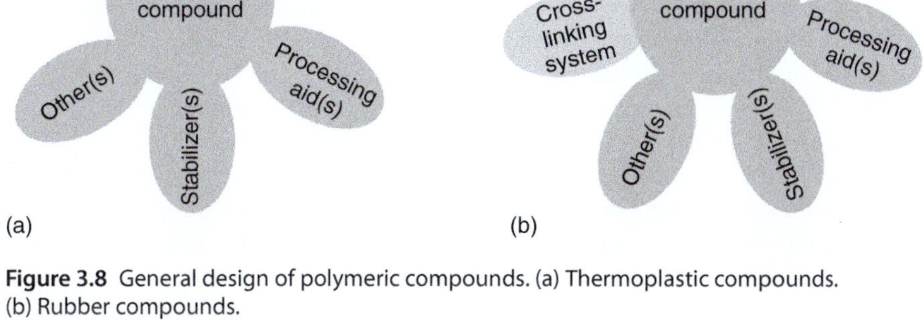

Figure 3.8 General design of polymeric compounds. (a) Thermoplastic compounds. (b) Rubber compounds.

Table 3.1 Abstract rubber compound recipe [13].

Ingredient	phr
Rubber(s)	100
Filler(s)	0–250
Plasticizer(s)	0–250
Stabilizer(s)	0–10
Processing aids	0–10
Others	0–10
Cross-linking system	0–10

(crystalline or amorph), short or long chain branching, and molecular weight distribution must be considered. In addition, beside many "normal" grades, also some specialties are commercially available:

- Oil extended grades
- Blends
- Carbon black batches
- Pre-cross-linked grades
- Carboxylated grades

If rubbers are modified, an additional letter describes the kind of modification:

- Highly hydrogenated nitrile rubber: H-NBR
- Carboxylated nitrile rubber: X-NBR

Different rubbers have different properties (Figure 3.9a, b). The choice of the rubber influences significantly the following:

- Mechanical properties
- Electrical properties
- Resistance against oils, liquids, and water
- Low/high temperature behavior
- Light and weather stability, outdoor behavior
- Fire resistance
- Corrosivity and toxicity
- Others

The selection of the rubber for a compound is influenced by the rubber properties as well as by the type of compound, cable specifications and standards, and finally by the final application and the customer.

In dependence of the manufacturing technology rubbers are delivered as powder (e.g. chlorinated polyethylene), as flakes (e.g. chloroprene rubber), as pellets (e.g. copolymers of ethylene and vinyl acetate) or as bales (e.g. nitrile rubber). These delivery forms have an impact on the compounding process.

The viscous behavior of the rubber grades can be characterized by measuring the Mooney viscosity (Figure 3.10). The Mooney viscosity test method is well established in the rubber industry, used for rubbers and rubber compounds as

Table 3.2 Rubber coding according ASTM D 1418 (selection) [14].

Last letter	Chemical characteristics	Examples	
M	Rubbers having a saturated chain of the polymethylene type	ACM	Copolymers of ethyl or other acrylates (polyacrylate rubber)
		AEM	Copolymers of ethyl or other acrylates and ethylene
		CM	Chloro-polyethylene
		CSM	Chloro-sulfonyl-polyethylene
		EPDM	Terpolymer of ethylene, propylene and a diene
		EPM	Copolymer of ethylene and propylene
		EVM	Copolymers of ethylene and vinyl acetate
		FKM	Fluoro rubber of the polymethylene type
R	Rubbers having a unsaturated carbon chain (carbon–carbon double bonds)	BR	Butadiene rubber
		CR	Chloroprene rubber
		IIR	Isobutene isoprene rubber (butyl rubber)
		IR	Isoprene rubber (cis-1,4-polyisoprene)
		NBR	Acrylonitrile butadiene rubber (nitrile rubber)
		NR	Natural rubber
		SBR	Styrene butadiene rubber
O	Rubbers having oxygen in the polymer chain	CO	Epichlorohydrin polymer
		ECO	Epichlorohydrin copolymer
Q	Rubbers having silicon and oxygen in the polymer chain	VMQ	Silicone rubber with methyl and vinyl substitute groups
		FVMQ	Silicone rubber with fluorine, methyl and vinyl substitute groups
U	Rubbers having carbon, oxygen, and nitrogen in the polymer chain	AU	Polyester urethane
		EU	Polyether urethane

In brackets are more common used names.

a) Wide spread also EVA.

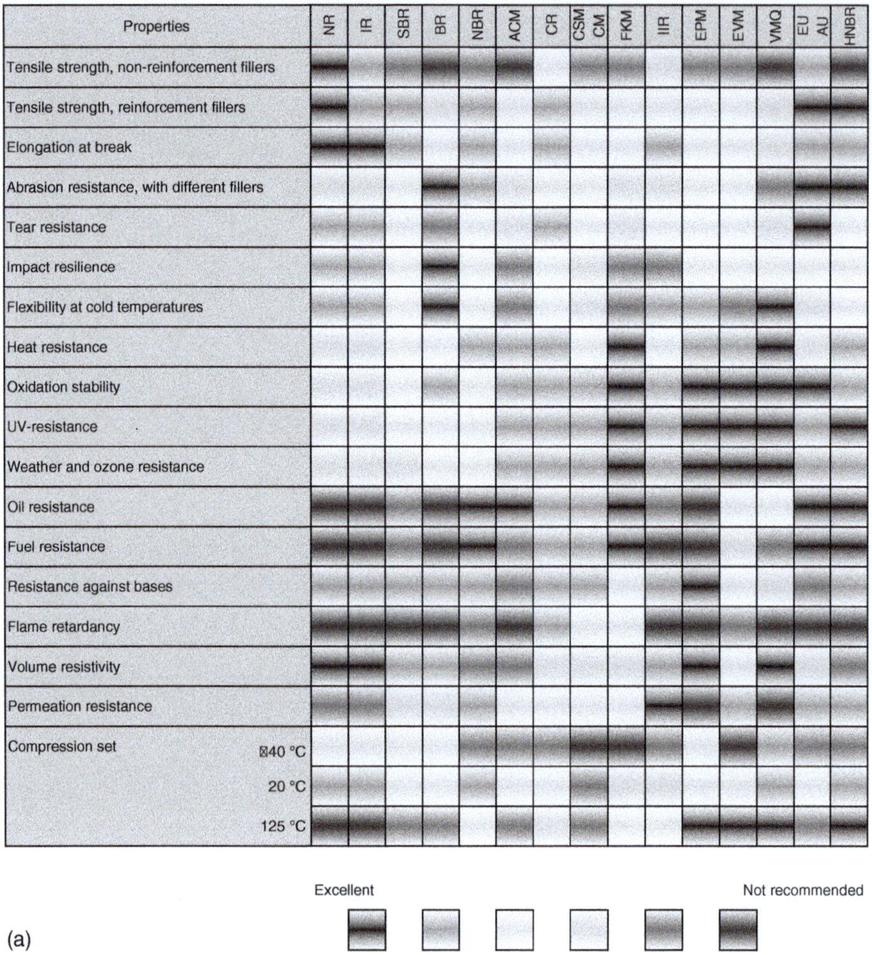

(a)

Figure 3.9 (a) Properties of rubber [15]. (b) Heat and oil resistance of rubber [15]. Rubber abbreviations are explained in Table 3.2.

well. The rubber sample is preheated for a defined time, and then sheared at a constant rate. Typically, the Mooney viscosity is measured at 100 °C: ML_{1+4} (100 °C), in Mooney viscosity units (MU).

3.2.2 Fillers

A lot of different filler classes, suitable for rubber compounds designing, are available on the market:

- Chalks
- Clays
- Carbon blacks
- Talc

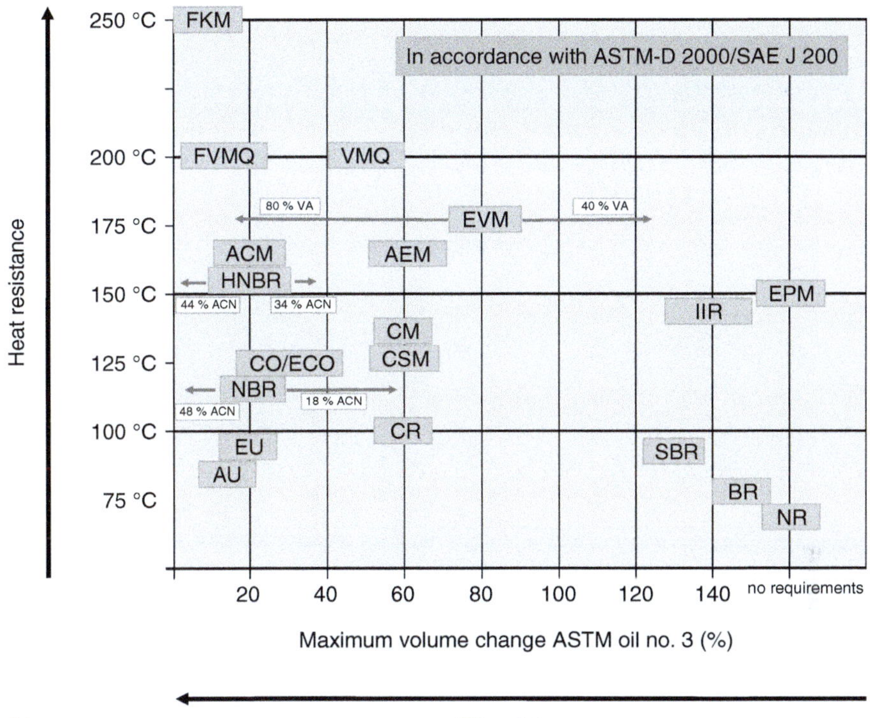

(b)

Figure 3.9 *(Continued)*

- Silicas
- Metal hydroxides such as alumina trihydrate (ATH) or magnesium dihydrate (MDH)
- Nano clays

Within these classes a lot of different grades are commercially available. A good example are the chalks, which can be grinded, whitewashed or precipitated. Among the clays there are soft, hard, or even calcinated ones. Furthermore, fillers can be surface treated.

The fillers can be classified in different ways:

- Natural and synthetic fillers
- Light and black fillers
- With emphasis on their reinforcement effect in the compound (Figure 3.11)
- With emphasis on special properties, e.g. flame retardancy

Fillers can also be classified according to their particle size (Figure 3.12). Important characteristic filler attributes include the following:

- Particle shape (e.g. round, plate-like, needle-shaped)
- Particle size and particle size distribution
- Surface morphology and specific surface (if porous)

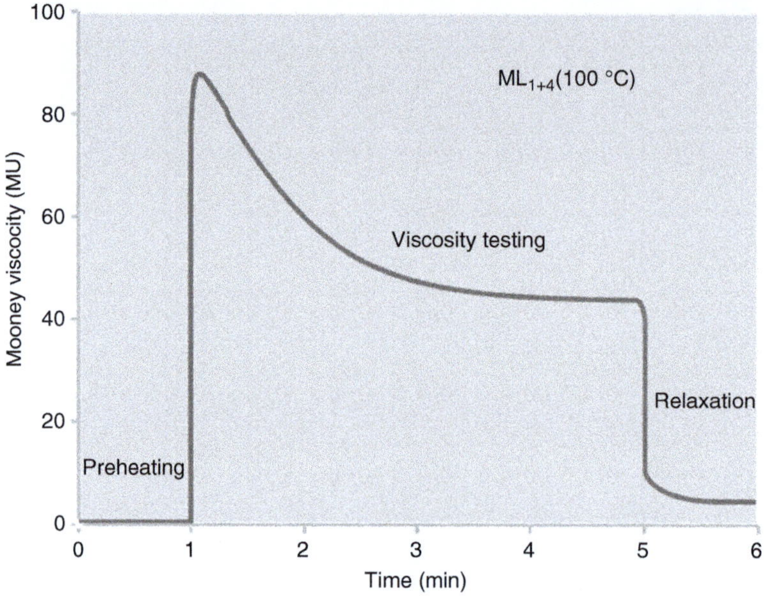

Figure 3.10 Mooney viscosity. Source: Ref. [13].

- Abrasive or non-abrasive (important for compounding and extrusion process)
- Special features (e.g. rheological behavior, optical parameters)

The fillers influence the mechanical, electrical, and flame-retardant behavior of the rubber compounds and, by extension, of the cable itself, significantly. State of the art for excellent electrical insulation compounds are calcinated clays, surface-treated ones, and for excellent flame retardancy, metal hydroxides such as ATH, MDH, and nano clays.

The filler load in a compound can vary between 30 and 90 wt%, strongly dependent on the kind of filler and the desired compound properties. Of course, fillers can also be combined with each other in one compound [1, 16].

Fillers are available in bags, big bags, and silos, depending on the local needs at the compounding site.

3.2.3 Plasticizer

Beside the fillers, the plasticizers belong to the quantitatively most important ingredients in rubber compounds. With respect to the compatibility between rubber and plasticizer, the plasticizer can be classified as

- Primary plasticizers
- Secondary plasticizers

Primary plasticizers are compatible with the rubber matrix. Those plasticizers migrate into the rubber and improve the viscous flow, reduce the viscosity of the rubber compound, and increase the elasticity of the elastomer as well. Different plasticizers can act as primary plasticizer in various rubbers. As a rough

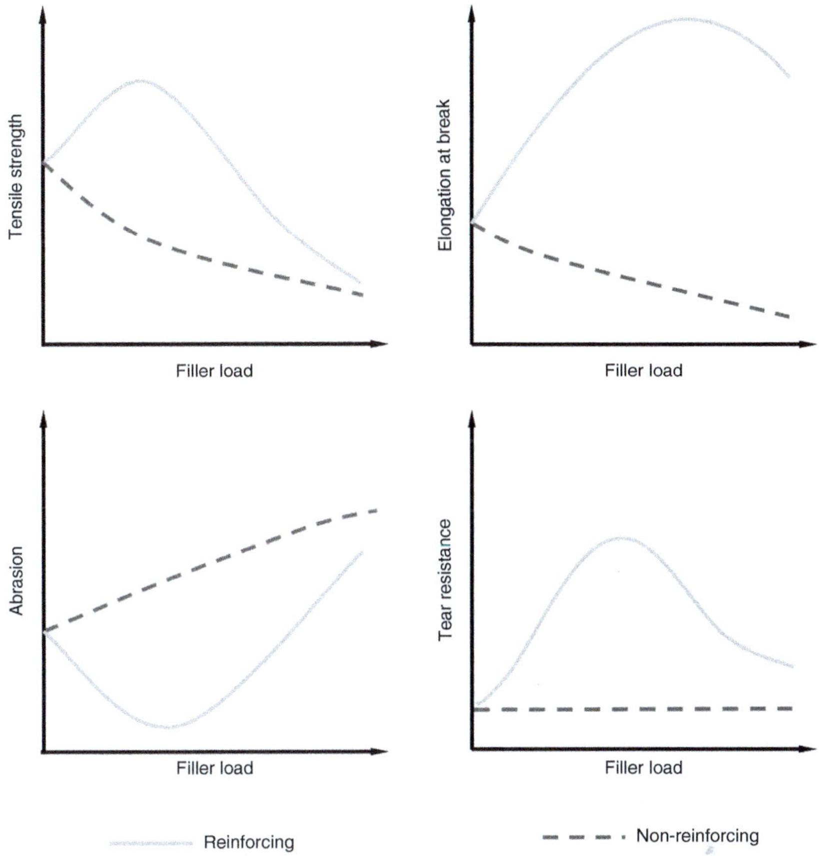

Figure 3.11 Reinforcement effects of fillers in rubber compounds. Source: Kleemann [15, 16].

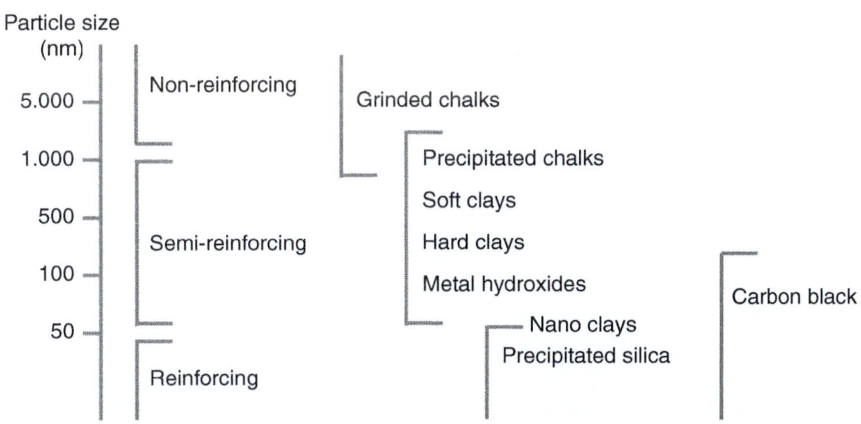

Figure 3.12 Classification of fillers according their particle size [15].

guide, polar plasticizers are used for polar rubbers and nonpolar plasticizers for nonpolar rubbers.

In the case of secondary plasticizers, the compatibility between plasticizer and rubber is either less or not given. The plasticizer acts as a lubricant between the rubber chains and can improve the filler dispersion. There is no influence on the viscosity of the rubber compound to register. In contrast to the primary plasticizers, the dosage of secondary plasticizers is much lower. Too high quantities migrate in a short time on the compound surface, becoming wet [1, 16].

As a rubber cable normally consists of different compounds, all containing different plasticizers in different quantities, usually undesired plasticizer migration can occur between neighboring compounds. The migration speed is dependent on the plasticizer itself, on the concentration gradient, and the ambient temperature. In the worst case, plasticizer migration leads to changes in the elastomer properties, e.g. discoloration of colored compounds, loss on electrical performance, or other performance characteristics.

Plasticizers influence not only the mechanical properties but can also improve cold temperature behavior significantly. Normally, cold temperature behavior is dependent on the used rubber itself. But it can be improved using a suitable plasticizer with a low pour point, e.g. ester plasticizers.

Some of the plasticizers such as chlorinated paraffins or phosphate ester can improve even the flame retardancy.

The plasticizers can be classified into

- Natural plasticizers
- Mineral oils
- Ester plasticizers
- Special plasticizers

Plasticizers are commercially available in barrels, intermediate bulk containers (IBCs), and tanks.

3.2.4 Stabilizers

Elastomers must be stabilized against long-term aging [1, 16]. Protection is required against

- Heat
- Oxygen
- Ozone
- Light (UV)
- Ionizing radiation
- Metal ions
- Humidity
- Others

Aging leads to structural changes. Aging means normally an attack on the rubber itself. Carbon–carbon bonds of the rubber chains are cracked. With the dynamic stress present in rubber cables, aging is intensified.

The difficulty is that aging in practice is a very complex matter. However, the different testing possibilities in one test procedure in the laboratory cannot be as complex as the practice is. Outdoor applications with a big variance in the temperature range in combination with mechanical stress constitute are the most challenging aging scenarios.

Stabilizers should be selected very carefully, keeping always an eye on the chosen rubber(s), the whole compound recipe itself and last but not least the final application.

From a chemical prespective, there is a very broad variety of different kind of stabilizers commercially available on the market, starting from pure single substances on the one side to combinations of different stabilizers in master batches on the other.

3.2.5 Classical Cross-linking Systems

The type of the cross-linking system derives from the chemical nature of the rubber.

3.2.5.1 Cross-linking with Sulfur Systems

Cross-linking systems based on sulfur, belonging to the oldest vulcanization techniques, going back to Charles Goodyear, are limited to unsaturated rubbers only (Figure 3.13).

The advantage of cross-linking with sulfur systems is the insensitivity to most of the other compound ingredients, humidity, and oxygen. Acid ingredients have a retarding effect while alkaline ingredients have an accelerating effect.

3.2.5.2 Cross-linking with Peroxide Systems

Peroxide based cross-linking systems can be used for almost all rubbers (Figure 3.14).

In the radical cross-linking process oxygen must be excluded. If not, a degradation process starts immediately.

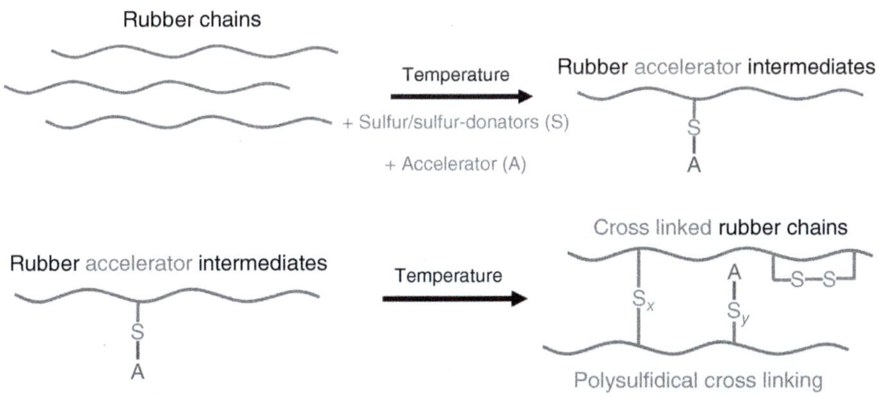

Figure 3.13 Cross-linking of unsaturated rubber chains with sulfur systems. Source: Kleemann, Schnetger [16, 17].

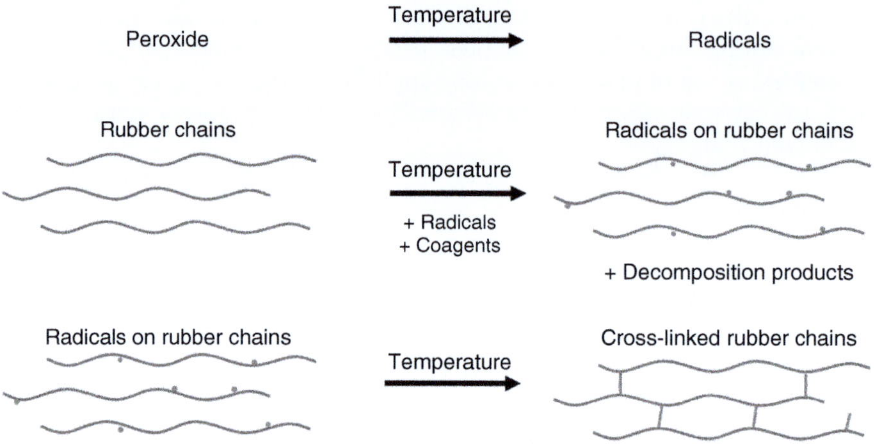

Figure 3.14 Cross-linking of rubber chains with peroxide radicals.

3.2.5.3 Moving Die Rheometers (MDRs)

The cross-linking behavior and the cross-linking degree of the rubber compounds can be determined with moving die rheometers (MDRs) (Figure 3.15).

Comparing the sulfur and peroxide systems with each other, sulfur systems are much more sensitive for scorch.

Cross-linking systems can be simple and complex as well. In any case, the cross-linking systems must be adjusted for optimal compounding and extrusion/vulcanization processes as well.

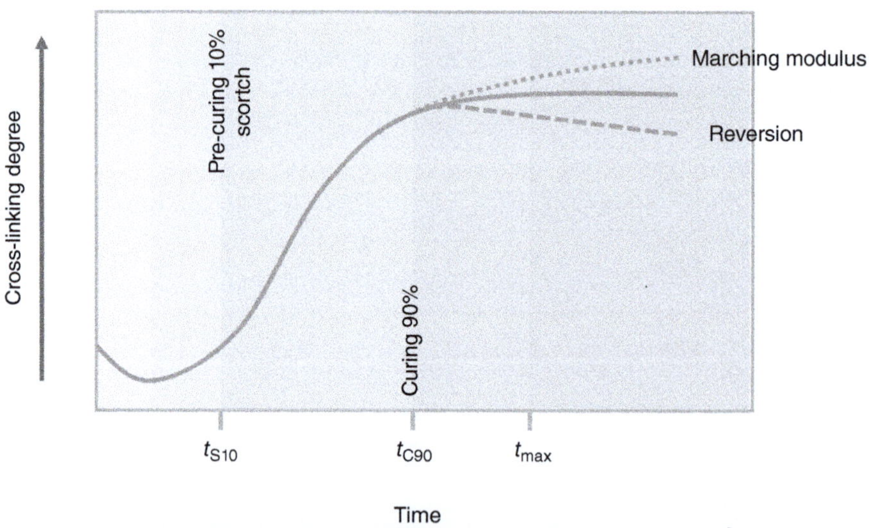

Figure 3.15 Cross-linking behavior measured with moving die rheometer (MDR). Source: Ref. [13].

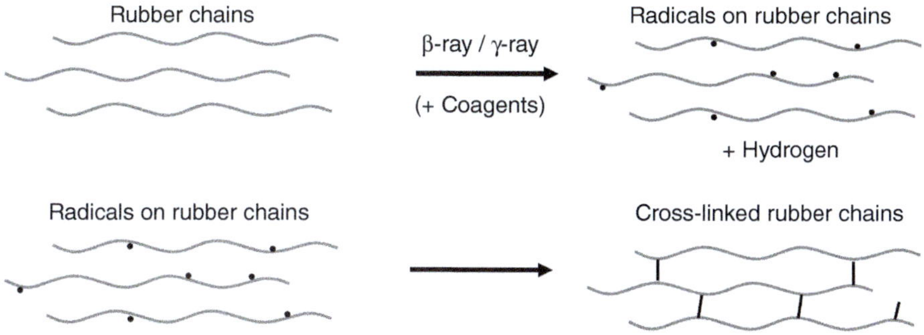

Figure 3.16 Cross-linking of rubber chains with radicals.

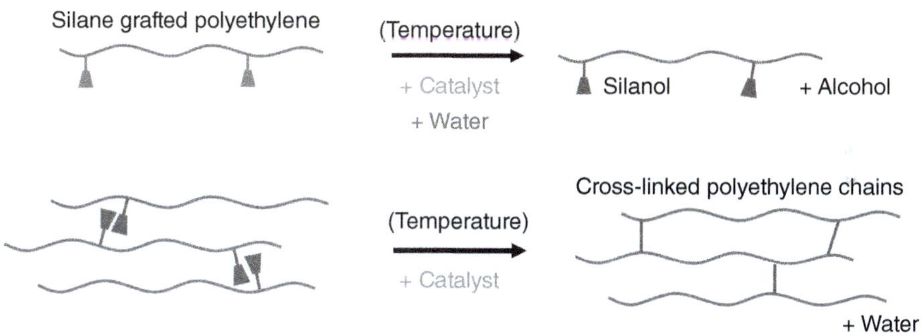

Figure 3.17 Cross-linking of rubber chains with silanes.

3.2.6 Other Cross-linking Systems

3.2.6.1 Cross-linking by Irradiation

Normally no cross-linking ingredients are necessary. Cross-linking follows a radical chain mechanism as with the peroxide-based systems. The only difference is the source of the initiating radicals. No limitations exist with regard to the rubbers (Figure 3.16).

3.2.6.2 Cross-linking with Silanes

Cross-linking techniques with silane systems are normally used today with ethylene-based polymers (Figure 3.17).

3.3 Compounding

The compounding of rubber compounds in the cable industry normally takes place in internal mixers discontinuously. Internal mixers consist of piston, hopper, mixing chamber, rotors, and folding saddle, whereby the geometry of the rotors significantly influences the intensity of the compounding. There are two different rotor systems (Figure 3.18):

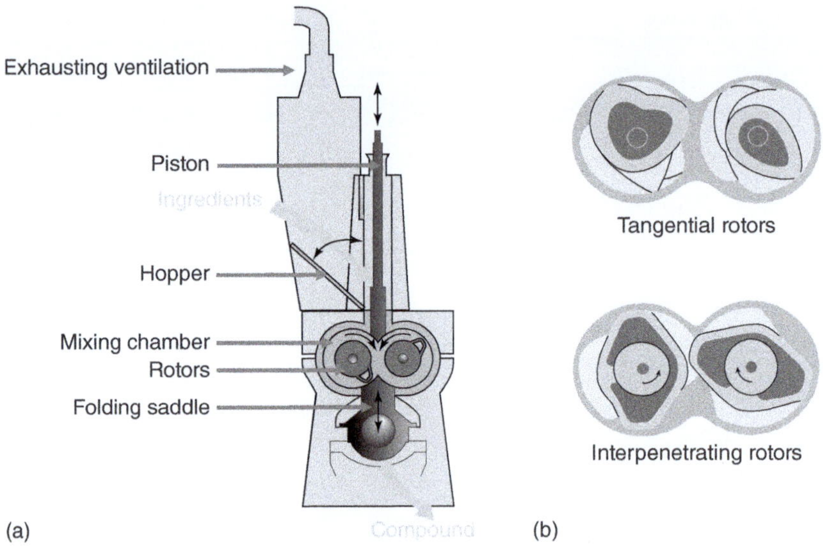

Figure 3.18 (a) Internal mixer. (b) Rotor systems. Source: Refs. [1, 15, 18].

- Tangential rotors
- Interpenetrating rotors

Comparing the two, the interpenetrating rotor geometry with the resulting enhanced friction allows a more intensive compounding.

In addition, the rotor speed should be stepless adjustable, and effective cooling/heating of the rotors and the mixing chamber brings benefits too.

The compounding can take place in one, two, or more passes. It depends on the kind of compound and compound recipe on the one hand and compounding machinery on the other.

The filling degree and the resulting batch weights depend on the mixing chamber volume and the filling factor. Normally the filling factors for tangential systems are between 0.7 and 0.9, and for interpenetrating ones between 0.6 and 0.7 [19].

$$\text{effective chamber volume} = \text{chamber volume} \times \text{filling factor}$$

$$\text{batch weights} = \text{effective chamber volume} \times \text{compounds density}$$

Of course, beside this theoretical approach, the batch weights must be optimized practically. A lot of experience is needed.

By compounding in one pass the cross-linking system is added into the mixer in the last third of the mixing cycle. Accurate temperature control is absolute required. The temperature of the compound batch must not increase over the pre-cross-linking threshold. Depending on the cross-linking system used the drop-out temperature is between 90 and 120 °C.

It is very important to have an excellent homogeneous dispersion of all ingredients during the compounding process in the mixer. For each compound recipe a compounding procedure should exist, a strict procedure for the order and the

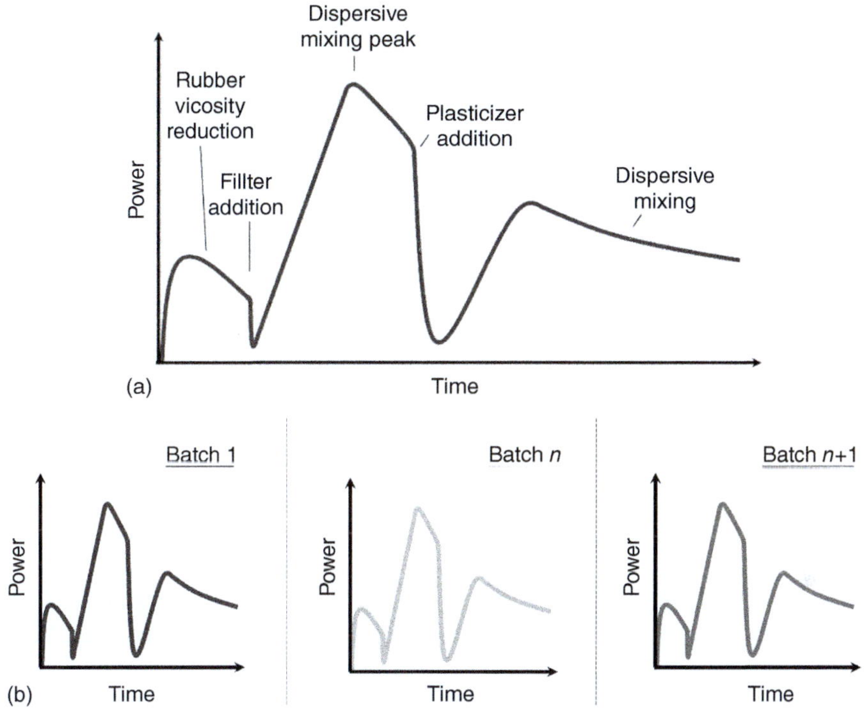

Figure 3.19 (a) Compounding diagram. (b) Optimal compound manufacturing. Source: Refs. [15, 17].

time of putting the ingredients into the dropper, the positions of the piston (up or down, piston pressure), rotor speeds, total cycle time, and drop-out temperature. The "classical" compound procedures start always with rubber. After a short plastification of the rubber the other ingredients are added according to the compounding procedure. Some rubber compounds can be manufactured with the so-called "upside down" procedure. In contrast to the "classical" procedures, fillers, stabilizers, and processing aids were put in the mixer first, pre-mixed, and shortly after, the rubber is added.

So, each batch of the same compound recipe should be produced identically. The quality of the mixing process, influenced by the filling degree, cycle time, mass temperature, rotor speed, and piston pressure, can be seen on comparing the compounding diagrams of each batch (temperature, cycle time, electric power consumption) (Figure 3.19).

After finishing the compounding process, the hot batch is removed by opening the folding saddle and goes directly to the next parts of the compounding line, getting formed and cooled down. Several compounding line arrangements are possible (Figure 3.20).

It is always advisable to be sensitive and keep an eye on the overall compounding process, starting from the ingredients themselves to the ready mixed compounds. Impurities can be introduced into the compounds, as well as with the ingredients and at the different stages of the compounding processes.

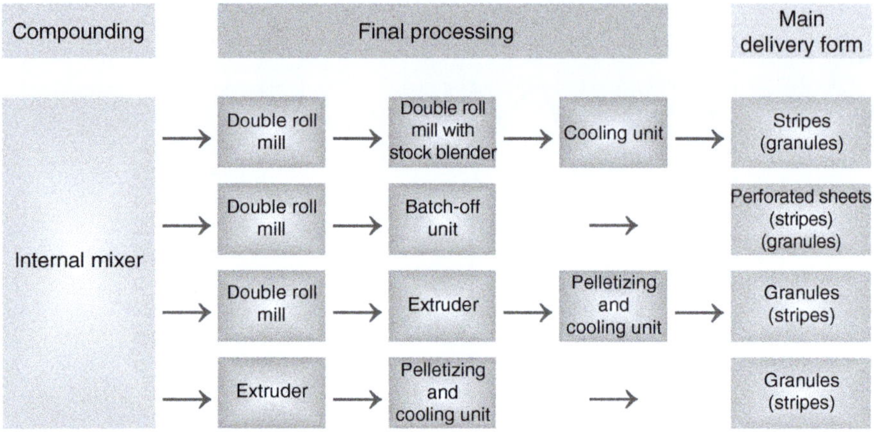

Figure 3.20 Compounding line arrangements.

In addition, as the compounding lines mainly consist of different kind of metals, alloys, and steel, a certain level of abrasion cannot be excluded. To ensure the compound quality at a high level, sieving systems, magnets, metal detectors, and strainers can be implemented at different parts of the compounding lines. State of the art for insulation compounds is the sieving over special sieve packages, where the smallest sieve size goes down to 75–80 μm. The sieving process can be controlled with backpressure. An increasing backpressure indicates an increasing sieve load with impurities. To avoid a sieve break through the change of the sieve package at the right time is important. Undoubtedly, these additional steps are necessary, and make the overall compounding process much more complex and challenging.

In order to avoid the sticking of the sheets, stripes, and granules, release agents are required. Depending on the process stearates, metal soaps and talc are used.

To validate the compound quality, density, Mooney viscosity, and cross-linking behavior can easily be used.

3.4 Extrusion

Rubber compounds are applied on the core or core assemblies by the use of single screw extruders (Figure 3.21). The extruder size is specified over the screw diameter d (common diameters between 60 and 300 mm) and multiples of the screw diameter as screw length l. The screw length corresponds to the barrel length. A very common ratio is $l{:}d = 16$ [1].

Granules or stripes – the two common delivery forms of the rubber compounds – have an influence on the feeding system of the extruder. In the case of stripes, a heated feeding roll is needed in any case. The screw transports and compresses the rubber compound to the extruder head. In the die the plasticized rubber compound is shaped on the core/core assembly. Problems exist in continuously feeding the extruder with the compound, such as the following:

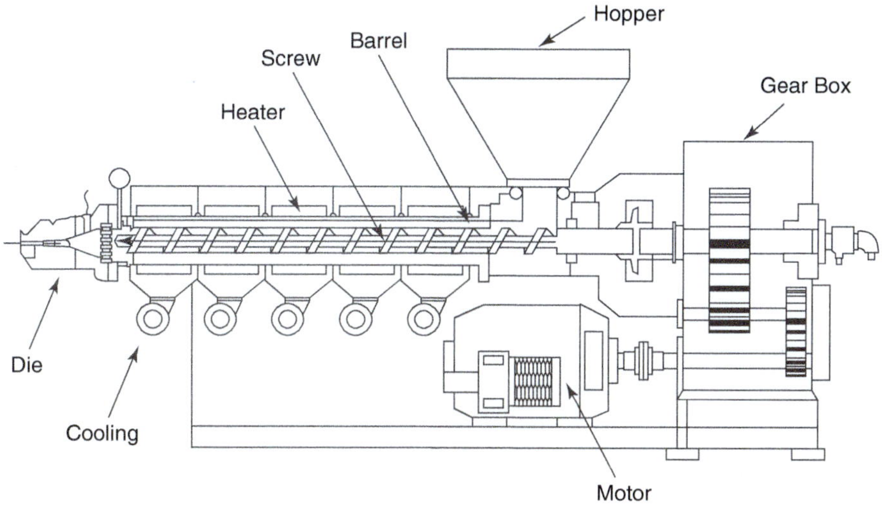

Figure 3.21 Extruder [1].

- Stripes sticking together due to irregular application of the release agent on the stripes
- Brittle or tight stripes because of exceeding the storage time
- Formation of granules bridges due to too little powdering of the granules
- Insufficient granules transport due to too much powdering of the granules
- Insufficient granules transport due to big variety in the granules size

These problems lead to pulsation of the extrudate and finally to unacceptable changes in the wall thicknesses.

During the extrusion process the mass temperature must be, on the one hand, high enough to grant an optimal excellent extrudability and, on the other, significantly lower than the starting temperature of the curing system. In addition, the line speed must be chosen carefully with the view on a possible friction resulting in an increase in the extrudate temperature leading to pre-curing in the extruder and/or die. By measuring the temperature of the heating zones, head and die, the extrudate temperature and pressure, the extrusion process can be controlled and is adjustable. It is always helpful to measure the compound output and the mass temperature depending on the number of screw rotations per minute to obtain information about the compound behavior on a certain extruder.

Different extrusion line configurations are possible:

- Single head
- Tandem single head
- Double head
- Tandem double head
- Triple head

So, up to three rubber compounds can be extruded at the same time. With complexity of the extrusion line the challenges for the measurement and control techniques are rising. The focus is on the wall thicknesses and the centricity.

Most challenging is the centering before starting the extrusion and vulcanization process. Especially in the case of the pressurized vulcanization techniques a lot of experience is required to adjust all process parameters accurately considering the diameters before and after the vulcanization line, the increasing of the line speed to the process line speed and a possible shrinkage of the wall thicknesses, which can occur.

3.5 Cross-linking/Vulcanization

To obtain the final properties the rubber compounds have always to be cross-linked after the extrusion process. This is the main processing contrast to most of the thermoplastics and TPEs. In the vulcanization/cross-linking process always chemical reactions take place. Energy is necessary.

The oldest vulcanization technique using hot air does not play any role in the cables industry today. The big disadvantage of this vulcanization technique, the unpreventable occurrence of porosities, led to the development of the vulcanization in autoclaves. In a first step the ready extruded cables had been jacketed with lead with thicknesses between 1 and 3 mm by use of lead presses or extruders [1]. The lead jacketed cables were put on drums. Afterwards these drums were put into the autoclaves and vulcanized in hot steam. After cooling the lead was removed and used again. The big advantages of this technology are the prevention of pores and deformations of the core assembly in multicore cables as well as an excellent cable surface finishing. On the other hand, there is a limitation to the production length and discolorations on the cable surface can occur. The vulcanization under lead sheaths had been well established in the cable industry. Today the vulcanization under lead is still used, but has no longer a significant importance, among others, especially because of the increasing environmental relevance of lead.

Since many years the continuous vulcanization (CV) techniques are state of the art in the cable industry. Cross-linking takes places immediately after the die of the extruder head. The vulcanization normally takes place in tubes. After vulcanization water cooling is always necessary.

The following vulcanization techniques are possible.

3.5.1 Vulcanization in Saturated Hot Water Steam

Developed already in the 1950s, CV in saturated hot water steam is to date the most widely used vulcanization technique in the cable industry [1].

Immediately after the extruder head the cable goes directly into a steam tube.

Based on the arrangement of the extruder and vulcanization tube the vulcanization process is distinguished as follows:

- Vertical continuous vulcanization (VCV)
- Horizontal continuous vulcanization (HCV)
- Catenary continuous vulcanization (CCV) (Figure 3.22)

Catenary lines are the most spread CV lines (Figure 3.22). With increasing cable weight, the risk of touching the cable on the tube ground is increasing.

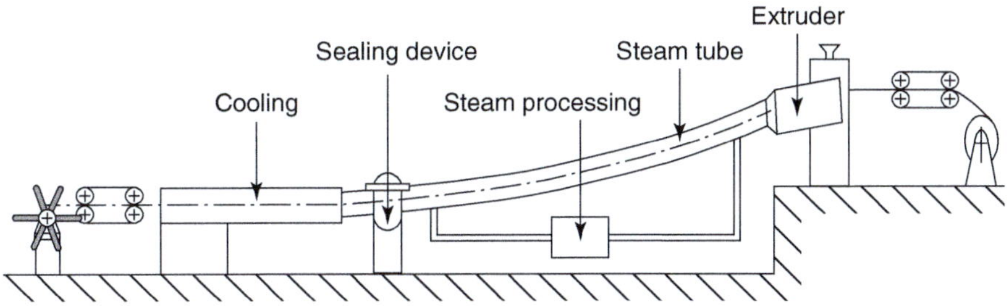

Figure 3.22 Catenary continuous vulcanization line [1].

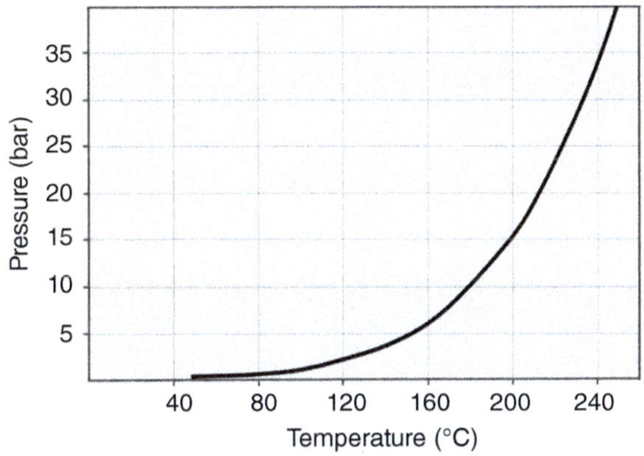

kp/cm²	bar	°C
6.0	6.1	164.0
8.0	8.2	174.4
10.0	10.2	183.1
12.0	12.2	190.7
14.0	14.3	197.4
16.0	16.3	203.4
18.0	18.4	208.8
20.0	20.4	213.9
25.0	25.5	225.0

Figure 3.23 Saturated hot water steam: correlation steam pressure and temperature [20].

Therefore, application of a certain tension load on the core assembly is necessary. This requires complex and sensitive monitoring, control and tuning systems, avoiding an extension of the conductors. Vulcanization tubes for smaller core or cable cross sections can be up to 150 m long and even longer, in contrast to bigger cross sections with shorter tube length, significantly below 100 m.

Vertical lines need high tower buildings. Therefore, there is a clear limitation due to costs and construction requirements. On the other hand, the tension load on the core assembly is much lower. Vertical lines are perfectly suitable for cables with large cross sections.

Horizontal lines are dedicated just for small cross sections. The arrangement is slightly sloped.

Heat transition occurs with saturated hot water steam. Owing to the condensation of the saturated hot water steam on the cable surface peroxide curing is also possible.

Normally steam pressures between 6 and 25 bars are applied. There is a direct correlation between steam pressure and temperature (Figure 3.23).

The applied steam pressures (and as result the temperatures) are strongly dependent on different parameters:

- The curing system of the rubber compound
- The cable design (e.g. wall thicknesses, core dimensions, metallic screens, bonding/non-bonding of layers)

- Tube length
- Required/desired line speeds

It needs no mention that with increasing wall thicknesses, increasing cross sections, and metallic contents there is always a temperature gradient inside the cores/cables during the time in the steam tube, even at low line speeds.

Especially for multi-core cable designs it should be clearly noted that with increasing steam pressure more deformations of the cores can occur.

The relatively high steam pressures and corrosive decomposition products are challenging the tube manufacturers.

3.5.2 Vulcanization in Liquid Salt Mixtures under Pressure

In the 1970s the pressurized liquid continuous vulcanization (PLCV) was developed in the United Kingdom. For the required heat transition a melted eutectic mixture of potassium and sodium nitrite/nitrates is commonly used. Avoiding pores coming from decomposition products or humidity, a system pressure of air or nitrogen is applied, normally around 6 bars. If a peroxide curing system is used, oxygen has to be avoided. Typical salt bath temperatures vary between 180 and 220 °C. The vulcanization takes places normally in tubes, sloped or catenary line arrangement [1].

The melted salt mixture runs in a closed circuit. The cross-linking occurs by passing the cable through the molten salt mixture in the pressurized tube, getting heated.

A big advantage of the salt bath vulcanization technique is the excellent heat transition. Furthermore, in multi-core cables the low system pressure helps to avoid deformations of the core assembly better. Salt bath vulcanization is especially suitable for large cross-sections with high cable weights. The cable is immersed in the salt bath.

Of course, there are disadvantages too; much focus on the compound ingredients is required – on the one hand, to avoid the formation of nitrosamines, and on the other to avoid the migration of ingredients into the salt, increasing the risk of fire.

The permanent risk of fire, which should not be ignored, can be reduced by the following:

- An effective temperature control of the molten salt bath, preventing localized spots and overheating
- Regular cleaning of the salt, removing impurities
- Avoiding direct contact between hot salt and water, generating steam
- Gentle compound recipes, preventing the migration of low melted ingredients into the salt.

Other important aspects needed to be respected:

- Much more focus on the compound ingredients is necessary to avoid the formation of nitrosamines.
- The cooling water must be cleaned regularly due to nitrite/nitrate environmental reasons.

3.5.3 Electron Beam Cross-linking

Cross-linking using radiation entered the cable industry in the 1960s [1].

In contrast to the other cross-linking procedures, heat transfer is not required. Therefore, no cross-linking agents are necessary usually.

In principle, there are two kinds of radiation possible [21]:

- Electron beam or β-ray
- γ-ray

The difference between both is the processing time and the penetration depth of the radiation.

As for cross-linking of polymers in general relatively high radiation doses are necessary, electron beams are mainly used for cross-linking cables.

The electron beams are generated by electron accelerators. Electrons, emitted from a heated cathode, are focused and accelerated in an electrical field along an evacuated tube. The speed of the electrons and with that the energy of the electrons are directly dependent on the voltage between the cathode and the anode. Then the electron beam passes through an alternative magnetic field. At the end of the scan horn, passing a thin metallic window, the electron beam is fanned out. The cables pass through this curtain of electrons and receive their required irradiation dose (Figure 3.24).

The electrons penetrate the rubber cable, which results in stimulation and ionization of rubber chains in the irradiated cable compound. Cross-linking occurs by a radical chain mechanism. The penetration depth is dependent on the accelerator voltage whereas the processing speed is dependent on the quantity of electrons generated by the accelerator amperage.

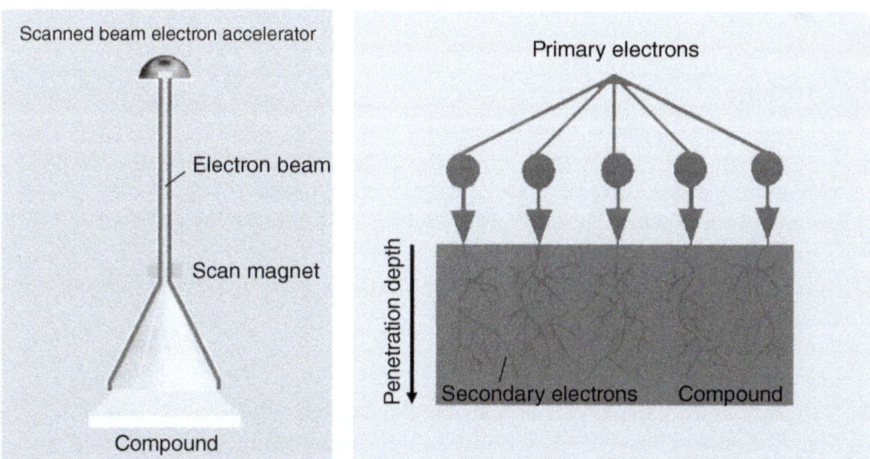

(a) (b)

Figure 3.24 (a) Electron accelerator. (b) Irradiation principle. Source: Refs. [21, 22].

The following are the main challenges for the electron beam cross-linking:

- Expensive irradiation protection is required. The irradiation part must be in a concrete bunker.
- Avoiding voltage flashovers, the electron accelerator must be in pressure tanks filled with an inert gas, normally sulfur hexafluoride.
- With increasing wall thicknesses and considering the low penetration depth of the electrons, for a consistent optimal irradiation over whole cable cross-section the cable has to pass many times the curtain of electrons, ideally turning during passing. So, special optimized core/cable handling designs are required. If it is possible from the equipment point of view, a two-sided irradiation gives also advantages.

Before starting a new product run, dose mappings ensure on the one hand an optimal cross-linking of the cable, and reducing the risk of depolymerization of the polymer(s) on the other side.

Comparing the different cross-linking techniques on identical cables with the same cross section, much higher line speeds can be achieved with electron beam cross-linking.

During the irradiation process the cable is heated up. Water cooling at and/or after the irradiation process is required. In addition, ozone and nitrogen oxides are formed.

In principle two different procedures are possible:

- Online procedure: extrusion and cross-linking in one process line
- Offline procedure: extrusion and cross-linking in separate process lines

The online procedure is done in house at the cable manufacturers and mainly dedicated for small cross sections. High line speeds can be applied.

For larger cross sections the offline procedure is the procedure of choice. Cable manufacturers who do not have any electron beam cross-linking facilities locally can go externally for the cross-linking.

References

1 Röthemeyer, F. and Sommer, F. (2013). *Kautschuktechnologie*. Carl Hanser Verlag GmbH & Co. KG.

2 Giersch, U. and Kubisch, U. (1995). *Gummi – Die elastische Faszination*. Dr. Gupta Verlags GmbH.

3 Köhler, F.E. (1897). Köhler's Medizinal-Pflanzen. https://de.wikipedia.org/wiki/Guttaperchabaum.

4 Fotos private.

5 Kausch-Blecken von Schmeling, H.-H (2012). Halle – Zürich – Freiburg Hermann Staudinger und das Konzept der Makromoleküle. Lecture at Martin-Luther-University Halle-Wittenberg, Jacob-Vollhard-Auditorium, GDCh-series of lectures, 19th December 2012.

6 Heimlich, S. (1998). *Portraits in Plastik – Pioniere des polymeren Zeitalters*. Hoppenstedt Publishing.

7 Holden, G., Kricheldorf, H.R., and Quirk, R.P. (2004). *Thermoplastic Elastomers*, 3e. Carl Hanser Verlag.
8 Morton, M. (1977). *Journal of Polymer Science*, Polymer Symposium 60: 1.
9 Wegner, G. (1978). *Kautschuk Gummi Kunststoffe* 31: 67.
10 Rader, C.P. and Stemper, J. (1989). *Progress in Rubber, Plastics and Recycling Technology* 6: 50.
11 Holden, G., Bischop, E.T., and Legge, N.R. (1969). *Journal of Polymer Science Part C* Part 26C: 37.
12 Holden, G. (1973). Properties and applications of elastomeric block copolymers. In: *Block and Graft Copolymerization*, vol. 1 (ed. R.J. Cesera). John Wiley & Sons.
13 Technical info brochure TA Instruments: Rubber Testing (2013).
14 ASTM D 1418 (2001). ASTM International, issued 1 May 2001.
15 Training documents "BAYER Rubber Course" (1995).
16 Kleemann, W. (1982). *Mischungen für die Elastverarbeitung*. VEB Deutscher Verlag für die Grundstoffindustrie.
17 Schnetger, J. (2004). *Lexikon Kautschuktechnik*. Hüthig Verlag.
18 DGUV (2008). *Regel 113-010: Sicheres Arbeiten in der Gummiindustrie*. DGUV.
19 Bloch, H.P. and Soares, C. (1998). *Process Plant Machinery*, 2e. Butterworth-Heinemann.
20 BAYER (1991). *Handbuch für die Gummi-Industrie*. Bayer AG.
21 Technical info brochure Beta-Gamma-Service: Strahlenvernetzung von Kunststoffen (10/1993).
22 Technical info brochure Isotron: Electron Beam Processing (11/2008).

4

Extrusion of Cables

Stéphan Puissant

Principality Building Society, Cardiff, Wales, UK

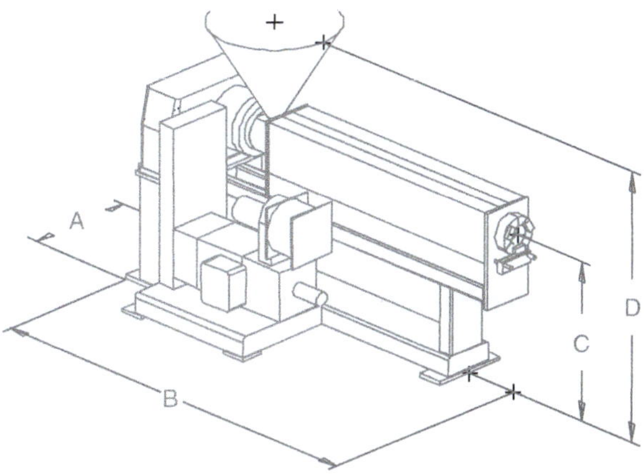

Example of extruder. Source: Maillefer SA.

4.1 Historical Introduction to Cable Extrusion

The first extruder (with a piston) was built by Joseph Bramah (1748–1814) from the United Kingdom to produce lead pipes. Gutta-percha, a gum produced from the milky sap of a tree from the south east of Asia, was discovered in 1843 [1]. Faraday found that this material has exceptional insulating properties. This material was used since then for the insulation of telegraphic cables. Nevertheless, Faraday was dissatisfied with the difficulty in obtaining continuous insulation without a weld line. He contacted his fellow citizen H. Bewley who invented the first extrusion line in 1845. These first extrusion lines already had the main components that are still in use today (Figure 4.1): extruder (letter A in Figure 4.1), cross head (letter B in Figure 4.1), and cooling trough.

[1]Currently Nexans Research Center, 17 rue Lortet, 69007 Lyon, France

The Global Cable Industry: Materials, Markets, Products,
First Edition. Edited by Günter Beyer.

Figure 4.1 One of the first extrusion lines from the middle of the nineteenth century. Source: Picture from the Porthcuno Telegraph Museum.

In the next century, during the 1930s new thermoplastics, PVC and PE, appeared. This last material, due to its exceptional dielectric characteristics, allowed the development of high frequency use, and of the radar in the United Kingdom, during the Second World War. Since then, the use of thermoplastics has continuously increased, until representing almost all insulation materials.

Even if the line construction may vary according to the different types of cables, one can recognize the main elements always constituting a cable extrusion line (cf. Figure 4.2). These elements are

* the pay off, followed, if necessary by the wire-drawing and annealing devices, or by a wire assembling machinery.
* the extrusion group composed of the extruder, and the cross head,
* the cooling trough,
* the pulling machinery, and
* the take-up device.

All cables, for electrical power or data transmission, have to be insulated electrically. Elastomer or thermoplastic materials that have good electrical properties are used for this insulation.

Extrusion, a manufacturing process to produce parts of constant straight section, is used to produce the insulation for cables. In the majority of cases, insulating material forming is carried out with the help of a single screw extruder. The extruder will have the function to melt the material in order to push it into the extrusion (or forming) head so that the insulating material is deposited on the conducting wire. The description of the single screw extrusion process, as also of the extrusion head, is also described in [2–4].

Nevertheless, the extruder alone would not be enough to guarantee product quality. Other elements placed along the production line play an important role. Besides the extrusion group (extruder and head), we will find in all lines winder, cooling troughs, diameter measurement devices, and reel wind up devices.

But the specificity of each line can be related to the three segments of the cable market. These are traditionally the energy transport segment, the copper wire telecommunications, and the fiber-optic telecommunications segment.

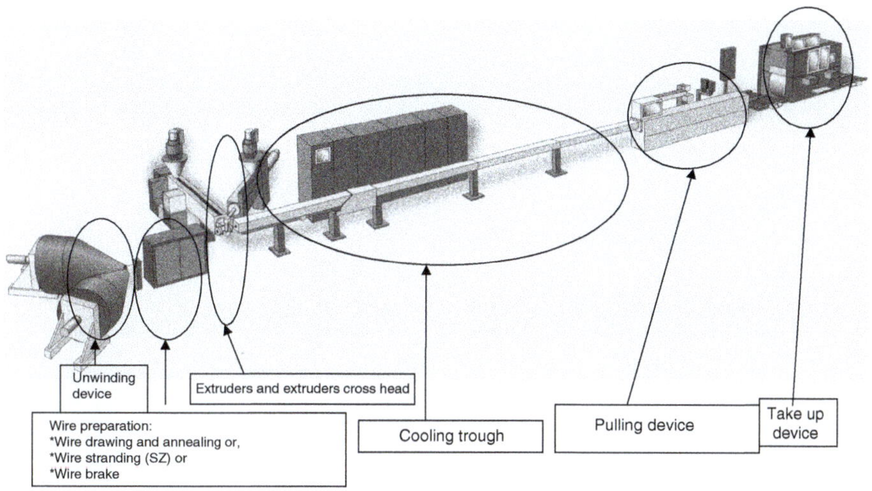

Figure 4.2 Figure presenting the different parts of a cable extrusion line. Source: Maillefer drawing.

In this part, we will propose a review of the cable line elements that have an influence on the melting, mixing, forming, and cooling of the insulation.

4.2 Extruder in Cable Lines

The extruder is an essential component of a cable line. In majority of the cases, it is a single screw extruder whose description is given below. It can be used in a classical manner or for reactive extrusion (foaming, cross-linking). During its "classical" use, the extruder has the purpose of conveying, melting, homogenizing, and putting under pressure the material.

Moreover, for two particular extrusion processes for cables, which are foaming and cross-linking, the extruder may be used as a chemical reactor. The reasons for the use of this particular process are related to the specific characteristics required of the cables. For data transmission, the signal propagation speed depends on the dielectric constant of the cable; for power transmission cables, a sufficiently high mechanical resistance at high temperature is needed.

4.2.1 Description of the Single Screw Extruder

The main function of an extruder is to melt, mix, and then generate the pressure necessary to pass the cross head. This pressure is needed to push the molten material through the extrusion die. The extrusion screw is always divided in three functional parts, the feeding, melting, and metering zones. Sometimes, a mixing zone may be added.

4.2.1.1 Different Functional Screw Zones

In order to realize the process in a continuous manner, a screw rotating inside a barrel (or cylinder) is used. The screw and barrel are specified by the ratio of

• Example of geometry: $D = 38$ mm screw for PE

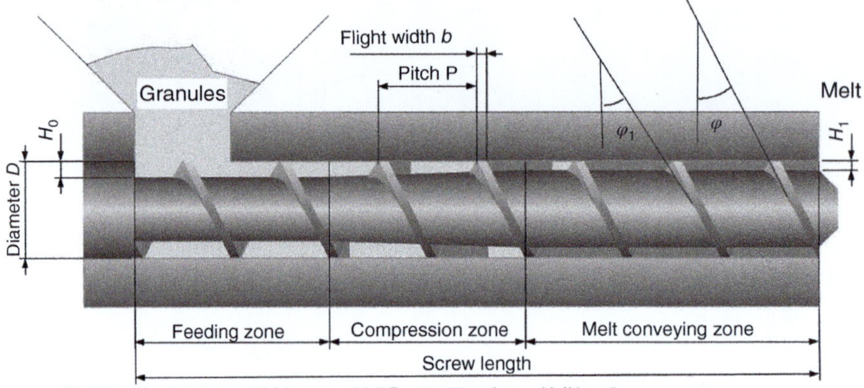

Real length of screw = 1000 mm = 26.3D, compression = $H_0/H_1 = 3$

	Feeding zone		Compression	Melt conveying
	Hopper	Feeding		
Length	1.3 D	14 D	3 D	8 D
Depth	6.5 mm	6.5 mm		2.1 mm

Figure 4.3 Sketch of the different zones of a compression screw $D = 38$ mm.

the working length to the diameter (L/D). The parameters describing a screw geometry are: D: nominal diameter, H or d: H (=channel depth) related to d (=root diameter) by: $H=(D-d)/2$, b: flight width, n: number of flights (here 1), P: flight pitch.The screw is divided into parts of different functionalities (Figure 4.3):

* *Feeding zone (beginning under the hopper)*: Helps to feed and compact the still cold granules.
* *Melting zone*: This zone is needed to melt the polymer.
* *Metering (or compression) zone*: It is needed to homogenize the melt and sometimes generate the pressure necessary to pass in the forming tool.
* *Mixing (or Melt conveying) zone*: In certain cases, it is necessary to mix color master batches or chemically active components in a homogeneous manner in the material; for this reason, mixing devices may be added.

Moreover, sometimes (foaming or cross-linking) the extruder has a role of initiating a chemical reaction. So, during chemical foam extrusion (telecommunication wires), or in the case of grafting silane groups on molecular chains (Monosil and Sioplast type extrusion), the end part of the screw must also play the role of a chemical reactor.

Diverse metering zones alternating with mixing zones are used. These metering and mixing zones are always positioned after the melting zone.

4.2.1.2 Description of a Screw Geometry

In Figure 4.3, a simplified screw geometry is represented for a screw of diameter $D = 38$ mm and length = 1000 mm = 26.3 × D, used for PE extrusion.

The geometry is described by the pitch of the flight(s) p, the number of flight(s) n, the width of the flight(s) b, and the depth H. On the nominal diameter D of the screw, the flight angle φ is related to the pitch p and the diameter by

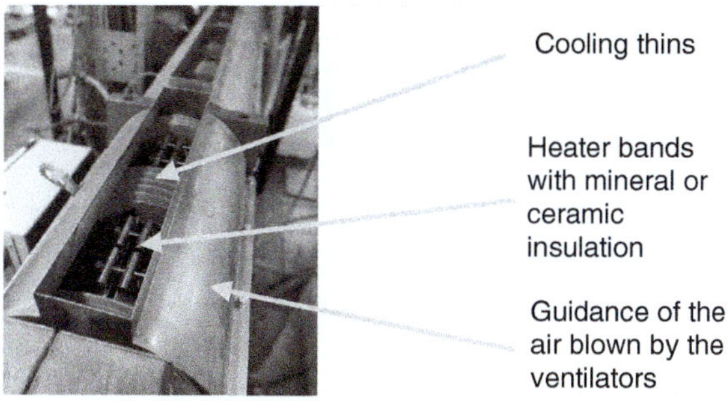

Figure 4.4 Heating bands and cooling thins. Source: Maillefer extrusion.

$\varphi = \arctan\left(\frac{P}{\pi \cdot D}\right)$. This flight angle is greater at the bottom of the flight than on top. The average angle φ_{av} (in the middle of the flight) is related to the average diameter $D_{av} = (D+d)/2$, so as the angle φ_1 (at the bottom of the flight) to the bottom diameter d. In a compression screw, the compression ratio of the screw is defined by the ratio of the height in the feeding zone to the height in the melt conveying zone.

The barrel (or cylinder) in which the screw is turning is regulated in temperature (Figure 4.4). The most common case is the use of electrical heating bands (with electrical resistances inside), and surrounding the external diameter of the barrel so as to ensure heating. To these bands, thins, or metallic stripes (good heat conductor such as copper, aluminum), are added. These thins, or stripes, are in contact with the external surface of the barrel, and are disposed radially to it. They ensure the cooling function under the action of air blown by the ventilators.

In the case of cross-linkable polymer extrusion, and in particular for high voltage energy cables, sometimes a precise temperature control of the barrel, based on the use of fluids (like oil or water), may be used. This point may be also justified for the extrusion of silicone, or rubber type materials, for which the work temperatures are quite low. A higher heat transfer coefficient of a liquid against air helps to better evacuate the dissipated heat, even at low processing temperatures.

The diameter of the extruder depends on the necessary output, and may vary from 20 mm (stripes on isolated wires) to 250 mm (insulation of a high voltage cable). The length of the extruder is expressed as a function of the diameter. It is, for common applications, 22–24 times the diameter. Nevertheless, in the case of reactive extrusion as also foaming or the production of cross-linked PE, the length may often be 30, and sometimes even 38 times the diameter.

More detailed information on the extruder may be obtained in [5–7], or [2, 3].

We will now describe in detail the three zones of the extruder: feeding zone, melting zone, and metering zone.

4.2.2 Feeding Zone of the Extrusion Screw

4.2.2.1 Friction-Based Feeding Mechanism

When granules slide on the metallic surface of the screw and the barrel, a friction force results. In his book, Ref. [5] proposes to use Coulomb's law, where the friction force is equal to the normal force or pressure (cf. Figure 4.5) multiplied

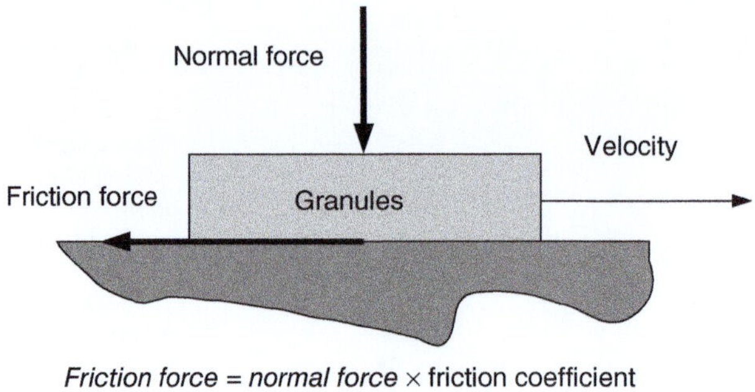

Figure 4.5 Friction coefficient.

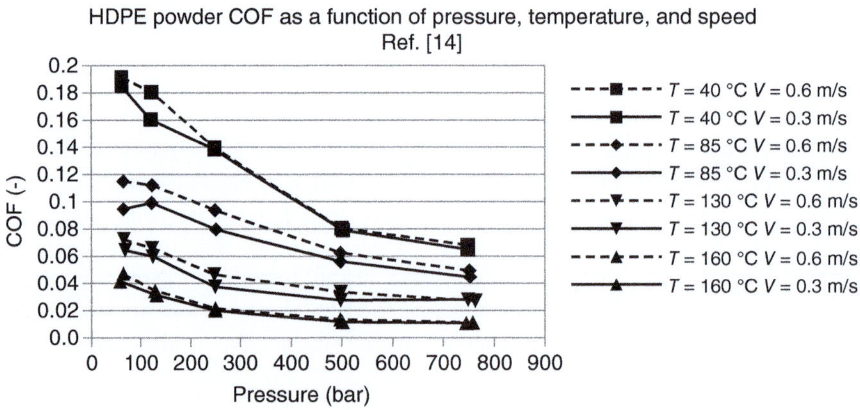

Figure 4.6 Evolution of the friction coefficient as a function of pressure for different sliding speeds and temperatures.

by the friction coefficient (COF).

$$F_{\text{friction}} = F_{\text{normal}} \times f_{\text{friction}}$$

As shown by [8], with HDPE (high density polyethylene) powder the COF is not dependent on the sliding speed but mainly changes with the interface temperature, as with the pressure (Figure 4.6 drawn with the data collected).

In the simple model presented below, the coefficient of friction is supposed to be independent of the sliding speed, but also of the pressure. And the friction coefficient on the screw is named f_s and on the barre f_b.

4.2.2.2 Simple Modeling of the Feeding Zone

In Figure 4.7, a ring of granules, considered as an incompressible solid, is shown. The screw is considered to be static and the barrel rotating. The speed of the barrel versus the screw $v_{b/s}$ (m/s) is related to the rotation speed N (rpm) and the diameter D (m) by $v_{b/s} = (\pi \cdot D \cdot N)/60$.

The balance of the forces on this ring induces a movement of the ring, movement with a direction imposed by the screw flights. The length of the ring is related to the peripheral length ($= \pi \cdot D$) diminished by the flight width ($= b/tg(\varphi)$).

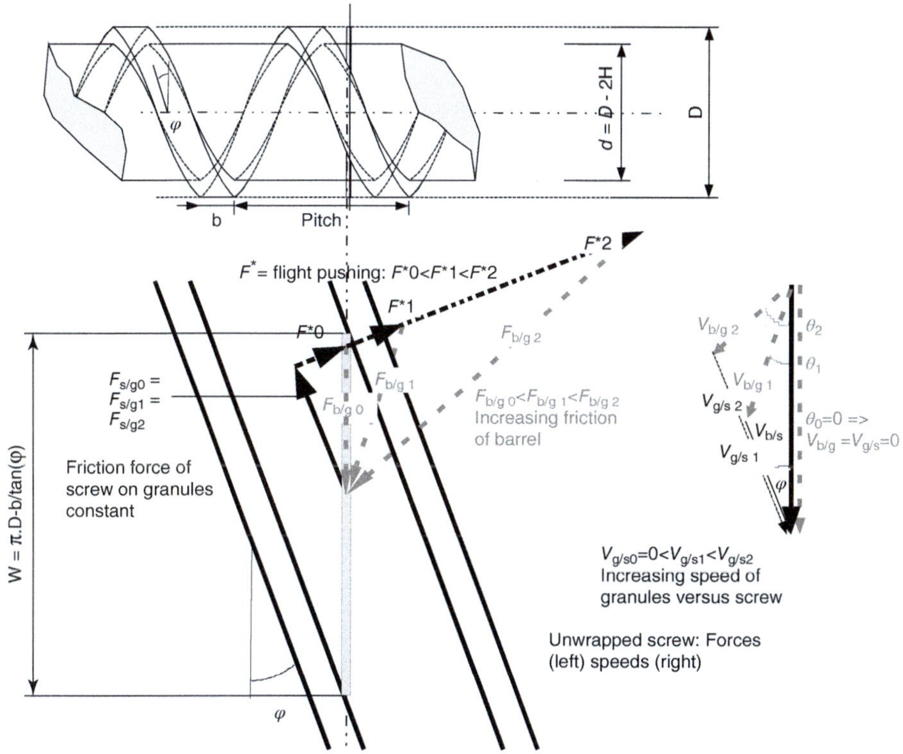

Figure 4.7 Simplified representation of the extruder feeding zone (a) screw feeding part geometry, (b) velocity diagram.

The ring length W_1 on the channel bottom is $W_1 = \pi \cdot d\text{-b}/\tan(\varphi_1)$, in the center $W_{av} = \pi \cdot D_{av}\text{-}b/\tan(\varphi_{av})$, and close to the barrel $W = \pi \cdot D\text{-}b/\tan(\varphi)$.

Balance of forces

The ring of solid material is represented unwrapped. The pressure is supposed to be the same in the whole ring. For simplification, in this figure the difference in pressure along the screw axis is supposed to be equal to 0. With this last hypothesis, only three forces act on the ring:

* On the external surface, the friction force of the barrel ($F_{b/g}$) has an unknown direction.
* On the internal diameter of the ring, the friction force of the screw ($F_{s/g}$) is oriented parallel to the flight, and its direction is opposed to the sliding movement of the screw.
* On the face on contact with the "pushing" flight, we have an unknown force F^* normal to the flight.

The balance of force gives $F_{s/g} + F^* + F_{b/g} = 0$.

The barrel friction force may increase ($F_{b/g0} < F_{b/g1} < F_{b/g2}$) due to different temperature settings, but the screw friction force is considered to remain constant. As the direction of $F_{s/g}$ remains constant and parallel to the flight and the direction of the unknown force F^* remains perpendicular to the flight, the angle θ between $F_{b/g}$ and the ring increases, $\theta_0 = 0 < \theta_1 < \theta_2$. The normal force F^* will also increase, with its direction remaining constant.

Velocities

The velocities considered are

* the speed of the barrel versus the screw ($V_{b/s}$, in vertical direction),
* the sliding of the ring against the screw ($V_{g/s}$, parallel to the flight), and
* the speed of the barrel versus the granules in the ring ($V_{b/g}$).

These three speeds verify $V_{b/g} = V_{b/s} + V_{s/g}$. The speed of the barrel versus the screw remains constant. The sliding of the ring of granules versus the screw has a constant direction. The frictional force of the barrel on the external surface of the ring gives the direction (with the angle θ) of the sliding speed $V_{b/g}$. From this, we can obtain the size of the velocity of the ring versus the screw.

If the friction force on the barrel $F_{b/g} \cdot \cos(\varphi) < F_{s/g}$, there is no movement of the ring inside the screw. By increasing the friction on the barrel, the velocity of the ring in the screw increases from $V_{g/s1}$ to $V_{g/s2}$.

This shows that the ratio of the COFs on the screw and on the barrel is very important, and that we should try to have the highest friction on the barrel, with the lowest on the screw.

4.2.2.2.1 *Volumetric Output Versus Screw Flight Angle* The balance of all forces and torques applied on the ring gives two equations. The unknowns are F^*, the pushing force of the flight, and θ the conveying angle. Using a geometry-dependent factor k, we can have an expression for the conveying angle:

$$\theta\,(rd) = \arcsin\left(\frac{-(M \cdot k) + \sqrt{k^2 - M^2 + 1}}{(1 + k^2)}\right)$$

with M and k defined through geometrical parameters by

$$M = \frac{1}{W \cdot f_b}\begin{pmatrix} 2 \cdot H \cdot f_s \cdot \left(k + \text{ctg}\,(\varphi_{av}) \cdot \frac{D_{av}}{D}\right) \\ + \\ W_1 \cdot f_s \left(k \cdot \sin\,(\varphi_{av}) + \cos\,(\varphi_{av}) \cdot \frac{d}{D}\right) \\ + \\ \frac{1}{z} \ln\left(\frac{P(z)}{P_0}\right) \cdot k \cdot W_{av} \cdot \frac{D-d}{2} \end{pmatrix}$$

$$\text{and } k = \frac{\left(\sin\,(\varphi_{av}) + f_s \cdot \cos\,(\varphi_{av})\right) \cdot D_{av}}{\left(\cos\,(\varphi_{av}) - f_s \cdot \sin\,(\varphi_{av})\right) \cdot D}$$

The output becomes

$$Q_v = V_{b/s} \cdot \left(\frac{\sin\,(\varphi_{av}) \cdot \sin\,(\theta)}{\sin\,(\varphi_{av} + \theta)}\right) \cdot \left(\frac{\pi}{4} \cdot (D^2 - d^2) - n \cdot \frac{b}{\tan\,(\varphi_{av})} \cdot (D - d)\right)$$

or

$$Q_v = V_{b/s} \cdot q \cdot \left(\frac{\pi}{4} \cdot (D^2 - d^2) - n \cdot \frac{b}{\sin\,(\varphi_{av})} \cdot (D - d)\right),$$

$$\text{with } q = \left(\frac{\sin(\varphi_{av}) \cdot \sin(\theta)}{\sin(\varphi_{av} + \theta)} \right)$$

4.2.2.2.2 *Influence of the Flight Angle* In the case of no pressure increase, (i. e. $P(z)/P_0 = 1$), M and k, and therefore the conveying angle θ are only dependent on the screw geometry, and the 2 friction coefficients f_s and f_b. As seen from the expression above, if the width of the flight b is neglected ($b \sim 0.1^*D$), the output Q_v changes with the peripheral speed (rotation speed), D (nominal diameter), d (root diameter), and factor q. At a constant rotation speed (i. e. peripheral speed) and fixed diameters d and D, the output depends only on q. Finally, q depends only on friction coefficients and flight angle (i.e. pitch), as represented in Figure 4.8.

By using the same set of coefficient of friction on the screw and on the barrel of $D = 100$ mm, and $d = 80$ mm (i.e. $H = 10$ mm), Figure 4.8, giving the evolution of q in function of the flight angle, is obtained. For the majority of polymers, the coefficient of friction is somehow between 0.2 and 0.5. For those values, an optimum flight angle of slightly less than 20° is obtained. This explains why in a majority of cases, the pitch of the flight is taken equal to 1D (or 17.65°).

4.2.2.2.3 *Bulk Density for Mass Output Calculation* The mass output inside the screw is related to the volume output of the feeding zone and the bulk density given by the granules. This bulk density is mainly dependent on the density of the material, the size and shape of the granules, and also on the material's ability to deform.

The size of the volume containing those granules may also play a role. Therefore, for granules with sizes around 2 mm, an extruder of diameter $D = 30$ mm, having channels with depth around 3–4.5 mm, we will have a lower bulk density than for a screw of diameter = 150 mm, with channels having a depth of 15–22.5 mm.

In Table 4.1, some indicative values of the ratio of bulk and solid density are given, in the case of an extruder of diameter 60 mm, for different materials.

4.2.2.2.4 *Influence of Channel Depth and Counterpressure on Output* The below graph in Figure 4.9 represents the calculation results for a 100 mm extruder

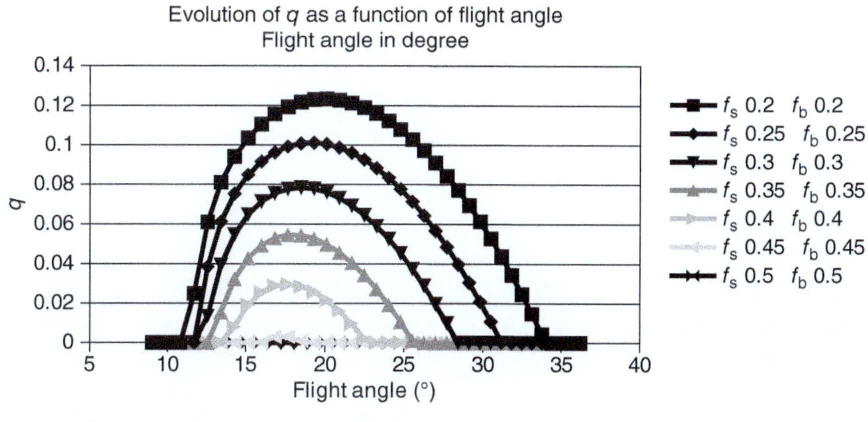

Figure 4.8 Influence of the flight angle on q.

Table 4.1 Some values of bulk densities.

Materials	FR, PUR, PVC	PVDF	PA, PE, PP	Polyester
(ρ_g/ρ_{sd})	0.52–0.54	0.56	0.58	0.60

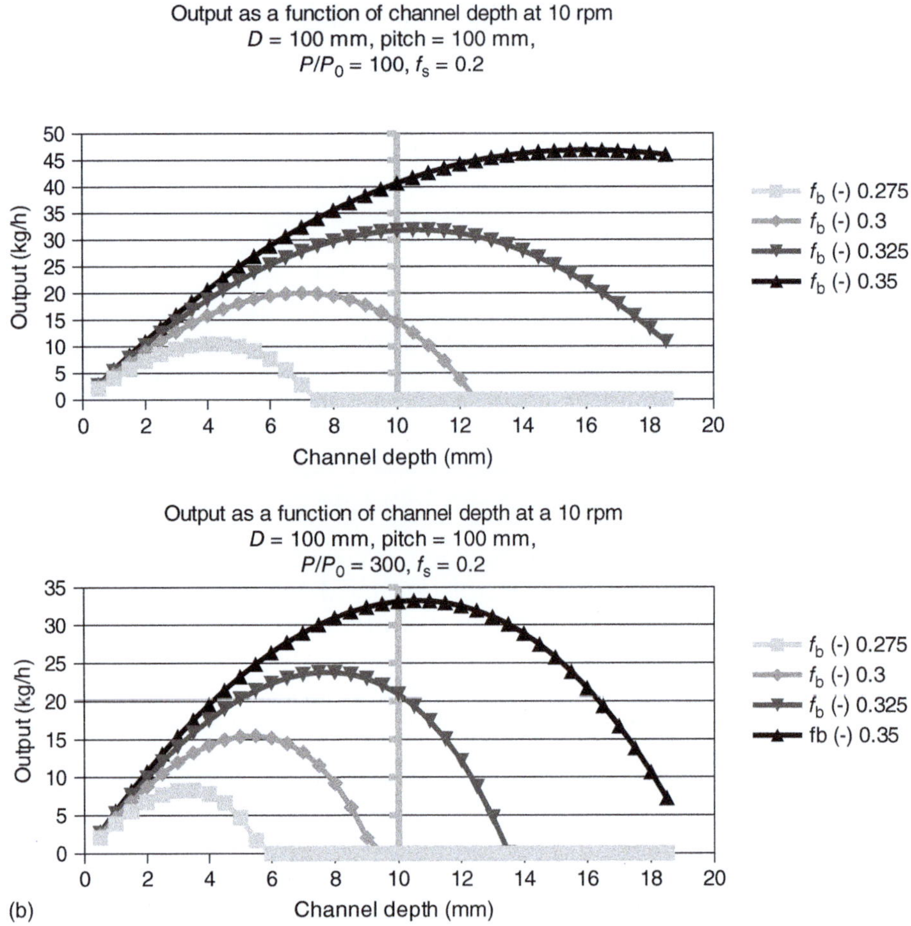

Figure 4.9 Output as a function of channel depth for different barrel friction coefficients. The top figure (a) is with a counterpressure = 100 bars, and the lower figure (b) with counterpressure = 300 bars.

at 10 rpm, with HDPE of density 0.96 kg/l, and a COF on the screw of 0.2. The output reaches the maximum for a given screw channel depth and then falls. This is due mainly to the fact that the surfaces in contact with the screw increase, as against the barrel, they do not change. Also, if the optimum lies at about 16 mm for $f_b = 0.35$ and a counterpressure of 100 bars (~45 kg/h), when a counterpressure of 300 bars is applied, the output diminishes to ~22kg/h. The standard value is therefore often taken at 10 mm. At this value, when the

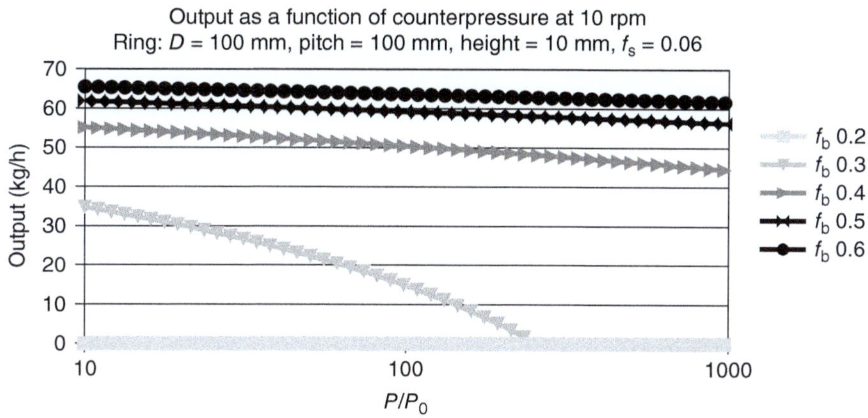

Figure 4.10 Output as a function of pressure at different coefficients of friction on barrel for an extruder of diameter 100 mm turning at 25 rpm.

counterpressure is increased from 100 to 300 bar, the output diminishes by about 33%, for friction coefficient of 0.35 and 0.325 on the barrel. For lower coefficient of frictions, there is no output anymore with a counterpressure of 300 bar.

4.2.2.2.5 Influence of Counterpressure on Feeding Zone Output Figure 4.10 shows the evolution of output (at constant screw rotation speed) as a function of counterpressure for different coefficients of friction. If the coefficient of friction on the barrel is lower than 0.3, there is no output.

The higher the coefficient of friction is on barrel, the less the output is dependent on the counterpressure. This explains the use of straight grooved type barrels.

4.2.2.3 Improvement of the Feeding Zone: Use of Helical Grooved Barrel

4.2.2.3.1 Balance of Forces and Velocities on Rings in the Screw and Barrel Helical grooves of inverse pitches to the screw are machined in the barrel in the entry zone. If we consider two rings, one in the screw and the other in the barrel, we may have a sliding between them. But in the case where the two rings do not slide against each other, we obtain an imposed output by a screw-nut mechanism, the output of which does not depend on the counterpressure.

Figure 4.11 shows an example of a helical grooved feeding zone. In the case of no sliding between the granules in the screw and in the barrel, the movement is specified by a conveying angle equal to the barrel flight angle.

Output calculation in the case of no sliding between the granule rings:

The forward movement of the granules in the screw axis direction (or advance) of the material by screw turn is known

$$\mathrm{Adv} = \pi \cdot D \cdot \left(\frac{\sin(\phi_{\mathrm{screw}}) \cdot \sin(\phi_{\mathrm{barrel}})}{\sin(\phi_{\mathrm{screw}} + \phi_{\mathrm{barrel}})} \right)$$

$$= \left(\frac{\mathrm{Pitch}_{\mathrm{screw}}}{(1 + \mathrm{Pitch}_{\mathrm{screw}}/\mathrm{Pitch}_{\mathrm{barrel}})} \right) \text{ in } \left(\frac{\mathrm{m}}{\mathrm{turn}} \right)$$

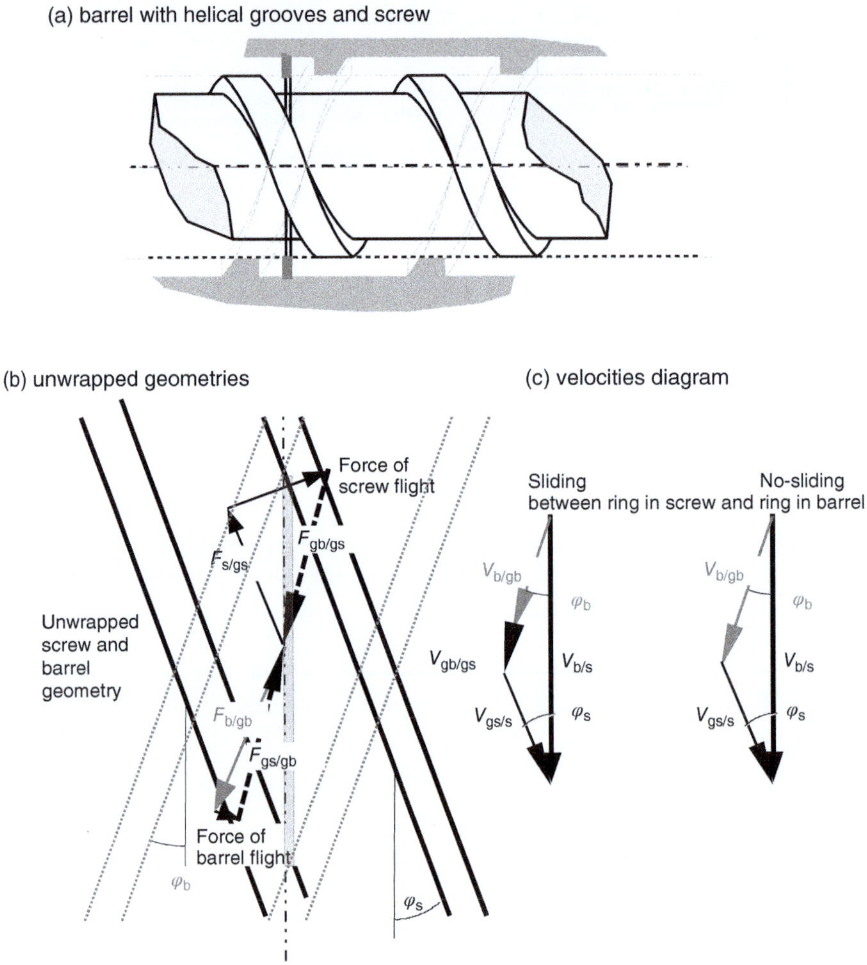

Figure 4.11 A helical grooved feeding zone. (a) barrel with helical grooves and screw, (b) unwrapped geometries, (c) velocities diagram.

The ring sections perpendicular to the screw axis are known, so the output per screw turn can easily be evaluated.

$$Q_{\text{vsp}} = (S_{\text{screw}} + S_{\text{barrel}}) \times \text{Adv in } (\text{m}^3/\text{turn})$$

This output depends on the geometry of the feeding zone of the screw. To obtain the mass output, we have to take into account the relative granular density (taking into account the empty spaces appearing between granules), which is always less than unity.

4.2.2.3.2 Output Order of Magnitude (Qualitatively) and Independence to Pressure The use of helical grooved feeding zones has two advantages, the first being that the outputs may be quite thoroughly estimated when knowing the diameter. The second is that the output is almost not sensitive to the pressure.

Table 4.2 Order of magnitude of specific outputs for different extruder sizes.

N (rpm) at V_{circ} = 30 m/min	*D* (mm)	*Q*_spec (l/h/rpm)
320	30	0.125
213	45	0.43
160	60	1.0
120	80	2.37
96	100	4.63
80	120	8
64	150	15.625

Concerning the qualitative order of magnitude of outputs for different size of machines, we use the specific outputs in l/h/rpm. These units are used because they are very convenient to calculate the output as a function of the screw rotation speed. We may observe that the specific output is proportional to the diameter in cube. It may be observed that the specific output is also equal to the output per screw turn multiplied by 60. By using this kind of feeding, the output is quite simply found when the diameter is known (Table 4.2).

It has to be noted that in the case of PVC, the circumferential velocity is diminished by 15%. For the foam PE, this maximum speed is roughly half of the maximum available speed. Finally, for the HFFR type (Megolon), it is the maximum acceptable temperature (in terms of product quality) that determines this speed ($T_m < 165\,°C$).

4.2.2.3.3 ***Example of an 80-24D: Comparison of the Model with Measurements*** A second advantage of this kind of feeding is that the output is almost independent of the counterpressure. In Figure 4.12, the measured specific output of an 80-24D are presented in the above graph, and the calculated ones are presented in the below one, by varying the internal coefficient of friction between granules. For an internal COF that is double of the friction of granules on steel, we observe the same behavior as measured.

4.2.3 Thin Film Plastification

4.2.3.1 Melting on the Backside of Flight

Maddock [9] described some experiments (screw pull out) in 1959, which showed that the melting began on the backside of the flight. Those observations are also described by Agassant [5].

Figure 4.13 represents a sketch of the mechanism that appears in the melting zone of the extruder. The picture is obtained after a pull out test. This test involves stopping the screw rotation and cooling the extruder violently in order to solidify the material.

A molten material zone appears in the back part of the channel, as the rest of the channel is occupied by granules.

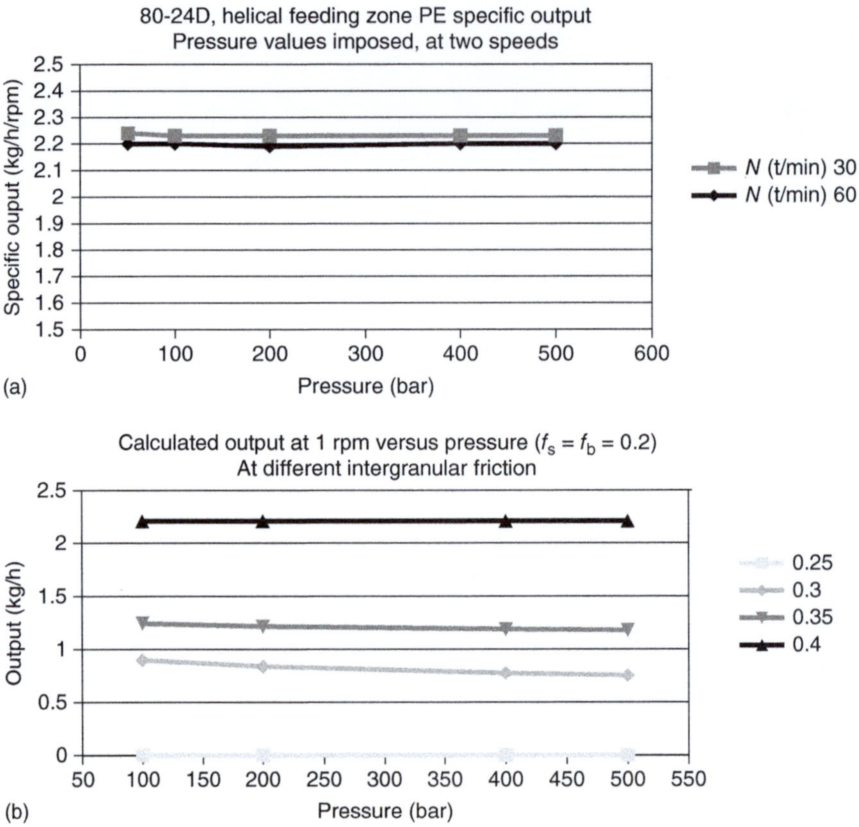

Figure 4.12 Examples of outputs obtained for an 80-24D with helical feeding (a) helical barrel feeding zone: specific output variation with counterpressure for two different screw speeds, (b)smooth barrel feeding zone: specific output variation with counterpressure for different barrel friction coefficient.

In order to better understand the melting mechanism, Figure 4.14 represents the melting zone in the same manner as was done for the feeding zone in Figure 4.7. The barrel rotates around the screw with a speed $V_{b/s}$ (velocity of barrel against screw). The solid material (solid bed) moves forward with a speed $V_{g/s}$ of the granules versus the screw. The angle θ of the velocity of the barrel versus the granules $V_{b/g}$ is the conveying angle calculated in the feeding zone.

4.2.3.2 Initiation of Liquid Film

A melt film is formed above the granules. When it becomes thicker than the flight clearance, the pressure builds up in the back flight. Above a critical pressure p_c, it has been shown that the solid bed begins to deform, creating a liquid pocket.

The pressure buildup is of the form : $$\Delta p = \frac{6 \cdot \eta \cdot v_{b/s}}{\sin(\phi)} \cdot \frac{\delta - \delta_f}{\frac{\delta_f{}^3}{b} + \frac{\delta^3}{W}}$$

Figure 4.15 shows the variation of pressure as a function of the ratio of melt film thickness to flight clearance for different flight clearances.

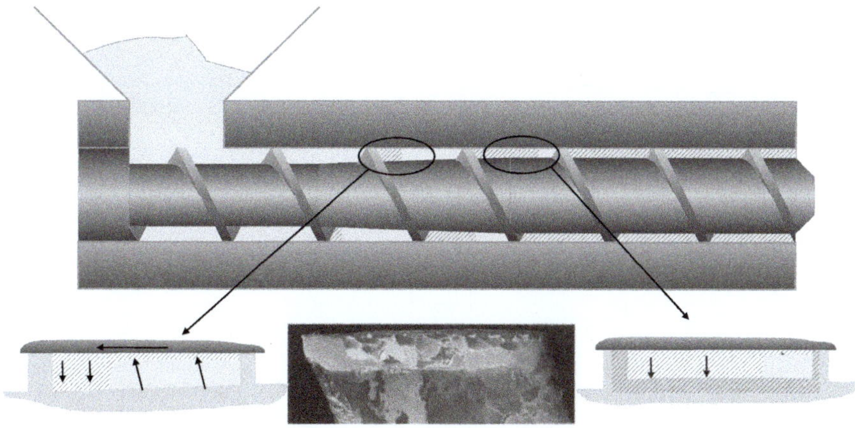

Figure 4.13 Representation of the melting mechanism in the melting zone and a picture of the experiment.

We can see that

* the pressure increases very fast for values of film thickness higher than the flight clearance, and
* the maximum value of pressure reached decreases very fast with increasing flight clearance.

From this graph, it can be seen that increased flight clearance may avoid reaching the critical pressure level, thereby preventing the formation of a liquid pocket.

4.2.3.3 Melt Flow Rate in the Liquid Film

From the equilibrium of forces and fluxes inside the liquid film, one may obtain the expression for the melt flow rate in down channel direction as a function of the channel width W_x, perpendicular to the channel direction:

$$\frac{d\dot{M}_p}{dz} = \Omega_1 \sqrt{W_x} \text{ with } \Omega_1 = \sqrt{\left(\frac{k_l(T_l - T_f) + \frac{1}{2}\eta(u_{b/g})^2}{(h_f + c_s(T_f - T_{s0}))}\right) \rho_l u_{b/s} \sin(\phi)}$$

In this equation, the melting output is dependent on the sum of conduction in the melt film $k_l(T_l - T_f)$ and viscous dissipation $\frac{1}{2}\eta(u_{b/g})^2$. If the energy by viscous dissipation begins to be higher than the one by conduction, the melting rate will be controlled by this first factor.

4.2.3.4 Influence of Different Parameters on Melting Length

The calculations presented below were done for an extruder of diameter 60 mm.

4.2.3.4.1 ***Influence of the Compression Rate in the Melting Zone*** The calculations were done for PE of specific heat 2500 J/kg/°C, with a melt enthalpy of 15 000 J/kg, a thermal conductivity of 0.2 W/(m °C), having a specific mass (non-molten) of 920 kg/m^3, and of 780 kg/m^3 when molten. The temperature of the barrel is

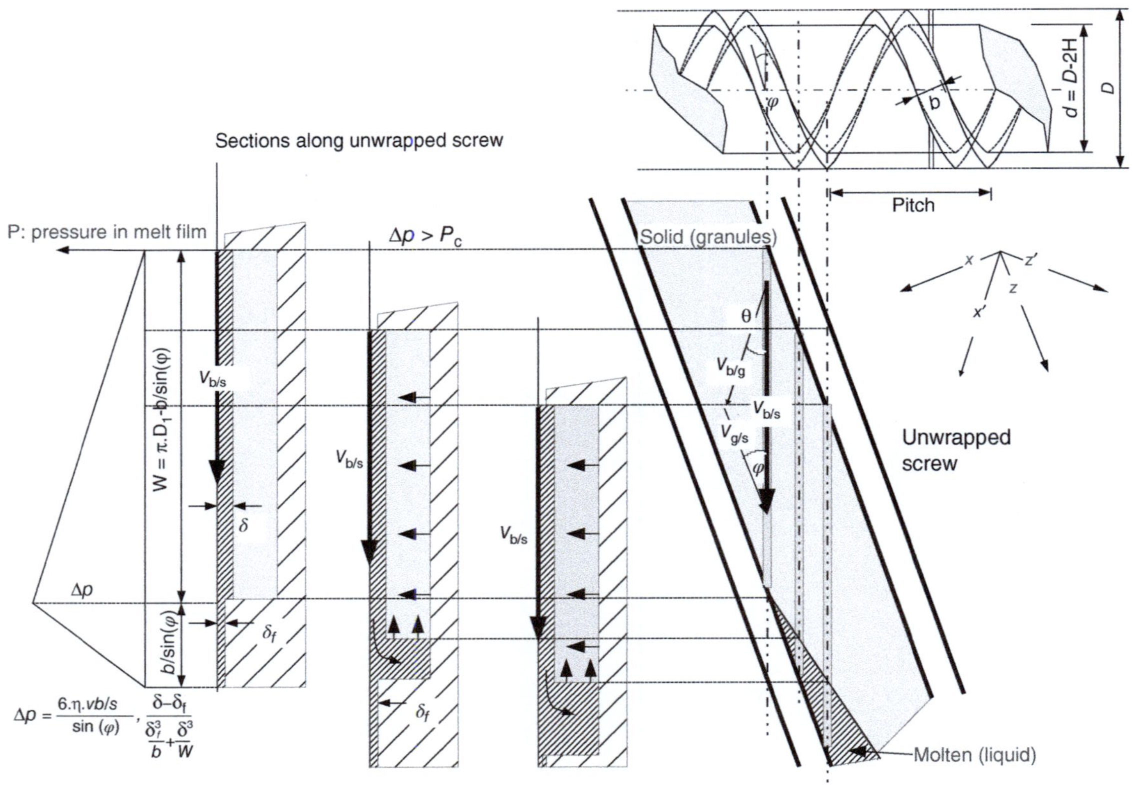

Figure 4.14 Melting zone mechanism. *Note*: θ = solid conveying angle.

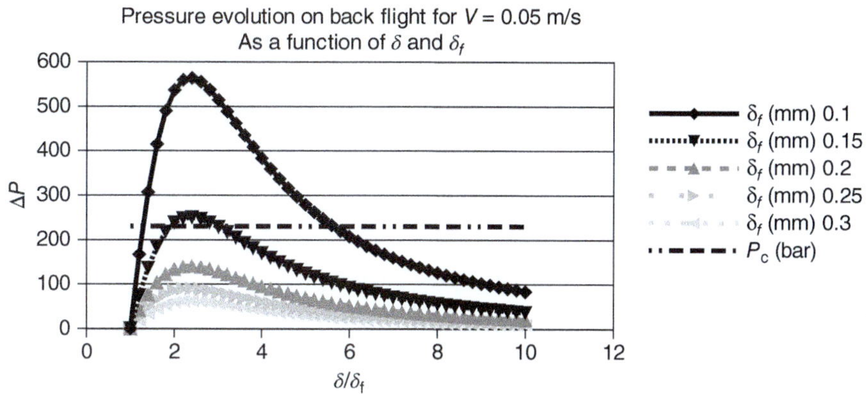

Figure 4.15 Evolution of the pressure along the back flight as a function of the clearance δ_f between the flight and the barrel and of the molten film thickness δ.

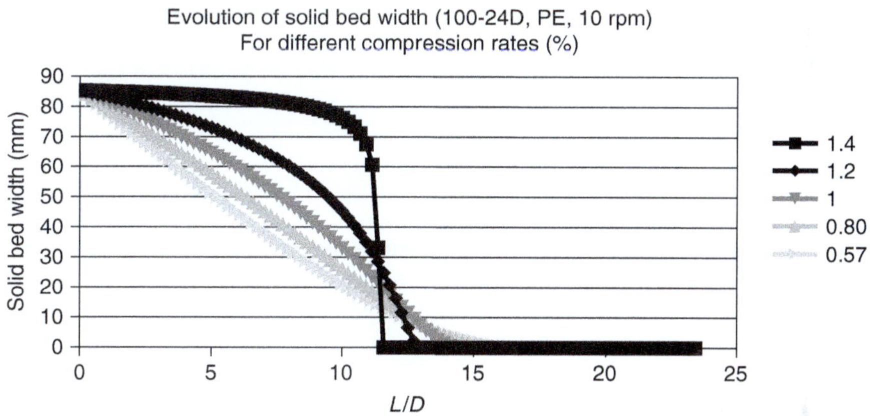

Figure 4.16 Influence of the solid bed width along the screw, as a function of the compression rate.

supposed to be 150 °C (Temperature of melting = 110 + 40 °C). The calculated output in the feeding zone is of 38 kg/h (Figure 4.16).

The curves of the solid bed width are shown for different compression rates:

$$\text{Compression rate in\%} = \frac{H_{\text{feeding}} - H_{\text{metering}}}{\text{Length}}$$

For a high compression rate (1.4%) the width suddenly drops to 0 at an L/D of 11. For lower compression rates, the change is slower at the beginning but an L/D of 16 is needed to completely melt the material.

In reality, two main types of screws are encountered.

The ones with a long compression zone (8–16 L/D) with low compression rates (0.15–0.6%) are used for polymers having limited stability (PVC, ABS), or high viscosity.

The ones with a short compression zone (from 1 to 3 L/D) and with a high compression rate (1–2%) are used for materials showing a very low viscosity in the molten state (PA…).

Screws are often characterized by their Compression ratio $= \frac{H_{\text{feeding}}}{H_{\text{metering}}}$. This value has to be handled carefully, as two screws having the same compression ratio but with very different compression zone length exhibit completely different behaviors.

4.2.3.4.2 *Influence of Rotation Speed and Material* The calculations were done for two materials:

* PE of specific heat 2500 J/kg/°C, with a melt enthalpy of 15 000 J/kg, thermal conductivity of 0.2 W/(m °C), having a specific mass (non-molten) of 920 kg/m^3, and of 780 kg/m^3 when molten. The temperature of the barrel is supposed to be 150 °C (Temperature of melting = 110 + 40 °C). The imposed output is of 3.8 kg/h/rpm.
* PBT of specific heat 1600 J/kg/°C, with a melt enthalpy of 48 800 J/kg, thermal conductivity of 0.25 W/(m °C), having a specific mass (non-molten) of 1310 kg/m^3, and of 1080 kg/m^3 when molten. The temperature of the barrel is supposed to be 265 °C (Temperature of melting = 225 + 40 °C). The imposed output is 5.5 kg/h/rpm.

The graph in Figure 4.17 shows the evolution of the melting length as a function of screw speed for two different materials (PE and PBT).

The melting length increases with increasing speed.

The length for melting the PBT completely is much higher than that for HDPE. At 40 rpm, the melting length will be of 10 L/D for the HDPE and about 17 L/D for the PBT. This means that a screw may be optimized for the melting of one material but cannot be the best for all kinds of materials.

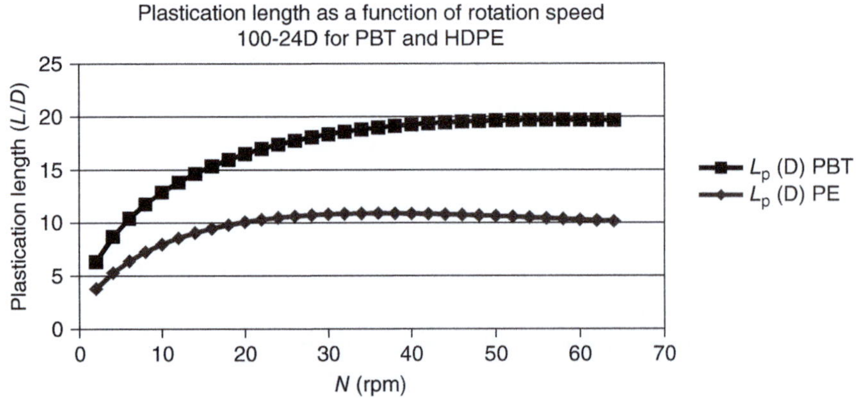

Figure 4.17 Influence of rotation speed and material on melting length.

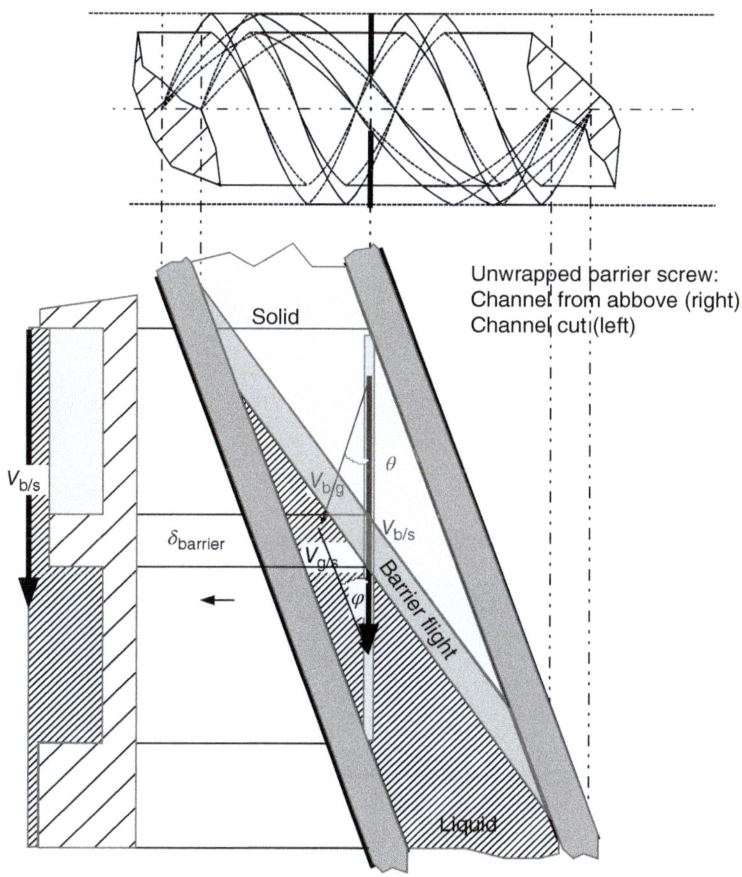

Figure 4.18 Functioning of the barrier zone.

4.2.3.5 Thin Film Melting with the Help of a Barrier Zone

As shown in Figure 4.18, the main function of the barrier zone is to separate the solid bed from the molten polymer.

4.2.3.6 Barrier Zone, Its Advantages, and Its Drawbacks

The goal of the barrier zone is to control the melting by separating the solid bed from the liquids bed. Also, the thickness of the film may be decreased, which increases the production of heat by shear in the film.

Advantages: Granules cannot pass above the clearance without being sheared, therefore heated up, it is better to melt the material.

Drawback: As mentioned earlier, the material has to pass through the clearance, and is therefore heated up by the shear of the material. This heating by shear has to be controlled not to be too high.

- If the length of the zone is too small, the pressure necessary to "push" the material through the clearance for a given output may also be too high.

Remark: This zone may also be used as a mixing zone.

4.2.4 Metering Zone

This last zone of the screw is named metering zone because in extruders without forced feeding, this zone helps to create the necessary pressure to go through the extrusion head. But it may also help homogenize the melt.

4.2.4.1 Representation of the Metering Zone

The metering zone is more often constituted by a simple flight with a pitch equal to the diameter. In Figure 4.19, the flow lines are represented. Also the flow in the

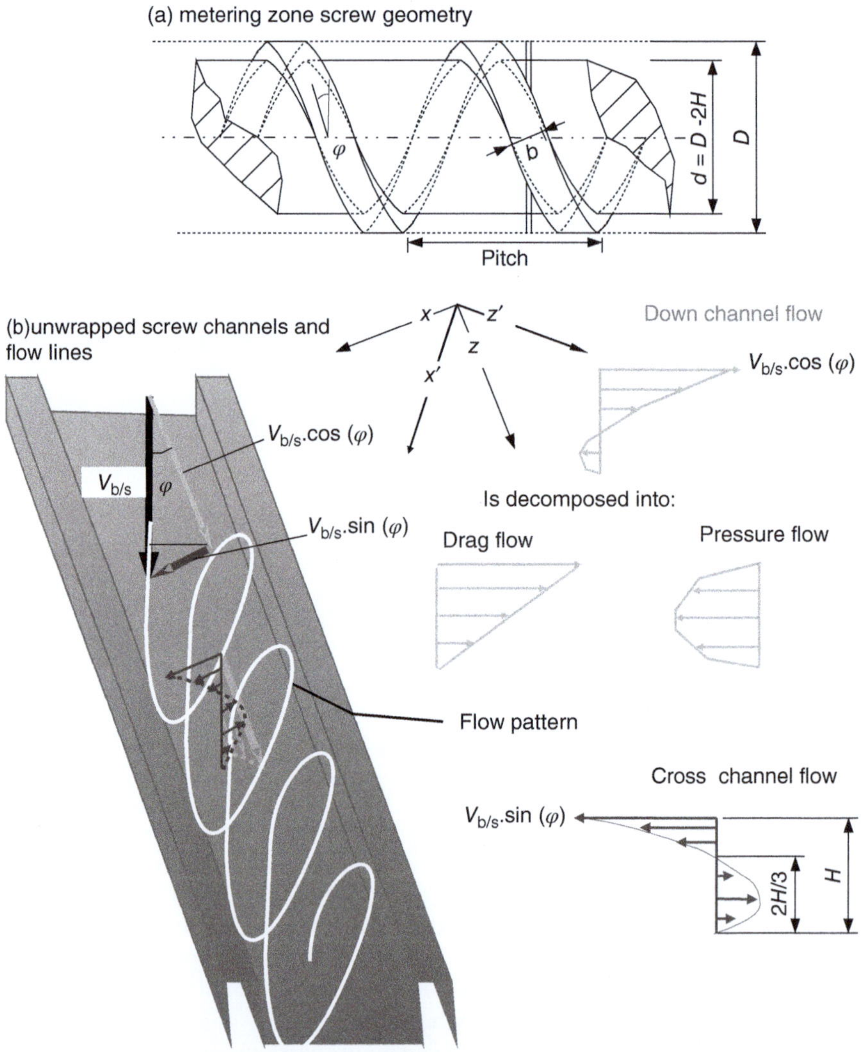

Figure 4.19 Presentation of the metering zone (a) metering zone screw geometry, (b)unwrapped screw channels and flow lines.

down channel direction is represented: it is the sum of a drag induced flow and pressure (or Poiseuille) flow.

4.2.4.2 1D Analysis

In the down channel direction, a combined shear and pressure (Poiseuille) flow exists, for which the speed and output expression can be written [6]: $u_z(y) = \frac{(p_1-p_2)}{L}\left(\frac{1}{2\eta}\right)(y^2 - hy) + \frac{u_{bz}}{h}y$ with the pressure induced flow being $Q = \frac{(p_1-p_2)}{L}W\left(\frac{1}{\eta}\right)\left(\frac{h^3}{12\eta}\right) + u_{bz}\frac{h}{2}W$

$Q_{\text{pressure}} = \frac{(p_1-p_2)}{L}W\left(\frac{1}{\eta}\right)\left(\frac{h^3}{12\eta}\right)$ and the drag flow $Q_{\text{drag}} = V_{b/s} \cdot \cos(\phi)\frac{h}{2}W$.

Perpendicular to the channel, the velocity is

$$u_x(y) = \left(\frac{V_{b/s} \cdot \sin(\phi)}{h} - \frac{\left(\frac{\partial p}{\partial x}\right)h}{2\eta}\right)y + \frac{\left(\frac{\partial p}{\partial x}\right)}{2\eta} \cdot y^2$$

The pressure gradient can be determined by supposing that the output perpendicular to the screw channel is equal to zero: $\frac{\partial p}{\partial x} = \frac{6 \cdot \eta \cdot V_{b/s} \cdot \sin(\phi)}{h^2}$.

The resulting cross channel velocity profile is

$$u_x(y) = \frac{-2V_{b/s} \cdot \sin(\phi)}{h}y + \frac{3V_{b/s} \cdot \sin(\phi)}{2\eta} \cdot y^2$$

In the down channel direction, the ratio of pressure to drag flow is determined by

$$r_d = \frac{(p_1 - p_2)}{L}\left(\frac{h^3}{6\eta V_{b/s} \cdot \cos(\phi)h}\right)$$

An r_d close to 0.5 allows obtaining a good mixing (and homogenization of temperature). This explains that for a given output (imposed from the feeding zone), we have to calculate the depth in order to be close to this value of 0.5.

4.2.5 Example of Results for 1D Model Including the Three Zones of the Screw

In reality, the three zones of the screw (feeding, melting, and metering) cannot work independently. With a simple model integrating these three parts of the screw, we can already analyze the extruder characteristics.

4.2.5.1 Influence of Friction Coefficients on the Screw Characteristics

The results presented below are for an extruder of 100 mm diameter working PE.

The characteristic curve of an extruder is the output obtained at a constant rotation speed by varying the pressure at the end of the screw. In Figure 4.20, we represent four curves for different frictions in the barrel feeding zone.

In order to obtain the functioning point of the extruder, the characteristic curve of the die, being the output obtained by varying the pressure in a Poiseuille flow for a given geometry (length and radius), is needed. In our simple model, this

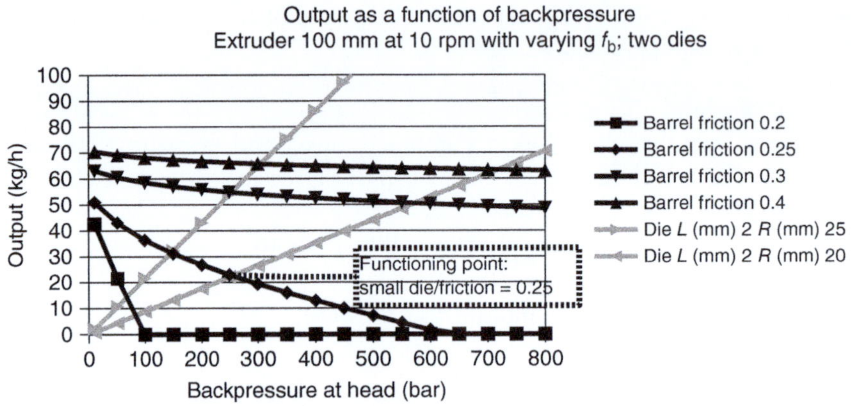

Figure 4.20 Characteristic curves of a 100 mm extruder at 10 rpm with PE for four different frictions in the barrel feeding, coupled with two dies.

curve is a straight one (Newtonian flow). The crossing of the extruder characteristic at a given coefficient of friction and the die characteristic gives us the functioning point. The coefficient of friction of the barrel feeding zone has a great influence.

For a given extrusion head, the output diminishes when the coefficient of friction diminishes (for example, because of wear in the barrel feeding zone).

4.2.5.2 Interaction Between Compression and Friction in the Feeding Zone

For a given set of friction and compression, the evolution of output and inside pressure (at the end of the feeding zone) as a function of the counterpressure are presented on a 100-24D turning at 10 rpm with PE.

In Figure 4.21, the influence of friction on the barrel, as also of compression are presented.

For low friction (smooth barrel, $f_b = 0.2$),

* high compression gives a low output (~20 kg/h) decreasing slowly when the head pressure increases and the inside pressure level stays low;
* low compression gives quite a high output (~45 kg/h) for low counterpressure but decreases very sharply with increasing head pressure (no output at 100 bar counterpressure), and the inside pressure stays low.
* For high friction (grooved barrel, $f_b = 0.3$),
* high compression gives an output around 45 kg/h quite independent of the head pressure, but the inside pressure at the end of the feeding zone is higher than 1000 bars (much too high);
* for a low compression screw, the output is the highest but diminishes with counterpressure (65 kg/h for no counterpressure to 50 kg/h for 600 bar counterpressure), as the feeding zone pressure reaches quite a high but acceptable value of 600 bars.

Therefore a smooth barrel is better with high compression screws, and a grooved barrel with low compression.

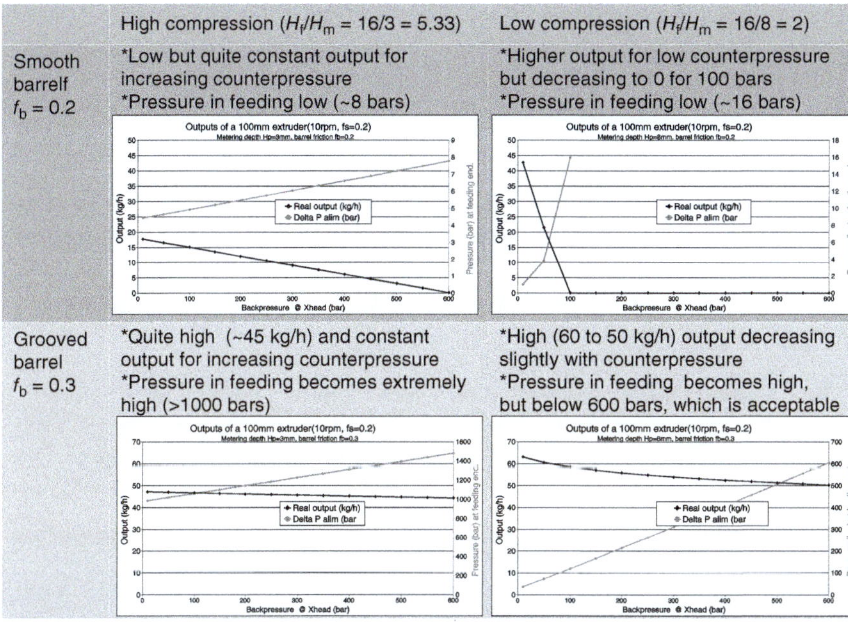

Figure 4.21 Output (left vertical axis) and inside pressure evolution (right vertical axis) as a function of pressure level at the die. The four graphs are with a constant screw coefficient of friction (0.2), for two different friction coefficients on the barrel ($f_b = 0.2$ and 0.3) and two compression levels.

4.3 Accessories for Extruders

4.3.1 Mixing Zones

At extruder exit, the material should be homogeneous. Homogeneity defaults of the material can in fact have a negative influence on the characteristics of the final products.

So, in the case of the production of tubes, the repartition default of black carbon in polyethylene influences the resistance to the sun's UV radiation. The mechanical resistance of the pipe to the internal pressure may be lowered in the case of the presence of homogeneity defaults. Also, during the production of electrical cables, these same defaults may induce insulation defaults. For the production of transparent films, the apparition of dots of different coloration is not acceptable.

4.3.1.1 Observed Defaults

The assessment of the quality level obtained with different mixing devices can be done by extruding a given base material to which a specified quantity of color master batch is added.

Then cuts in the final products are done. An observation under a binocular microscope allows to obtain an image of the number and the size of the color clusters. The method to obtain these samples and also to quantify their mixing level is described in some different standards.

* The international standard ISO 13949 (Method for the assessment of the degree of pigment dispersion in polyolefin pipes, fittings, and compounds)
* The British Standard BS 2782: Part 11: Method 1106A: 1983
* The Australian Standard AS 1685–1984.

4.3.1.1.1 Homogenization Mechanism of the Material The mixing of the material can generally be defined as the capacity to diminish the nonuniformity of the extruded form. We can distinguish two main types of defects or nonuniformities in relation with the type of defects. We consider either aggregates (color aggregates, for example) or local smooth variation of the color component's concentration inside the main matrix.

4.3.1.1.2 Aggregates Type of Defects This type of defect corresponds to the fact that the aggregates (of different components find inside the main matrix) have to be sheared with significant shear stress in order to suppress the aggregation between them. This kind of problems can be solved by bringing an added shearing work.

This type of mixing base on shearing is named dispersive mixing in the following.

4.3.1.1.3 Smooth Color Concentration Type of Defects If we consider that the material agglomerate added in the matrix has been sheared, this second type of defect corresponds to a lack of random physical movement of the particles. We can define three different mechanisms that can allow this kind of random movement:

* Molecular diffusion
* Mixing due to turbulent flow. These two phenomena apply more to gases and to low viscosity liquids, which the molten polymers are not.
* Convective diffusion, applicable to high viscosity liquids like polymers with laminar flow, and therefore presents laminar type of flow.

The last kind of random repartition of particles is the one that applies in the case of high viscosity polymer extrusion. In the following, it will be named distributive mixing.

From what we have seen, we can understand that a good mixing behavior can only be provided in the presence of these two types of mechanisms: dispersion and distribution.

4.3.1.1.4 Example of a Lack of Dispersive Mixing When the components of a different material than the basis matrix stay agglomerated together (Figure 4.22), this indicates a lack of shear in the dark part in order to disperse it in the main material. We can therefore obtain dark or colored dots distributed randomly (good distributive mixing), but separated by large noncolored base material (matrix) parts. The necessary mixing type, which associates mechanical shear actions on aggregates with a good particle distribution, is named dispersive type of mixing.

To solve this kind of defects, special screw tips that shear the material have to be used.

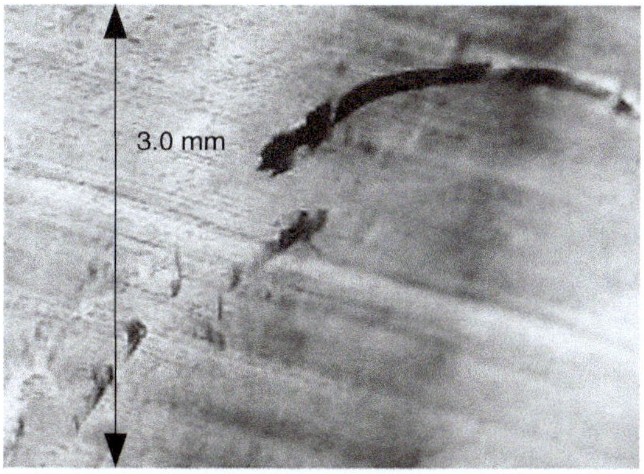

Figure 4.22 Defect due to lack of shear rate. Source: Maillefer extrusion.

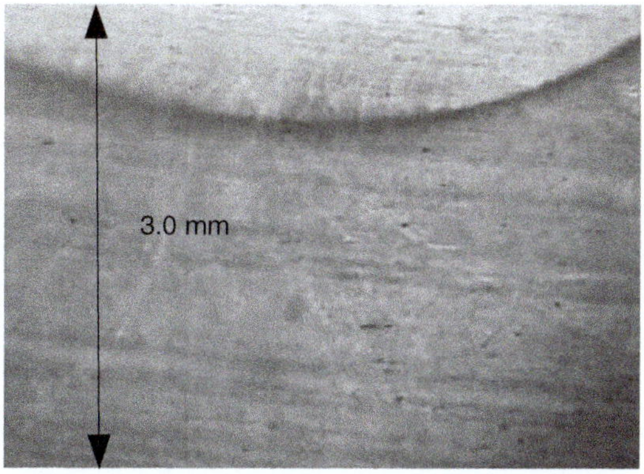

Figure 4.23 Example of default due to lack of distribution. Source: Mailefer extrusion.

4.3.1.1.5 Example of a Lack of Distributive Mixing In the case of an extruder, we are transforming high viscosity materials. So a good distributive mixing implies that the flow lines allow a random repartition of particles that were grouped together before. On the contrary, progressive but localized color changes point to a lack of distribution (Figure 4.23).

In order to solve the problems due to a lack of material distribution, screw tips that have the goal to generate "random" flow lines are used.

4.3.1.2 Devices Selection Criteria

A significant number of mixing devices have been developed along the years.

The main criterion for device selection is the obtained mixing quality, but other criteria have also to be taken into account, and according to the application, one criterion may be more important than the other.

The mixing device should present a minimal, and if possible a positive pressure gradient.

The stagnation points have to be banished in order to avoid dead spots that will induce material degradation.

Finally, the mixing device should be easily dismounted by operators so that their cleaning is done easily.

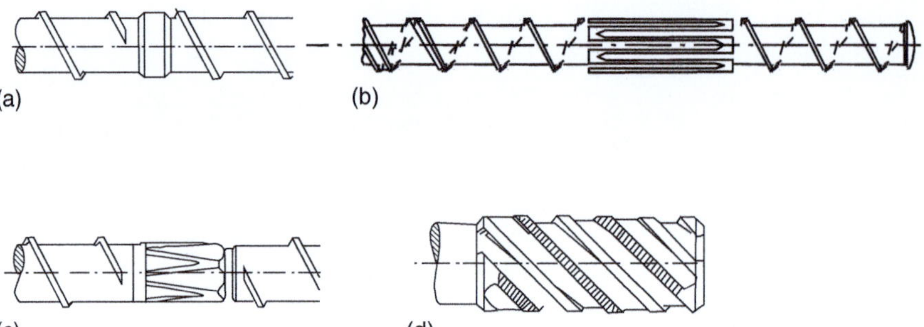

(a) (b) (c) (d)

Figure 4.24 Different dispersive type mixing devices. (a) Mixing device with shear ring. (b) Maddock type mixing device. (c) Dray type mixing device. (d) Egan type mixing device.

4.3.1.2.1 Examples of Dispersive Mixing Devices. A screw tip device for dispersive mixing should have the following characteristics.

The part corresponding to the screw tip should avoid high shearing of the polymer.

The design of this kind of screw tip should be done so that the time during which the material undergoes a high shear rate is relatively short. This is to avoid excessive heating of the polymer.

All fluid parts have to undergo the same level of shear during the same time interval in order to allow uniform action.

Among the different solutions proposed that verify the above criteria, modifications of the barrier kind of geometry have been proposed. A forced flow above a shear ring (Figure 4.24a) or a barrier flight (Dray, Egan, or Maddock type mixing devices) insures a significant shear during short times. The first proposed was the Maddock type, with flights parallel to the screw or device axis (Figure 4.24b). Also some z-shaped flights (Figure 4.24c). An interesting evolution of the Maddock device was proposed by Egan (see Figure 4.24d) who used helical flights, which is therefore very close to a barrier with multiple flights.

4.3.1.2.2 Examples of Distributive Mixing Devices In order to allow a good distributive mixing, the device must exhibit the following characteristics:

High shear rates must be applied to the polymer fluid in order to insure the dispersive function. Then, the flow must be such that the flow lines are randomly oriented ("distributive" flow).

The first kind of mixing devices presenting the above characteristics are the devices with round or diamond shaped pins (pineapple type device), which are there to induce a perturbation of the flow lines. But their drawback is the creation of dead zones, which are prone to induce material stagnations.

Another kind of device is the Rapra type device with semi-spherical holes situated in two half cylinders (so that the whole device can be dismounted) and on the device itself. This configuration helps attain a good level of homogeneity but the high shear rates obtained result in a high temperature increase.

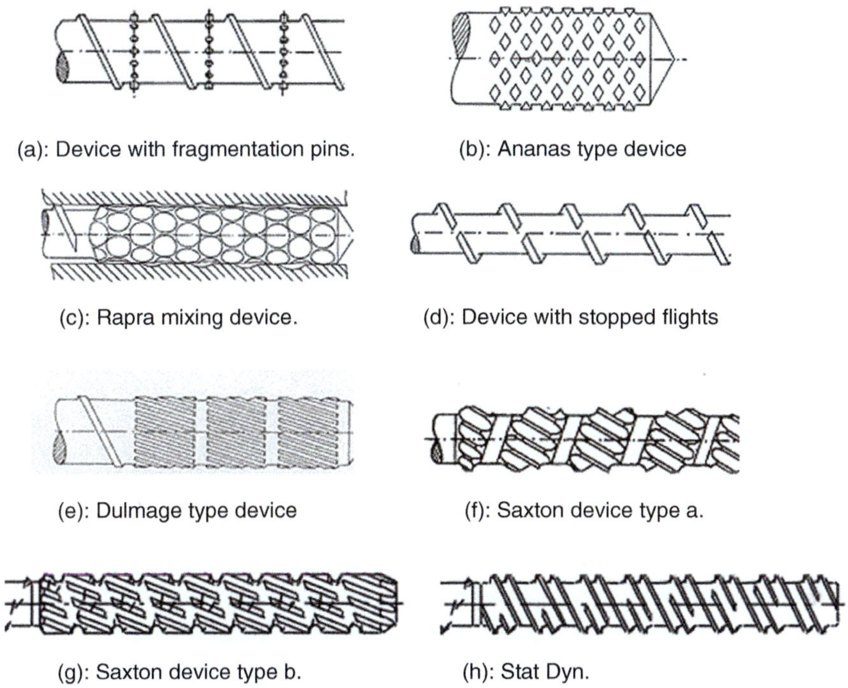

Figure 4.25 Different kinds of distributive mixing devices. (a) Device with fragmentation pins. (b) Ananas type device. (c) Rapra mixing device. (d) Device with stopped flights. (e) Dulmage type device. (f) Saxton device type a. (g) Saxton device type b. (h) Static dynamic.

The last kind of device consists of one or more interrupted flights. When a certain number of these flights are used, the devices are called Dulmage, or Saxton type. But the presence of numerous flights reduces the available space for the material flow, which gives high shear rates and therefore induces a dispersive action.

The Saxton mixing device consists mainly of a device with multiple flights with low cross section channels. This induces high shear rates. The interrupted part of the flights is obtained by grinding a wide channel with a pitch either in the opposite direction to the main channels (type a), or in the same direction (type b): the main channels are therefore regularly interrupted perpendicular to their direction (see Figure 4.25). Stagnations are therefore avoided.

4.3.1.3 Example of Results for Finathene HDPE

In order to have a better idea of how the different types of devices react, tests have been done at Maillefer extrusion to be able to quantify their mixing behavior.

Two different devices have been placed successively at the tip of an 80 mm diameter screw. The same material (HDPE Finathene 3802 from Fina) has been extruded. All other parameters that may have an influence on mixing quality (just to name two: the extrusion cross head, and the kind of mesh) remain the same for the two devices. The first device used was an Egan type device (dispersive,

Figure 4.26 Egan mixing device. Source: Maillefer extrusion.

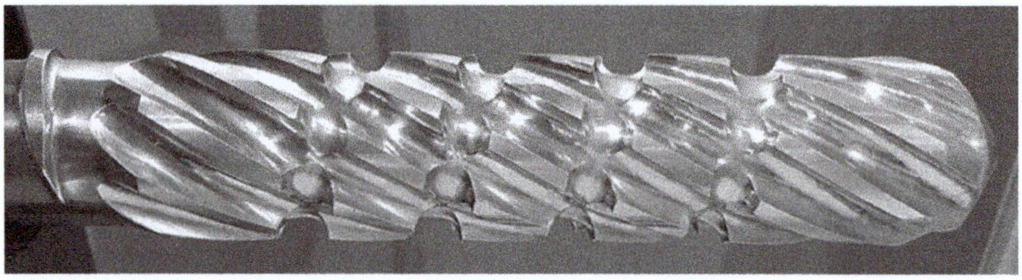

Figure 4.27 The Saxton type mixing device. Source: Maillefer extrusion.

see Figure 4.26) whose barrier flight has been designed with significant width by purpose in order to increase the shear rate.

The second device tested consists of a Saxton type mixing device (Figure 4.27) whose channel cross sections have been reduced in order to increase the effect of shearing (dispersive effect).

Qualitative results on mixing for the two kinds of mixing devices presented above are presented in Figure 4.28. This figure represents the evolution of the quality level (0 = bad evaluation, 10 = perfect mixing) as a function of extruder rotation speed (in rpm). All parameters are the same for the two different devices (each curve represents the result for one device). The criteria used are those based on the international standard ISO 13949. Only the evaluation level of the standard (7 = bad evaluation, 0 = perfect mixing) has been changed in a scale of 0–10, with 10 being the best score. Each point of the graph (for a given screw rotation speed) has been obtained by calculating the average of each of the five criteria chosen, on six samples obtained under identical conditions.

These criteria help quantify the amount and the size of agglomerates, as also their form (more or less pronounced Vortexes), intensity, and the eventual appearance of non-colored parts.

We observe that the Saxton device, which enhances dispersion and distribution of the material, has a much better note than the Egan device. But, it is more sensitive to the screw speed increase.

A simple explanation for this result is that the Egan device insures only dispersive type of mixing, as the Saxton device has a dispersive and distributive action. When we analyze the average mass temperatures (average of mass temperatures for all screw speeds) for the two devices (see Table 4.3), we can see that the Egan

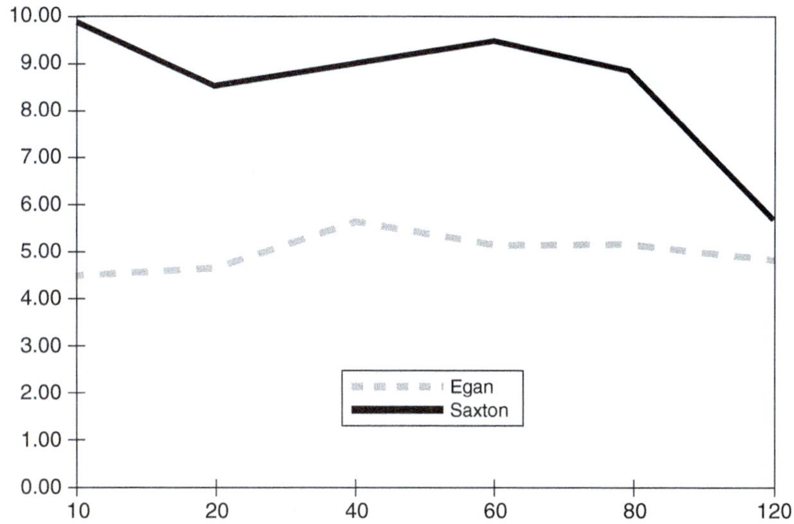

Figure 4.28 Evolution of the average evaluation of quality as a function of screw speed for two different devices.

Table 4.3 Mass temperature as a function of the device used.

Mixing element	Mass temperature (°C)	Average notation
Saxton HDPE	241	8.6
Egan HDPE	245	5.0
Saxton PVC	186	4.9
Egan PVC	187	3.2

Source: From Maillefer document.

device exhibits a higher mass temperature, and therefore generates more shear (dispersion). But the bad notation obtained can be explained by the fact that the random distribution of particles and flow lines (distributive mixing) is not obtained by this device.

4.3.2 Melt Filtration Systems

At the screw tip, we place, before the die, a hole plate (or breaker plate) and metallic canvases (filters), which have the function of filtering the molten mass.

In the most simple version, it is composed of a disk, or a circular breaker plate, of the screw diameter where numerous holes of smaller diameter (1–3 mm) are drilled, and at the front of which more refined woven metallic screen canvases (named screen-packs) are placed (see Figure 4.29). These screen-packs are in general in stainless steel. But we can also use micro-perforated metal sheets instead of these canvases.

(a) Breakerplate (b) Meshes

Figure 4.29 (a) Breaker plate; (b) Meshes.

The role of the filters is not limited to purifying the molten mass of all impurities. But the filters can also enhance homogenization by increasing the pressure at the end of the screw, in the case of a cross head generating low counterpressure. This helps increase the back pressure Poiseuille flow in the metering zone (see metering zone part).

The clogging of the filters can be controlled continuously by a pressure transducer situated upstream or even better by two transducers (one upstream from the screen-pack, the other downstream).

The filtration surface is also an important parameter because it influences the periodicity of screen-pack change. That is why, in certain cases, we replace the circular breaker plate of screw diameter by bigger systems. For example, by using circular filtering systems situated in the feeding channels of the extrusion cross head, we can make use of a filtering surface three to five times bigger.

4.3.3 Melt Gear Pumps

Sometimes, a regular output is necessary to insure a high quality product. For foamed coaxial cables with big diameters (2–4 mm) used in cellular transmission towers, and obtained by physical foaming with expansion rates of 80–85%, a regular output is very important to insure good transmission. Therefore, melt gear pumps may be used. They contain two counter-rotating, inter-meshing gears that produce a positive volumetric displacement of the melt. One of the two gears, driven by an electric motor, brings the other gear in rotation (Figure 4.30). The polymeric material flows in the space between the teeth of the gears and the

(a) Drawing of a gear pump

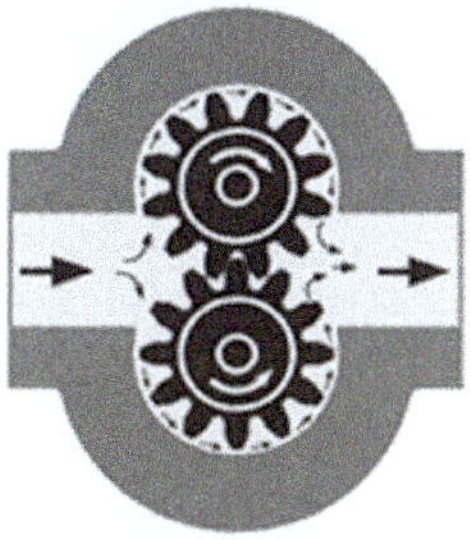

(b) Figure obtained by cutting prependicular to the gear axis,

Figure 4.30 Principle and drawing of a melt gear pump (a) drawing of a gear pump, (b) figure obtained by cutting prependicular to the gear axis.

periphery. It is then pushed inside the extrusion cross head. The output given by the volume between the teeth is precisely controlled by the rotation speed of the pump.

4.4 Extrusion Heads or Dies

The circular extrusion cross heads (for pipes, cables, blown film) have to distribute the melt flow identically across a circumference.

4.4.1 Description of the Extrusion Head

4.4.1.1 Extrusion Head Function

The plastic material exits the extruder in the form of a cylindrical extrudate. The extrusion head is necessary in order to conform the material to the form of a tube disposed thereafter on a conductor (to isolate a wire) or on an assembly of wires (in order to produce the sheathing).

Furthermore, the flow at the exit of the extruder is parallel to its axis and the head has to be perpendicular in order to be able to pass the conductor (or wire or wire assembly). The function of changing the flow orientation is insured by the assembly of the cross head and the connecting flange.

The extrusion head (Figure 4.31) is bound to a flange fixed on the extruder. This flange will diminish the diameter of the material flow from the extruder diameter until the head entry diameter (which explains its internal conical form). This connecting flange is fixed on the barrel by means of a clamp.

The extrusion head itself is made up of an external part, the body, in which one or more distributor(s) and tooling are positioned.

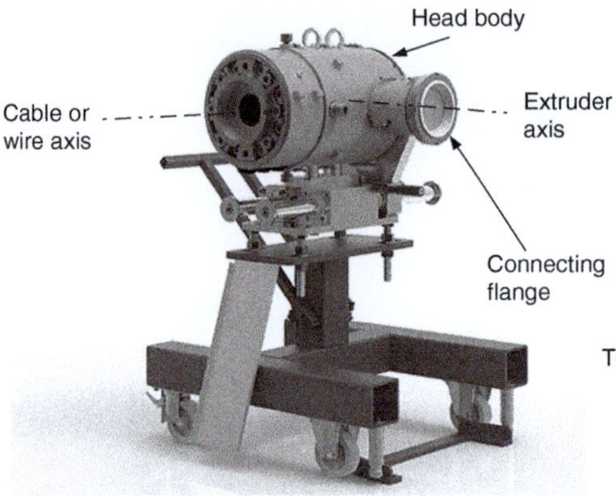

(a) Extrusion head 115/130 with its connecting flange

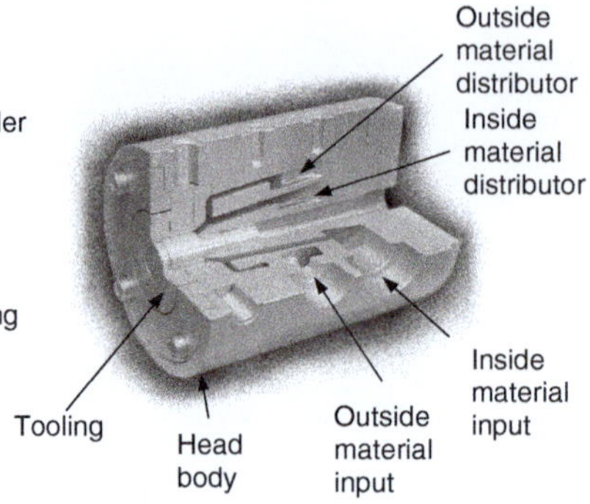

(b) Cross section of a head with two distributors

Figure 4.31 (a)Single layer sheathing extrusion crosshead. (b)Twin layer insulation extrusion crosshead. Source: Maillefer extrusion.

The extrusion head helps to conform the polymeric material, which has been plasticized before in the extrusion screw, and shape him to a constant straight section. The head should distribute the material in the flow channels in such a way that the material exits with a uniform speed. The distribution of the flow will be a function of the polymer flow properties, of the channel geometries, of the output, and of the temperature imposed on the die or head.

The geometry of the head and of the distribution channels being optimized for certain conditions, a change of melt temperature, output, or polymeric material may therefore change the repartition of flow at the die exit.

In order to obtain an extrusion head that allows the use of different materials at different output rates, we can adjust the flow pattern by changing the setting of temperatures on the different cross head zones, or by changing the geometry of the final part of the head (tooling). These adjustments through temperature settings or final flow geometry render the design of this head complex (for example, a certain amount of thermal regulation zones have to be planned, or the possibility to introduce different tooling in the case of a round extrusion), but allow flexible use of the extrusion head or flat die.

The following are the general design rules for extrusion heads or dies:

- There should be no dead zone (stagnation zone) in the flow channel.
- There should be constant increase in the average flow speed of the material.
- The assembly should be such that dismounting of the head or die should be easy.
- The length of the final part (die lips or tooling) should be at least equal to 10 times the thickness of flow.
- All steep section changes in the geometry should be avoided.
- Small angles should be used.

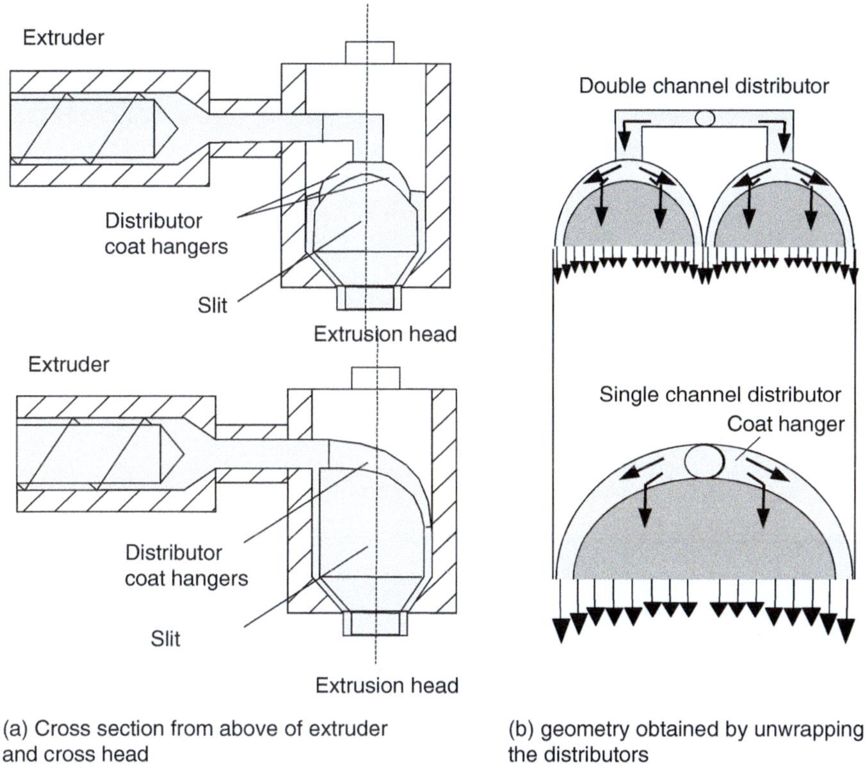

Figure 4.32 Presentation of an extrusion head with single or double distribution channel. (a)Cross section from above of extruder and cross head, (b)geometry obtained by unwrapping the distributors.

4.4.2 Distributors

One of the most important goal of the extrusion heads for wire insulations or cable sheaths is to give a regular thickness across the circumference of this insulation or sheath. That means the most even distribution of output across the circumference has to be achieved. Geometrical designs close to the ones used in flat dies (coat hanger type distribution channels) are the more usual. In some cases, distribution with the help of helical channels (developed in the beginning for blown film manufacturing) is used for the production of tubes and cables.

4.4.2.1 Head with Coat Hanger Type Distribution Channels

Heads with distributors are directly inspired by the geometries of flat dies with coat hanger channels (Figure 4.32). The unwound geometry of this kind of distributor corresponds in fact to the flat die geometry.

The distributors use one or more distribution channels (named also coat hangers) whose geometry is similar to the one used for the production of flat films or plates. The distribution function is guaranteed by a distribution channel having the shape of a coat hanger, and for which the flow arrives in the center. This geometry is wound around the head axis in the case of an extrusion head for cables

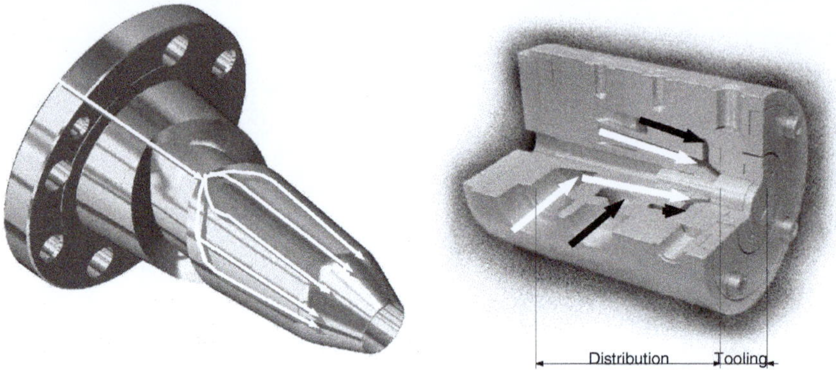

(a) Coat hanger type distributor (b) complete extrusion head (two layers)

Figure 4.33 Coat hanger type distributor, and complete extrusion head (two layers). (a)Coat hanger type distributor, (b)complete extrusion head (two layers). Source: Maillefer extrusion.

Table 4.4 Comparison of distributors with single or double distribution channels.

Distributor type	Ovality	Excentricity
Single channel distributor	+	−
Double channel distributor	−	+

(Figure 4.33). In order to obtain a regular distribution of the flow, the pressure loss along each flow line must be the same. Numerical optimization, which takes into account the rheological behavior, is done with the help of software of the same kind as those described in papers [4, 10, 11]. A given distribution geometry depends therefore on the extruded material.

The use of a distributor with only one channel helps correct certain eccentricity defaults by acting on its centering inside the head body. But it has the disadvantage of being more sensitive to the rheology of the polymer processed. Another drawback exists that corresponds to the generation of a weld line, which will be later a weakness line for pipes.

Because of the sensitivity to the material processed, distributors with two distribution channels have been developed. They allow a better distribution, but we can no more correct the errors by de-centering the distributor die against the body of the extrusion head. Also, if distribution problems appear, the final product may become oval. This kind of geometry will also generate two weld lines, thus weakening the lines in the tube or the insulation. Table 4.4 summarizes the advantages and drawbacks of these two kinds of distributors.

Based on the coat hanger design, different variations have been proposed. As seen above, we may have single or double channel distributors. Then, the distributor may be cylindrical or conical. The conical version has been developed to facilitate the dismounting of the head in order to clean it. The direction of extrusion

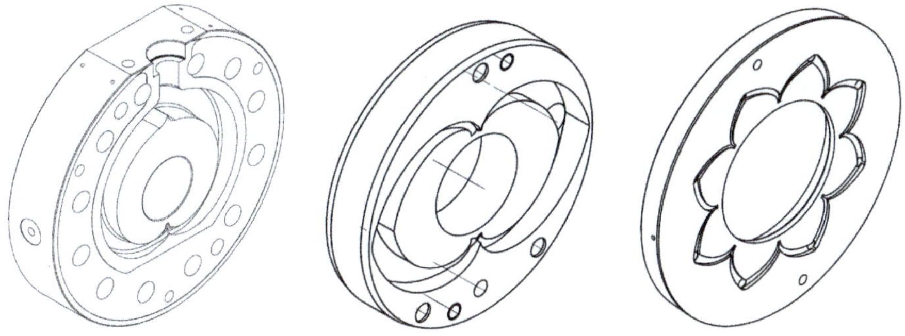

Figure 4.34 Flow in flat distributors with a (a) single distributor, (b) double distributor, and (c) multiple distributor.

being perpendicular to the extruder axis, this kind of head is used mainly for the production of cables (insulation of sheathing) because for cables, the wire has a perpendicular direction to the extruder axis.

4.4.2.2 Distribution Function Through Flattened Distribution Channels

A last variation of coat hanger type distributors is obtained with the design of a flat distributor. Use of this kind of flat distributor allows to decrease the size of the extrusion head (see Figure 4.34). In this figure the radial flow generated by the geometry is represented, as also its use in the production of a three-layered pipe (one central layer and two identical external layers). We can observe that the place occupied by the whole system is quite small.

As for the cylindrical distributor, evolutions with two distribution channels have been developed in order to enhance a better symmetry, and also to have a repartition less dependent on the material rheology.

4.4.2.3 Distributor with Helical Channels

To avoid weld lines parallel to the radius, heads with helical distributors have been developed [12]. These are used since long in the case of blown film extrusion (allow an enhanced distribution), but recently they are also used for pipes.

The advantage of these kinds of distributors is that the weld lines between the different flows from different channels are not radial. We can see this in Figure 4.35 representing the flow lines of one of the entry channels obtained by numerical simulation. The pressure resistance of the pipe is therefore increased.

A version of helical grooved distributors has also been created by using flat distributors. This idea has been developed principally for the production of multilayered films, with five layers or more. The extrusion head is therefore smaller.

4.4.3 Diameter Adaptation Function (Tooling)

This function is insured by the tooling [13]. These are of two types. If we have a cable, we can use pressure tooling, which will press and apply the material on the conductor, or we may also use tube tooling (Figure 4.36).

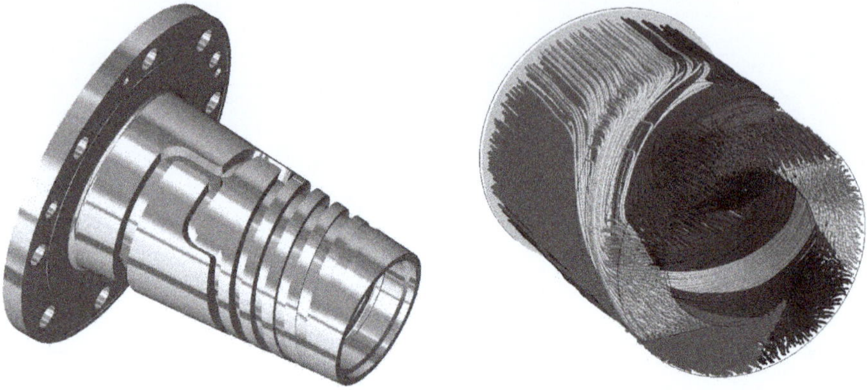

(a) distributor geometry

(b) Flow around the distributor: The flows from the different entry channels being at different colors, weldlines at the interface between 2 colors are easily visualized.

Figure 4.35 Helical distributor: (a) (a)distributor geometry; (b) flow around the distrobutor: weldlines between the flows from the different entry channels. Source: Maillefer extrusion.

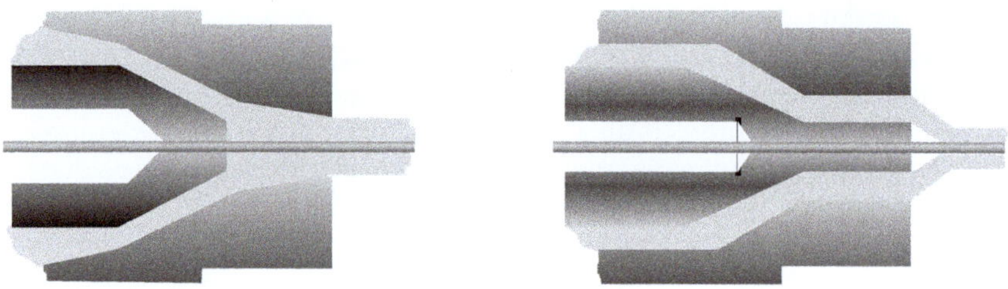

(a) Pressure tooling

(b) Tube tooling

Figure 4.36 (a) Pressure tooling and (b) tube tooling. Source: Buluscheck [13].

$$\mathrm{DDR} = \frac{D_e^2 - D_i^2}{d_e^2 - d_i^2}$$

$$\mathrm{DRB} = \frac{D_e}{D_i} - \frac{d_i}{d_e}$$

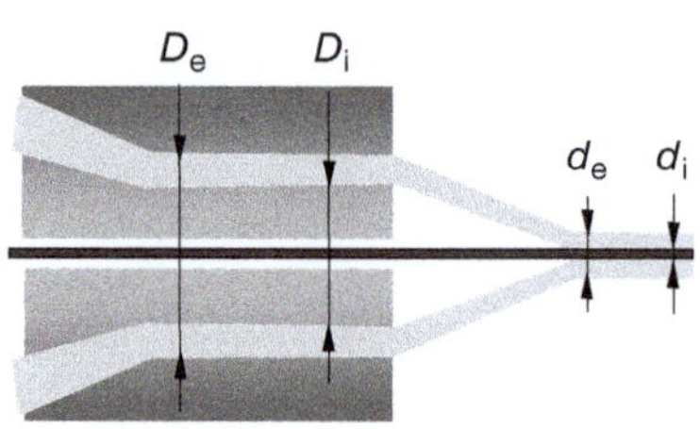

Figure 4.37 Tube tooling design parameters (case of cables).

4.4.3.1 Tube Tooling, DDR

The setting parameters for tube tooling are the section ratios (Draw Down Ratio [DDR]), and the Draw Balance (DRB). DDR is dependent on the material used and can take important values (200 for the Teflon 340). As for DRB, it stays close to 1 (Figure 4.37).

Some indicative values of DDR [13] for different materials may be used as guidelines. They are given in Table 4.5.

Table 4.5 Indications of values of DDR for some materials.

Material	PVC pure	PVC with 50% chalk	PE, all types	PVC with 50% ATH	PP	PA12	PA6	PBT	TPE	PVDF	ETFE	FEP	PFA
DDR<	4	2	4	2	10	10	40	16	20	25	25	150	200

Source: Buluscheck [13].

The following points have to be considered:

* Draw down ratio takes into consideration only the elongation inside the cone existing at the exit of the tooling. It is important to keep in mind that the more significant the elongation is (variation of elongation speed in the extrusion direction), the more the acceptable DDR is reduced.
* Higher material temperatures imply a higher acceptable DDR.
* Tooling with big diameter associated to materials of low viscosities may lead to gravity-based deformation of the cone, which means that the acceptable DDR is lower (this is no more true in the case of vertical extrusion).
* A higher DDR means a more significant molecular orientation in the extrusion direction, which indicates mechanical properties (of the formed pipe) different in the extrusion direction or in the perpendicular direction. Finally, this molecular orientation in the extrusion direction will mean a significant shortening in the same direction (a PE tube or pipe of 1 m length may even diminish to 0.8 m after a few weeks). In trying to reach the smallest shortening possible, we may have to diminish the DDR.

4.4.3.2 Tube Tooling, DRB

In reality, a DRB of 0.9–1.05, which allows good adhesion and low ovality, is used. For a value below 0.9, we may have a nonsymmetrical insulation, or even folds. For a DRB of more than 1.05, the material hangs around the central conductor.

4.4.4 Relation Between Pressure and Average Temperature Increase in an Extrusion Head

The high pressure necessary at the entry of the extrusion head will mean an increase in the mass temperature. This can be shown in a simple manner by using thermodynamics. The relation obtained relates the material density, calorific capacity, and pressure to the temperature increase:

$$\Delta T = \frac{\Delta P}{\rho \cdot c_p}$$

This simple relation obtained in the case of an incompressible and adiabatic fluid helps link the temperature increase to the pressure measured in the fluid.

Knowing the basic thermal properties of polymers, one can deduce the temperature increase occurring at a pressure drop of 100 bars (see Table 4.6, with some indicative values).

Table 4.6 Temperature increase for 100 bars counterpressure for different materials.

Polymer	PS	PVC	PMMA	SAN	ABS	PC	LDPE	LLDPE	HEPE	PP	PA6	PA66	PET	PBT	PVDF	FEP
ΔT/100 bars (°C)	7.86	6.49	5.84	6.61	7.00	5.95	4.73	4.73	4.68	5.23	4.12	4.08	4.78	5.93	4.12	3.94

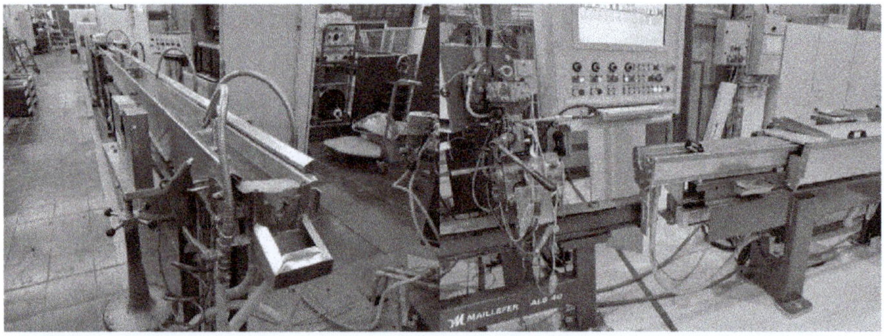

(a) Fixed first cooling trough (b) mobile trunk

Figure 4.38 Cooling trough: left fixed; right with mobile trunk (a) fixed first cooling trough, (b)mobile trunk. Source: Maillefer extrusion.

4.5 Cooling

Following its deposition on the conducting central wire, the insulation or sheathing material has to be cooled in order to be able to wind the cable. This cooling is done by passing the wire (or cable) inside cooling troughs (Figure 4.38). In the majority of cases, the cable is cooled through water aspersion.

Aspersion is often used instead of immersion for wires and cables of small sections (diameter < 2 mm) because it exhibits a better cooling efficiency. Also, a cable immersed in water undergoes higher drag forces, which may lead to significant plastic deformation of the wire (wire elongation).

During the design of the line, it is necessary to measure the length of the cooling troughs because too short a cooling trough would not cool sufficiently the wire or cable for the next step (winding on reel), and during this step permanent deformations could occur. On the other hand, too long a trough would mean an elongation of the conductor (due to the high drag forces in water) so as unnecessary costs (longer troughs mean more place used, and also higher water consumption).

For cooling calculation, it is necessary to take into account three terms in the heat equilibrium equation [14, 15]: forced convection with water, conduction inside the cable, and finally, the heat stored in the material (Figure 4.39).

4.5.1 Cooling Length Analytical Calculation for Fine Wires

For a small radius, corresponding to the case of telephone wires (of diameter 0.5–0.7 mm), the temperature may be replaced by the average temperature. With

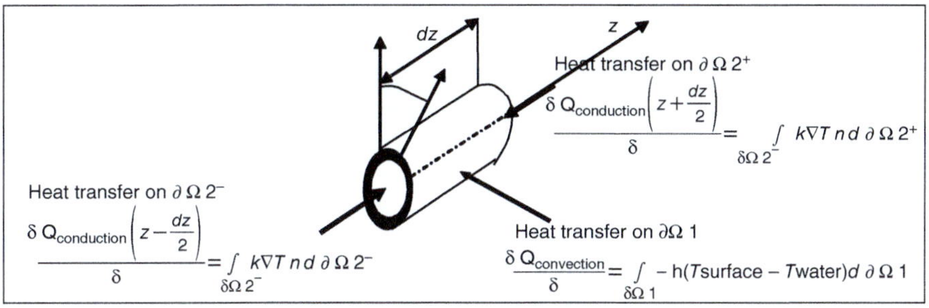

Figure 4.39 Active heat transfer in a cooling cable.

the conduction along wire or tube length being neglected in comparison to convection terms, we can easily obtain the necessary cooling length expression from the heat balance equation:

$$L = \frac{(R_1{}^2 \cdot \rho_{\mathrm{Cu}} \cdot C_{\mathrm{pCu}} + (R_2{}^2 - R_1{}^2) \cdot \rho_{\mathrm{Insul}} \cdot C_{\mathrm{pInsul}}) \cdot w}{-2 \cdot h \cdot R_2} \cdot \log\left(\frac{T_m(L) - T_{\mathrm{water}}}{T_m(z=0) - T_{\mathrm{water}}}\right)$$

With the help of this equation, we can determine the minimum trunk length for which the insulation temperature will be low enough before it will go on the capstan. The criteria chosen will therefore be a temperature of 10 °C lower than the melting temperature, in order that the insulation material does not deform under the action of the capstan.

Numerical application:

Wire with cupper conductor and polyethylene insulation:

Conductor radius $R_1 = 0.15$ mm,

Insulation radius $R_2 = 0.3$ mm, $h = 1000\ \mathrm{W}/\left(\mathrm{m}^{2\circ}\mathrm{C}\right)$

For a line speed of $w = 1\mathrm{m/s}$, $T_m\,(z=0) = 200^\circ\mathrm{C}$, $T_{\mathrm{eau}} = 20^\circ\mathrm{C}$

To obtain a cable temperature of 100°C, we need $L = 6.6$ m

4.5.2 Cooling Length Finite Difference Calculation for a Wire of Radius >1 mm with Copper Core and PE Insulation

In real cases, the wire insulation is built with different layers of different physical characteristics. Also, the radius is often bigger than 0.7 mm, so the *Bi* number is greater than 1, which means that the temperature change along the radius cannot any more be neglected. The heat balance equation in a wire element in cylindrical coordinates, neglecting the conduction in the axis direction against convection in the same direction, gives

$$\rho \cdot C_p \cdot w \cdot \frac{\partial T}{\partial z} = k\left(\frac{1}{r}\frac{\partial}{\partial r}\left(r\frac{\partial T}{\partial r}\right)\right), \text{with boundary conditions :}$$

$$\text{on the } \partial\Omega 1, \int_{\delta\Omega_1} -h(T_{\mathrm{surface}} - T_{\mathrm{water}})d\partial\Omega_1 = k_{\mathrm{Insulation}}\int_{\delta\Omega_1} \nabla T\vec{n}\,d\partial\Omega_1$$

$$\text{at the Copper} - \text{insulation interface}:$$
$$k_{\text{Cu}} \int_{\delta\Omega} \nabla T\vec{n}\, d\partial\Omega = k_{\text{Insulation}} \int_{\delta\Omega} \nabla T\vec{n}\, d\partial\Omega$$

This equation is solved with the help of an implicit finite difference method, in slices [16, 17]. The comparison between the evolution of average temperatures obtained analytically or by a finite difference method is shown in Figure 4.40. The parameters are the same as the ones used before. By changing the wire diameter, the Biot number increases. We can observe that the analytical calculation is not valid for Bi numbers greater than 1, or a radius greater than 0.33 mm.

The results of Figure 4.40 are obtained for wires with copper conductor and PE insulation of

* 0.1 mm insulation radius, 0.05 mm conductor radius ($Bi = 0.6$)
* 0.3 mm insulation radius, 0.15 mm conductor radius ($Bi = 0.9$)
* 1 mm insulation radius, 0.5 mm insulation radius ($Bi = 3$).

The convection coefficient, water temperature, initial wire temperature, and line speed are

$$h_{\text{water}} = 1000\,\text{W}/(\text{m}^2\ {}^\circ\text{C}),\ T_{\text{water}} = 20^\circ\text{C},\ T_m(z=0) = 200^\circ\text{C},\ w = 20\text{m/s}$$

We observe that the average constant temperature approximation is not valid for $Bi > 1$, which corresponds to a radius of 0.33 mm.

The analytical calculation allows evaluating the cooling length only for fine wires. In other cases, it will be necessary to use a finite difference or finite element model.

4.6 Quality

4.6.1 The Quality Parameter and Its Measurement

In order to produce according to specified characteristics, it is necessary to proceed to different measurements during the diverse steps of the insulation and sheathing processes. The cold insulation diameter, after the passing of the product in the cooling tanks or troughs, is found on all lines. According to the application of the line, the following measurements are also done:

- Diameter circularity measurement
- Insulation and conductor concentricity
- Insulation continuity testing (by spark tester)

Two diameter measurement devices may be used, for the conductor and for the insulated wire diameter.

Also, a device positioned at the exit of the cooling trough (or tank) may control the concentricity between the conductor and the insulation in the cold cable. The association of these different measurements allows to control continuously the excentricity, the diameter, and the thickness of the insulating wall.

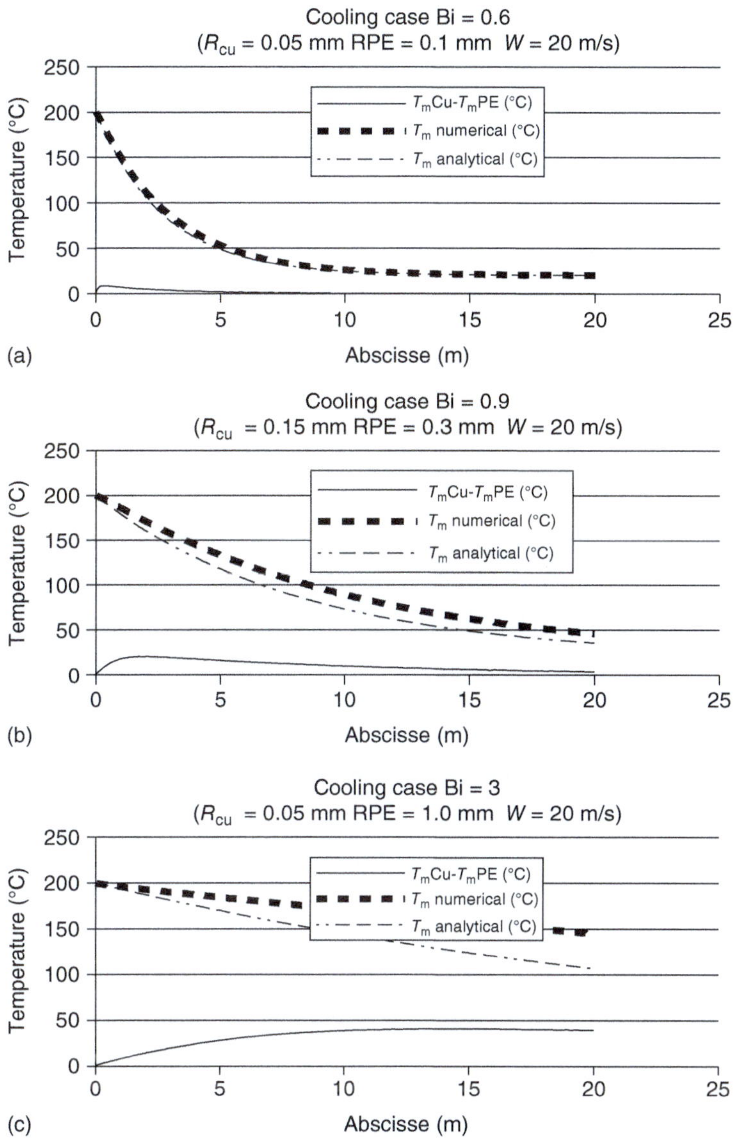

Figure 4.40 Evolution of average temperature for a *Bi* number of 0.6, then 0.9 and 3 (copper conductor, insulation PE) (a)cooling for Bi = 0.6: R_cu = 0.05 mm, R_PE = 0.1 mm, W = 20m/s, (b)cooling for Bi = 0.9: R_cu = 0.15 mm, R_PE = 0.3 mm, W = 20 m/s, (b)cooling for Bi = 3.0: R_cu = 0.5 mm, R_PE = 1.0 mm, W = 20m/s.

4.6.1.1 Diameter and Product Circularity Measurement

At least one diameter measurement is available on all lines. This measurement is done on the insulation, on the cold product (after the cooling trough or tank, on the final product). For cable extrusion lines with a quite low production speed, it may also be interesting to proceed to a diameter (and/or circularity) measurement on the hot cable at the exit of the extrusion head. Because if the diameter is

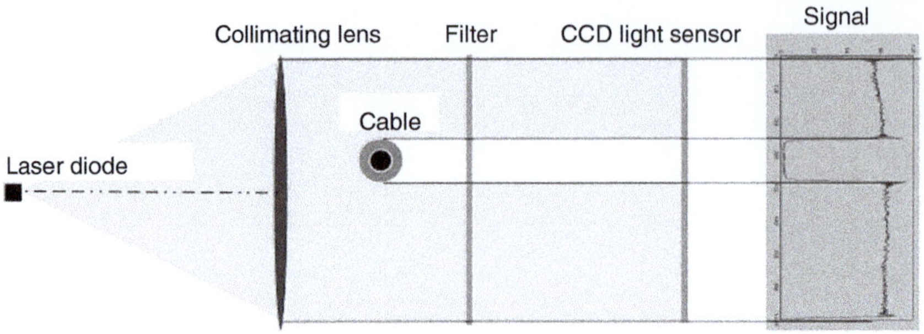

Figure 4.41 Use of the shadow of a cable to measure its diameter.

only taken on the cold cable, this corresponds to quite a long time after it passes through the extrusion head. In case of a default, all the length of the cable or pipe in the cooling tank is not at specification and the correction time is too long.

Diameter measurement is mainly done with the help of optical devices that measure the size of the shadow given by a lightened cable (Figure 4.41). Three different kinds of diameter measurement technologies can be differentiated:

- Lighting time measurement on a diode by first rotating a laser beam with a mirror, and then using a collimation optical system.
- Measurement of a surface lighted by a beam perpendicular to the wire and collimated by an optical lens.
- Measurement of a surface lighted and analysis of the wire shadow side fringes: in this case, no collimation optical system is necessary.

In order to quantify the circularity defects, diameter measurements are done along two axes included in the plan perpendicular to the product (pipe or cable). To obtain more precision, three different measurement heads positioned each at an angle of 60° may be used.

4.6.1.2 Concentricity Measurement

The most common concentricity measurement (for cable or wire) is done with the help of high frequency magnetic fields. A high sensibility bobbin rotating around the cable surface at high speed (up to 1200 rpm) detects the position of the conductor relative to the insulation. The system measurement principle uses different component induction.

In contrast to the preceding device, which has to measure by being in contact with the product insulation, other types of measurements, based on X-ray or electromagnetic frequency generator, without contact, have been developed.

4.6.1.3 Insulation Defects Measurements (Cable or Wire)

Insulation defects as also tearing or micro-holes are controlled by systems having electrodes (metallic chains) in contact with the insulation. These electrodes are put under a high voltage potential (from 0.5 to 50 kV) according to the type of cable tested. When insulation defaults appear, the system (the spark tester) detects a current.

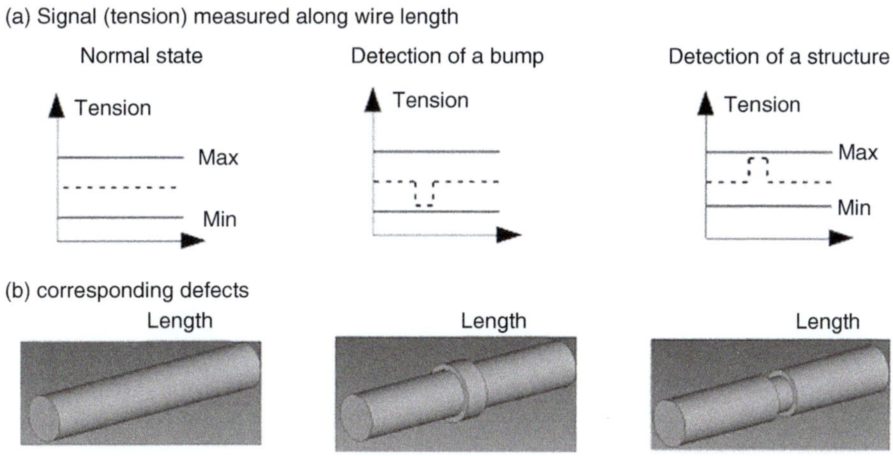

Figure 4.42 (a) Different types of defects (a) signal (tension) measured along wire length, (b) corresponding defects. Source: (a) Based on continuous detection of surface defects, Zumbach Electronic, (b) Based on diameter measuring devices, Beta LaserMike Product.

The origin of these defects may also be the polymer particles cross-linked too early (in the case of cross-linking processes), or degraded polymers, which have stagnated in the screw or head, or the beard formed at the exit of the tooling, which detaches itself to go on insulation. These defective particles are harder than the molten polymer, and are dragged by the wire and create a bump followed by a hole. This kind of phenomenon may be solved by using heads specifically designed to avoid stagnation (in the case of cross-linking) or through the use of a gas flame directed on the tooling and burning this beard.

Beyond a certain number of defects or failures by length unit (for example, 32 km for communication cables) the cable is rejected.

Insulation defects such as bumps are not detected by the spark testing unit. This explains why an infrared beam sensing system is used (Figure 4.42).

An infrared light plane is generated by a light source and different optical parts. This light plane is also perpendicular to the wire. A receptor disposed symmetrically to the wire gives an electrical tension as a function of the quantity of light received. For the nominal diameter, this tension remains constant. In the case of the arrival of a bump, the tension diminishes, and in case of a hole, the tension increases. A minimal tension is specified, corresponding to a maximum diameter, and also a maximal tension (corresponding to the minimum acceptable diameter).

4.6.1.4 Capacity Measurement (Telecommunication Wire)

For fine telecommunication wires the main measurement criterion is the coaxial capacity. This is a function of the internal diameter (conductor), the external diameter (on insulation), and of the dielectric constant. This last constant is dependent on the expansion level. By measuring the internal and external diameters, as also the capacity, we can calculate the expansion rate.

A capacitive measurement device is used. It is generally situated in the cooling tanks, with water playing the role of the external conductor. This measure is more important for foamed polymers than for solid polymers.

The capacitance and diameter measurement helps regulate the process parameters, which are the line speed, the extruder rotation speed, the position of the first mobile cooling trough, and the nitrogen injection pressure.

Concerning the line control, we know that by fixing all the line parameters, apart:

* The line speed: its increase will lead to a diminishing diameter.
* The screw rotation speed: its increase will lead to greater diameter.
* The mobile cooling trough: its movement toward the head will lead to a diminishing diameter and expansion.
* The nitrogen injection pressure: its increase will lead to a higher diameter and expansion rate.

For fine-tuning during production, a closed loop regulation module helps maintain the capacity and diameter values set by the operator. The line control system acts automatically on the extrusion rotation speed, the line speed, the mobile cooling tank, and the injection group in order tomaintain the quality parameters at the desired level. But the regulation cycles are quite complex because they influence themselves mutually.

4.6.1.5 Sheathing Wall Thickness Measurement

This measurement is done in the case of cables such as optical cables with a tube structure for which it is difficult to know the internal diameter of the tube.

In this case, ultrasonic waves reflecting on the internal and external surfaces of the tube are used. The echoes of these two surfaces are re-transformed into an electrical signal by transducers. The phase difference between the two echoes helps obtain the wall thickness. This system is positioned inside the cooling tanks, in water, for which the sound propagation velocity is known.

By disposing, in relation to the cable or tube axis, two diametrically opposed transducers with a known distance separating them, we can deduce the diameter and also the two wall thicknesses.

4.6.1.6 Periodicity of Measurement Analysis

The use of FFT (Fast Fourier Transform) helps identify the problems appearing in cycles on the line. This analysis can be done on the measures described above or on the end wire or pipe [18].

4.6.1.6.1 Analysis of Inline Measurements An analysis of the periodicity of the measurements helps detect the deterioration problems on the different parts of the extrusion line. The signals obtained from the different measurement devices are analyzed with the help of Fourier transforms. By using an analysis by Fourier transform, the periodicity of the parameters can be found, and therefore the root cause of the phenomenon can be found. This online measurement allows also to detect, during production, defects that would be visible only after the end production of the cable is complete.

Table 4.7 Common defects and possible remedies.

Defects	Main causes	Possible remedies
Irregular dimension along the produced product	* Irregular feeding at the level of the hopper * Irregular output of the extrusion group	* Use of a vibrating hopper * Extruder feeding zone temperature stabilization * Output regulation may be obtained by increasing barrel pressure with some screen-packs * Synchronization of haul-off and extruder rotation speeds
Thickness of sheathing or insulation varying along cross section	* Product too hot (material creep) * Head tooling bad centered * Geometry of distributor inadequate (symmetrical default)	* Diminish the head and extruder last zone temperatures * Centering of tooling * Choice of a new distributor
Wire or cable surface rough or with regular waves	* Visco elasticity of the material (shark skin) * Mass temperature too high in the tooling (orange skin) * Not enough mixed material (surface mat)	* Increase of temperature and length of tooling * Decrease mass temperature, increase tooling temperature * Increase of head and extruder end zones extruder temperature, increase of the material mixing by adding a filtering mesh between the extruder and the head
Surface with bubbles or blistering	* Humidity in the extruded material * Maximum admissible temperature of the material reached * Trapping of air in the extruder feeding zone	* Drying of raw material before transformation * Decrease of the head and extruder temperatures * Use of an inverse temperature profile (feeding zone warm, end zone of extruder cold)
Particles or lumps	* Unmolten particles in the raw material * Inclusion of heterogeneous parts (wood, paper)	* Adding of a screen between extruder and head. If use of a barrier zone, decrease of the barrier clearance * Control of the raw material introduces in the hopper
Product with low mechanical characteristics	* Lack of material mixing * Material too hot and degradated * Material batch different from the reference	* Adding of a screens * Diminution of extrusion temperatures * Return to the reference material
Apparition of bumps and lumps	* Presence of degraded or burned material grains	* Avoid stagnations in the extrusion head of the extruder

By using this kind of analysis, defects due to mechanical elements and also pulleys, capstan, and caterpillar are found by analyzing the diameter. Another type of periodical defect such as a cyclical variation of the preheating temperature can be detected with the help of the capacitance measurement. In fact, conductor temperature variation before the extrusion head will lead to an adhesion variation of the insulation.

4.6.1.6.2 SRL Analysis for a Telecommunication Cable The SRL (Structural Return Loss) measurement is done by injecting a high frequency signal in the conductor and measuring the reflection of this signal. This measurement is in general done after the production of the cable is complete. We can therefore obtain the periodicity of reflection of this signal, which corresponds to a regular capacity defect on the cable. With the high frequency signal being reflected, the cable acts as a filter and impedes good transmission of the signal. Specific norms to the produced cable specify the acceptable default.

Defects detection during insulation of the conductor may be obtained with FFT analysis and SRL simulation. This detection is done before the assembling of the diverse wires. It guarantees that each wire used in the next production step possesses the minimum necessary quality for the final cable. In case of non-conformity, the wire can be rejected before the final assembly of the cable, thereby decreasing production costs (when a completely assembled cable fails because of a single wire). In telecommunication fields, where the signal frequencies attain 3 GHz, this FFT analysis is a necessity to enhance the technical and economic performance of the production line.

4.6.2 Common Production Defects, Causes, and Remedies

Some defects are associated with the extrusion group or with the pull-through units. Table 4.7 summarizes the main problems with the extrusion group or the line as the defects origin. For more extensive information, the technical booklet of Maillefer [19] may be interesting for the reader. Those are more extensively addressed in the short paper [20], and also in the chapter defects and remedies of the article [2].

References

1 Salvador, R. (1991). Câbles sous-marins de télécommunications. E 7 550. Traité Réseaux et Télécommunications.

2 Vergnes, B. and Puissant, S. Extrusion monovis (partie 1). AM 3 650. Traité Plastiques et Composites.

3 Vergnes, B. and Puissant, S. Extrusion monovis (partie 2). AM 3 651. Traité Plastiques et Composites.

4 Vergnes, B. and Agassant, J.F. Modélisation des écoulements dans les filières d'extrusion. A3 655. Traité Plastiques et Composites.

5 Agassant, J.F., Avenas, P., Sergent, J.-P. et al. (1996). *La mise en forme des matières plastiques*, 3ème. Lavoisier Technique & Documentation.

6 Rauwendahl, C. (1994). *Polymer Extrusion*, 3èmee. Munich, Vienna, New York, NY: Hanser Publishers.

7 Martin, G.A. (2001). *Der Einschneckenextruder: Grundlagen zur Optimierung*. Düsseldorf: VDI-Verlag.

8 Bonvin, P.Y. and Schläfli, D. (2003). *Hochgeschwindigkeitsextrusion: Aspekte des Extrusionsprozesses und der Regelung de Kabelextrusion*, 241–255. Düsseldorf: VDI Verlag.

9 Maddock, B.H. (1959). A visual analysis of flow and mixing in extruder screws. *Society of Plastic Engineers Journal* 15: 383.

10 Schlaefli D. Analysis of polymer flow through coathanger melt distributors.-pp. 195-199 *International Polymer Processing* n°3, 1995.

11 Schlaefli D. Design and analysis of spiral mandrel die for wire coating application.-pp. 251-260 *Journal of Reinforced Plastics and Composites*, Volume 18, n°3, 1999 251-260.

12 Puissant, S. and Bonvin, Y. Lignes d'extrusion en câblerie: Étapes de fabrication. AM 3 641. Traité Plastiques et Composites.

13 Buluscheck, B. (2003).The Art and science of extrusion for wire and cable 1. *52nd IWCS/Focus Conference*, International Wire and Cable Symposium, Philadelphia Marriott Downtown, Philadelphia, Pennsylvania (17–20 November 2003).

14 Lienhard, J.H. IV, and Lienhard, J.H. V, (2004). *A heat transfer textbook*, 3e. Cambridge, MA: Phlogiston Press.

15 Schlaefli, D. (2000). Plastics extrusion: a short overview. *Nextrom Technical Symposium 2000* (10, 11 April 2000), Ecublens, Switzerland.

16 Pham Q.T. A fast, unconditionally stable finite-difference scheme for heat conduction with phase change *International Journal of Heat and Mass Transfer*, Vol. 28, n° 11, pp. 2079-2084, 1985.

17 Schlaefli, D. (2004). Méthode de calcul par différence finie implicite du refroidissement d'un fil. Communication écrite.

18 Blohm W. FFT spectrum analysis and structural return loss (SRL) prediction using SIKORA's INLINE 2000 series of measuring devices *EuroWire Journal*, May 2000, pp. D15-D18.

19 Maillefer, C. *Petit guide pratique de l'extrusion*, Bulletin technique. Nokia-Maillefer.

20 Koopmans, R. Défauts d'extrusion. AM 3 657. Traité Plastiques et Composites.

5

Foam Extrusion

Horst A. Scheid

Siebe Engineering GmbH, R&D, Am Eisenstein 23, Neustadt (Wied) 53577, Germany

5.1 Motivation

Consumer and business demand for high speed communication increases the market for high frequency cables in computer networks, CATV, and mobile phones. There are two main drivers for high bandwidth:

- *High quality video content for mobile phones*: According to an Ericsson mobility report [1], the total monthly mobile traffic was growing from around two exabytes in 2014 to 27 exabytes in 2018 (one exabyte is equal to one quintillion bytes or one billion gigabytes). In their latest forecast, they expect an increase to some 130 exabytes, where 74% of this amount will be video data.
- *Cloud traffic and workloads*: According to the Cisco global cloud index [2], the yearly cloud traffic increased from 1.6 zettabytes in 2013 to 11.6 zettabytes in 2017 (one zettabyte is equal to one sextillion bytes or 1000 exabytes). By 2021, a further increase to 20.6 zettabytes is forecasted.

This can only be accomplished with (backbone-) cables that have excellent radio frequency (RF) bandwidth characteristics, mainly realized by foam insulation. The optimization of electrical and mechanical characteristics of cable insulation by usage of foam techniques requires a good understanding of thermodynamical effects and their influence on the production process. Other than production of foamed injection molding parts, the main focus in foamed cable insulation is not on saving weight or material cost (even both are welcome as well) but an improvement in terms of electrical behavior.

5.2 Physical Basics

Figure 5.1 shows the generalized construction of coaxial cables, where several layers are optional and not every speciality (like slitted outer conductor) is mentioned. Within this chapter, the production of foam cables is treated as being coaxial – even for twisted pair constructions, the single wire can be described during extrusion as coaxial insulation.

The Global Cable Industry: Materials, Markets, Products,
First Edition. Edited by Günter Beyer.

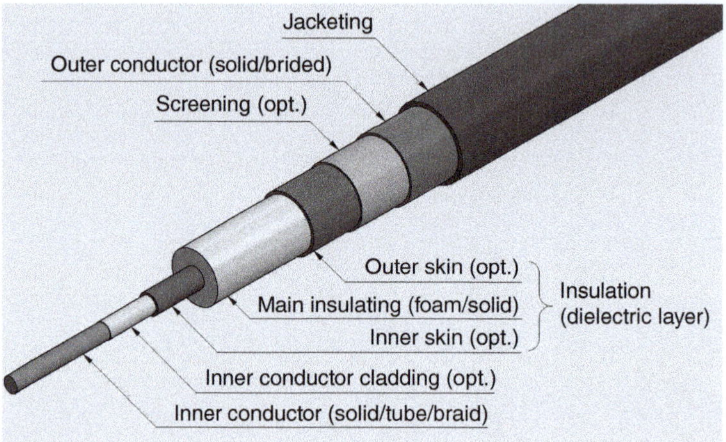

Figure 5.1 Cable construction of RF coaxial cable types.

Dielectric constant ε_r (or relative permittivity) is the ratio of capacitance measured with the insulating material compared with that of vacuum. To obtain the absolute permittivity, multiply by the absolute permittivity of vacuum: $\varepsilon_0 = 8.854 \times 10^{-12}$ F/m. Dielectric constant is slightly temperature and frequency dependent, but this is not considered here.

Deriving equations from cylinder geometry can be found for example in [3].

With some formulae derived from cylindric geometry shown in Table 5.1, electrical parameters can be calculated. A replacement of ε_r (= foam permittivity) by the permittivity of the solid polymer simplifies the expressions to the standard solid cable production terms. μ_r (i.e. the permeability) can be neglected for calculations as being almost equal to 1. Another important cable parameter is propagation speed. This value is given in absolute speed or in percentage of speed compared to vacuum light speed c_0:

$$v_{\text{p\_abs}} = \frac{c_0}{\sqrt{\varepsilon_r \mu_r}} \quad \text{in} \quad [\text{m/s}]; \qquad v_{\text{p\_rel}} = \frac{100}{\sqrt{\varepsilon_r}} \quad \text{in} \quad [\% \text{ of } c_0] \tag{5.4}$$

Finally, the permittivity of the foam layer can be described by Eq. (5.5):

$$\begin{aligned} \varepsilon_r &= \frac{\varepsilon_P}{2}\left\{(1+S_{\text{mf}})\cdot\left[(1-\text{vr})+\frac{\text{vr}}{\varepsilon_P}\right]+\frac{(1-S_{\text{mf}})}{(1-\text{vr})+\varepsilon_P\cdot\text{vr}}\right\} \\ &\approx \frac{\varepsilon_P}{2}\left\{1-\text{vr}+\frac{\text{vr}}{\varepsilon_P}+\frac{1}{(1-\text{vr})+\varepsilon_P\cdot\text{vr}}\right\} \end{aligned} \tag{5.5}$$

where

ε_r	permittivity of foam
ε_P	permittivity of solid polymer
Vr	void ratio (values from 0 to 1, representing 0% void to 100% void)

The foam model factor (S_{mf}) is used to describe the cell shape, while values in the range of [0 to −1] describe bubble elongation in production direction and

Table 5.1 Equations for calculation of some electrical cable parameters.

	Coaxial (accurate)	Coaxial (simplified)	Twin axial (accurate)	Twin axial (simplified)
Capacitance Accurate in (F/m) Simplified in (pF/m)	$C' = \frac{C}{l} = \frac{2\pi\varepsilon_r\varepsilon_0}{\ln\left(\frac{R}{r}\right)}$ (5.1a)	$C' = \frac{55.6\cdot\varepsilon_r}{\ln\left(\frac{R}{r}\right)}$ (5.1b)	$C' = \frac{C}{l} = \frac{\varepsilon_r\varepsilon_0\pi}{\ln\left(\frac{2a}{r}\right)}$ (5.1c)	$C' = 27.8\cdot\frac{\varepsilon_r}{\ln\left(\frac{2a}{r}\right)}$ (5.1d)
Inductance Accurate in (H/m) Simplified in (μH/m)	$L' = \frac{L}{l} = \frac{\mu_r\mu_0}{2\pi}\ln\left(\frac{R}{r}\right)$ (5.2a)	$L' = 0.2\cdot\mu_r\cdot\ln\left(\frac{R}{r}\right)$ (5.2b)	$L' = \frac{L}{l} = \frac{\mu_r\mu_0}{\pi}\ln\left(\frac{2a}{r}\right)$ (5.2c)	$L' = 0.4\cdot\ln\left(\frac{2a}{r}\right)$ (5.2d)
Impedance In (Ohm)	$Z = \sqrt{\frac{\mu_r\mu_0}{4\pi^2\varepsilon_0\varepsilon_r}}\cdot\ln\left(\frac{R}{r}\right)$ (5.3a)	$Z = 60\sqrt{\frac{1}{\varepsilon_r}}\cdot\ln\left(\frac{R}{r}\right)$ (5.3b)	$Z = \sqrt{\frac{\mu_r\mu_0}{\varepsilon_0\varepsilon_r\pi^2}}\cdot\ln\left(\frac{2a}{r}\right)$ (5.3c)	$Z = \frac{119.9}{\sqrt{\varepsilon_r}}\cdot\ln\left(\frac{2a}{r}\right)$ (5.3d)

l, unit length (:=1 m); r, inner conductor radius; R, outer conductor radius (coaxial); a, wire pair distance (twisted pair).

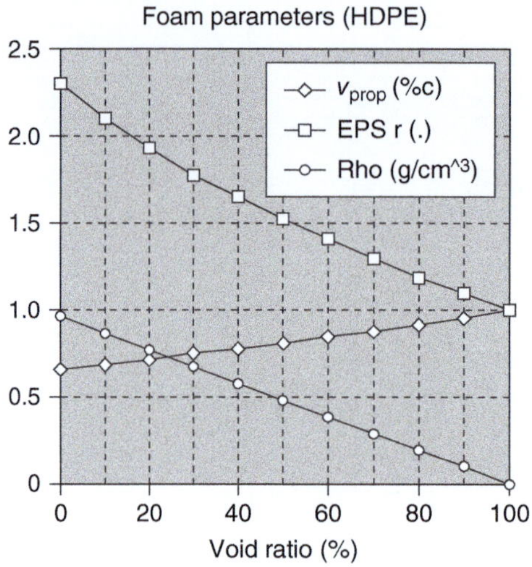

Figure 5.2 Relationship between void ratio and permittivity/propagation speed. Calculation based on HDPE with a solid density of 0.96 g/cm³ and $\varepsilon_{solid} = 2.3$.

the range [0–1] is the perpendicular direction. Assuming an ideal spherical cell shape, S_{mf} is zero and Eq. (5.5) simplifies a little.

Figure 5.2 shows the dependency of propagation speed and permittivity to the void ratio based on a HPDE grade.

Other than in solid production, ε_r is now no longer a constant due to its dependency on void ratio. Therefore capacitance and impedance have no longer a strict relationship to the diameter of the inner conductor and insulation. In a so-called "combined plot" the tolerance field of capacitance and diameter is shown. Additional characteristic curves display the ideal working point for mass output, gas flow, and permittivity. Adjusting the difference from actual workpoint to those curves is quite easy as the necessary mathematics could be provided by the line HMI system (Figure 5.3).

5.3 Selection of Polymer

Even though many polymers are foamable, only a certain number of polymers is used for cable applications. The criteria for the choice are the electrical properties and the rheological characteristics. Considering the molecular structure, foaming crystalline materials such as HDPE is more difficult than amorphous materials such as polystyrene. The reason for this can be seen in Figure 5.4. For polycrystalline polymers, the process temperature range is smaller, which means that the production of fine cell-structured foams requires a more exact process control.

One additional aspect is the electrical damping, particularly in RF applications. When using an insulating material as dielectric in a capacitor, this results in an adjustment of the phase displacement between current and voltage. The displacement from the normal angle of 90° is known as the loss angle δ. The loss factor (or dissipation factor) is defined as the tangent of the loss angle. Derived

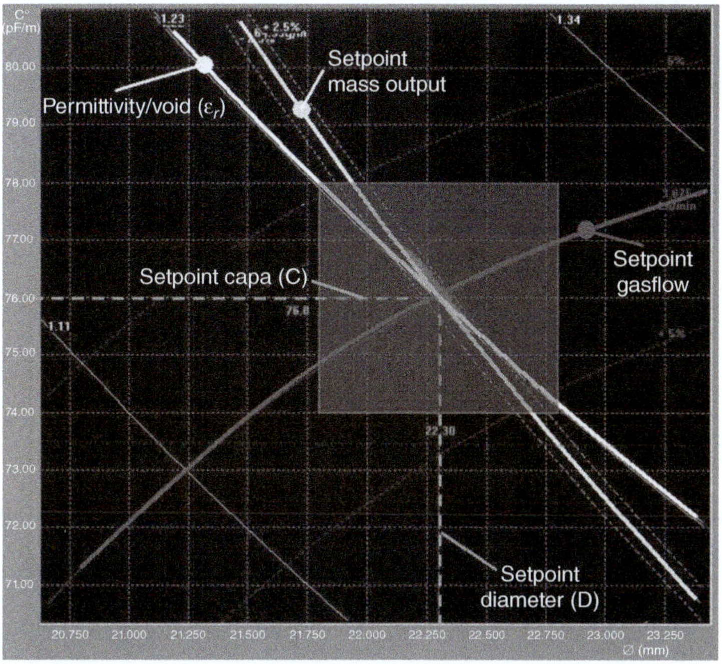

Figure 5.3 Combined plot of capacitance, diameter, output rate, permittivity, and gas flow.

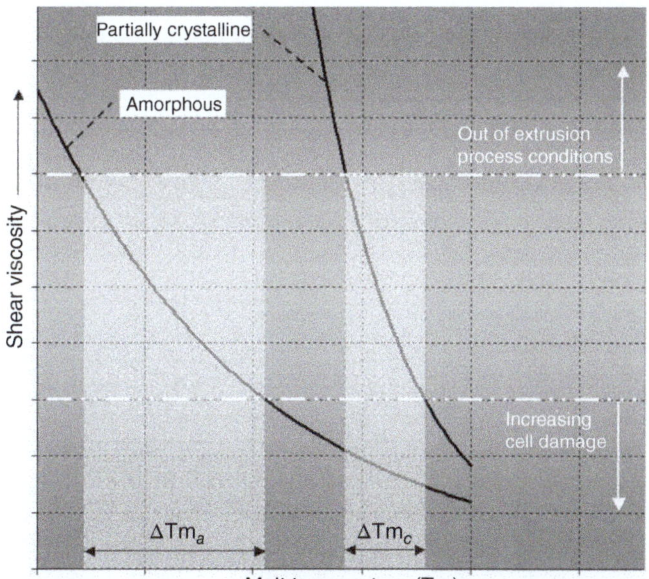

Figure 5.4 Processing window (not to scale) for amorphous and semicrystalline thermoplastics.

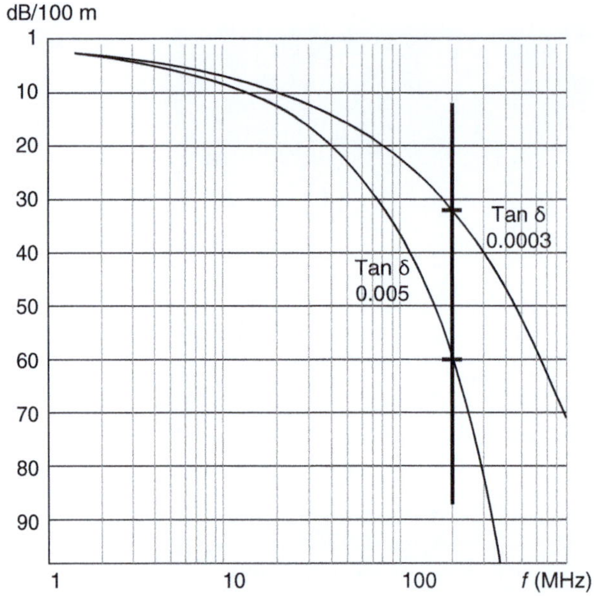

Figure 5.5 Signal attenuation due to different tan δ.

from the dielectric constant, tan δ varies slightly with temperature and significantly with frequency [4]. The exact calculation uses an equivalent circuit with the telegraphic equation and is a function of resistivity, dissipation (loss) factor, and permittivity, but will not be treated here (for details, see for example [5, 6]). Tan δ can be considered as the relationship between regainable energy (= real component of complex permittivity, $\varepsilon_r(\omega)'$) and absorbed energy (= imaginary component of permittivity, $\text{i}\cdot\varepsilon_r(\omega)''$). The reason for energy absorption is the increasing interaction of materials with the electrical field with increasing frequencies. In the GHz range, the most common dissipation mechanism is molecular dipole oscillation (rotation and precession) in the alternating electrical field [7].[1]

The relationship tan $\delta = \varepsilon_r(\omega)''/\varepsilon_r(\omega)''$ should be as small as possible and typical values are in the range of 10^{-4} for nonpolar polymers up to some 10^{-2} for highly polar materials. The higher the applied signal frequency on the cable is, the more the polar flipping increases damping as shown in Figure 5.5, where at a moderate frequency of 200 MHz, the same cable geometry has almost double the signal attenuation just caused by different tan δ values. Polymer data sheets sometimes list dissipation factors at different frequencies for the same grade. So care should be taken when comparing tan δ to get values at the same or similar frequencies.

Table 5.2 lists a selection of polymer types with typical values of ε_P and tan δ. Also mentioned in the middle row is the minimum foam permittivity that can be achieved at a high void ratio for the concerning polymer. Obviously,

1 In the literature, the expression tan δ is also used to represent the ratio of the viscous to elastic response of a viscoelastic material. So care should be taken to separate the electrical and mechanical values when looking for tan δ.

Table 5.2 A selection of polymers with permittivity and attenuation values.

Polymer	ε_p solid (typical value)	ε_r foam at void ratio (%)	Tan δ (10^{-4}) at 1 GHz
PFA	2.0	1.3 at 60	3
PS	2.2	1.2 at 80	2
FEP	2.1	1.35 at 60	4
PE	2.3	1.15 at 85	3
PP	2.3	1.35 at 65	2
ETFE	2.6	1.5 at 60	55
PVC	3.1	1.4 at 70	210
PVDF	4.5	1.9 at 60	150
TPU	4-6	2.0 at 60	190

grades such as PVC or TPU are not the right choice for high quality RF signal cables. Same care should be taken when selecting skin polymers, as they are part of the insulation and are active in the electrical field. So the abovementioned explanations are valid for skin materials as well. This concerns mainly the inner skin layer due to the highest electrical field strength near the inner conductor.

5.4 Selection of Blowing Agents

Foam processing can be done with several types of blowing agents (BAs). The type and amount of BA determine the maximum void ratio and the processing equipment needed. Most common blowing agents are chemical propellants (Chemical Blowing Agent [CBA]) or physical gases (Physical Blowing Agent [PBA]). Cell creation is almost the same for both types, but processing is different. While PBAs are needed for higher void ratio and RF cable types, CBAs are used for keeping machinery expenses low. But CBAs cannot cover high void and low loss applications. There are also other ways of foam creation, such as gas filled microspheres [8] or even releasing gas by partially decomposing the polymer. As these are rarely used for cable production, they are not considered here.

Chemical Blowing Agents are often separately compounded into the base polymer, but can also be admixed by dosing masterbatches or powder at the feeding hopper. With increasing temperature profile along the barrel, gas is released by decomposition of the CBAs. PBAs are injected into the barrel at a section where the polymer is completely molten and under pressure. They mix and dissolve within the melt. The mixing and diffusion process can be described by Fick's law [9], but more interesting is the solubility of gases in the melt according to Henry's law [10]

$$c_A = p_l \cdot S(T) = p_l \cdot S_0 e^{-\frac{\Delta H}{RT}} \tag{5.6}$$

where

c_A	blowing agent mass concentration
p_l	partial pressure
S_0	gas solubility coefficient at 0 K
ΔH	molar solubility enthalpy of the gas
R	gas constant (from ideal gas law $pV = nRT$)
T	temperature

While S_0 is a constant of the gas used, the maximum concentration is proportional to the pressure and reduces with increasing temperature T. For both blowing agent types the resulting one-phase mixture goes into oversaturation when leaving the cross head (= pressure drop), which initiates the bubble creation.

Table 5.3 shows some typical chemical blowing agents (In the last row, the normally unwanted gases from degradation side reactions of the blowing agents are underlined). The basic selection of CBAs is done according to their gas yield and the decompositon temperature range, which has to fit to the polymer melt temperature.

The already mentioned tan δ statements are valid for blowing agents and their by-products as well. As many CBAs create more or less polar residues while decomposing, high percentages of these additives cannot be used for low attenuation cables. This affects especially sodium bicarbonate types, as they release water as gas, but water is a polar molecule par excellence.

Table 5.3 Typical CBAs and some of their properties.

Name (chemistry name)	Decomposition type	Temperature range (°C)	Gas yield (cm³/g)	Released gases
ADC (azodicarbonic acid diamide)	Exothermic	200–215	150–300	N_2, $\underline{CO}$, CO_2, $\underline{NH_3}$
Sodium bicarbonate	Endothermic	100–160	140–185	CO_2, $\underline{H_2O}$
TSH (*p*-toluene-sulfonylhydrazide)	Endothermic	110–160	115–140	N_2, $\underline{H_2O}$
OBSH (4,4-oxybisben-zenesulfonylhydrazide)	Endothermic	140–160	120–150	N_2, $\underline{H_2O}$
THT (2,4,6-tri hydrazino-1,3,5-triazine)	Endothermic	245–285	185–210	N_2, $\underline{NH_3}$
TSS (*p*-toluene-sulfonylsemicarbazide)	Endothermic	210–230	120–140	N_2, $\underline{NH_3}$, $\underline{CO}$, CO_2
DNPT (dinitrosopen-tamethylenetetramine)	Endothermic	195	190–200	N_2, $\underline{NH_3}$, HCHO
5-PT (phenyltetrazol)	Exothermic	240–250	190–210	N_2
PPSO (polyphenylene sulfoxide)	Exothermic	300–340	80–100	$\underline{SO_2}$, $\underline{CO}$, CO_2

For a controlled and uniform bubble growth, nucleating agents[2] are used as "seeds" or low-energy wells to which gas can migrate and in which the bubbles begin to grow. Those nucleating agents are either incorporated in the polymer as a ready-mix or have to be added in the dosing station as powder or preferably as a masterbatch. As CBA residues are already acting as nucleating agents, there are no additives necessary here. This changes when processing physical blowing foam. Seeds can be either active or passive particles. When they are active CBAs are used in very small amounts whereas passive nucleants (organic or inorganic) do not modify during the extrusion process. Common passive particles can be, for example, talc, calcium carbonate, TiO_2, or even fine milled high temperature polymer powder that can be used in low temperature melt. In the case of foaming high temperature polymers (such as PFA or FEP), boron nitride is a good choice as well. The mean nucleant particle size ranges typically from 0.2 to 20 μm. Unfortunately, very fine powder tends to agglomerate and quickly results in large voids and/or poor cell size uniformity. Usage of active nucleants creates a small intrinsic foaming ratio (around 5% void) in the melt without adding extra gas. During the decomposition, the original particles often break into fragments, where each fragment acts as a nucleating hot spot. Therfore, active nucleants are preferred when fine cell size is required.

Some important properties of gases used for physical foaming are listed in Table 5.4 to give a comparison between (H)CFCs, organic and inorganic types. Inert gases have better properties as PBAs. In most cases, nitrogen is in use. The significance of CFCs almost vanishes despite its high solubility because of the ozone layer depletion. Water steam as blowing agent cannot be used in cable manufacturing because of its bad electrical properties; pentane and butane are combustible. Carbon dioxide is not simple to handle but its higher solubility in comparison to nitrogen promotes its use especially for higher void rates. The noble gases argon and helium could be used in special applications, but they are expensive and especially helium is difficult in handling. So most physical insulation foam for cables is based on nitrogen or carbon dioxide.

Contrary to some publications, the high void of foams when using CO_2 does not happen due to its overcritical (or "supercritical") phase. When looking at the second and third row of Table 5.4, all listed inert gases are overcritical at normal extrusion conditions ($T > 120\,°C$, $p > 10\,MPa$). Water, CFCs, and organic gases can switch between undercritical and overcritical while changing the process temperature or pressure. Such phase transitions should be avoided because the flow measurement at the critical point using parameters such as heat capacity c_p or density is not well defined [10].

5.5 Extrusion Equipment

The physical foaming process can be separated into several steps (Figure 5.6).

---

2 Also called "kicker," but the chemistry used to shift decomposition temperature of CBAs is sometimes named "kicker" as well.

Table 5.4 Gases for physical foaming with some relevant properties.

Gas formula	Tc (°C)	Pc (MPa)	bp (°C, 1013 mbar)	Permeation in PE (10^{-12} mol/msPa)	Solubility compared to nitrogen	Usability
Argon (Ar)	−122	4.91	−186	220–1100	2	+
i-Pentane ($(CH_3)_2CHCH_2CH_3$)	197	3.37	28.0	n.a.	1000	–
Carbon dioxide (CO_2)	31	7.38	−78.4	2900–3800	10	+
Nitrogen (N_2)	−147	3.39	−195.8	30–225	(:=) 1	++
CFC 11, $CFCl_3$ (trichlorfluormethane)	198	4.37	23.8	n.a.	n.a.	—
HCFC 22, CHF_2Cl (chlordifluormethane)	96	4.93	−40.8	50–60	30	—
Water (H_2O)	374	22.0	100.0	n.a.	n.a.	—
Helium (He)	−268	0.23	−269.0	1300–2000	0.25	O

n.a., data not available.

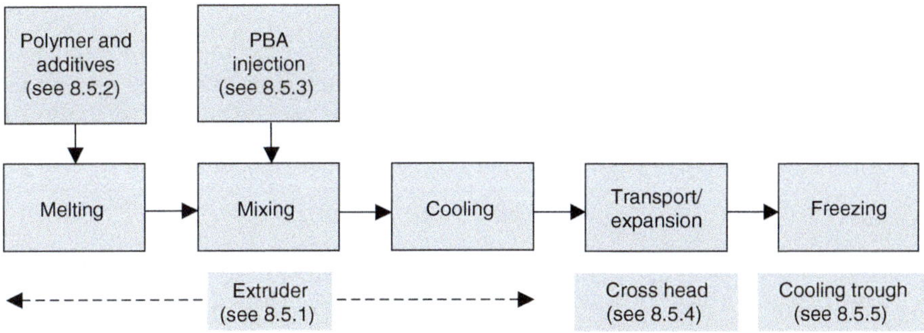

Figure 5.6 Simplified steps of physical foaming (numbers refer to sections in the chapter).

Only components that are different from solid extrusion will be mentioned in the following. In general, foam extrusion lines have higher requirements on pressure stability, temperature management, and line speed accuracy as variations affect the foam layer significantly because foam expansion acts as a magnification of instabilities.

5.5.1 Extruder and Screw

In the case of chemical foam processing, extruders for solid production might be used, if the L/D ratio is at least 27 and the temperature control is stable enough to keep temperatures within ± 0.5 to 0.8 °C of setpoint, because void ratio directly depends on the temperature profile affecting the amount of decomposed CBA. In most cases, the solid screw has to be replaced by a screw with additional mixing elements for improved homogenizing of the released gas.

For physical foaming, formerly used cascades (one extruder for melting the polymer plus gas injection feeding and a second one for homogenizing and cooling the melt) are meanwhile substituted by single screw extruders with an L/D ratio between 30 and 40, where a value of 30–33 fits most applications. The barrel has an injection point for the gas (typically a 1/2″-Unified National Fine [UNF][3] thread between half the length and 2/3 of the length). The screw needs a design so that the polymer melts completely before reaching the injection point. Otherwise, the injected gas might drain through the unmolten pellet channel backwards to the feeding section of the extruder. In the metering zone additional mixing elements help in gas distribution and homogenizing. In first order, the efficiency of the mixing process is related to the geometry of the screw and the mixing elements. But it is important not to think only in terms of mechanical mixing – the process is not based on a two-phase (emulsion) system, but on a single-phase (solution) system with a solubility gradient. Mixing elements that are too long or static mixers with high shear rates can even deteriorate a homogeneous solution.

3 UNF (Unified National Fine) thread is a very common thread type defined in the Unified Threads Standard (UTS) form and series. The 1/2″-20 UNF is used for example for most pressure transducers in extruders and it has a nominal diameter of 1/2″ with 20 threads/in.

Foam insulation consists in most cases of several layers as depicted in Figure 1.1. Inner skin is a thin solid bonding layer between the inner conductor and foam. Especially physical foam polymers have poor direct adhesion to metal surfaces and tend to lift off while cells are growing. Inner skin material should be metal stabilized, processible into thin wall thickness (typically approximately 50 μm), and compatible to the foam layer. Outer skin can be applied for several reasons:

- *Surface smoothness:* Especially for small cable constructions, where cell size is a significant portion of the insulation wall thickness. The growing foam cells can make the surface bumpy without the outer skin.
- *Protection against water entering into the foam*: When processing high void foam or using polymers with low melt strength, foam consists not completely of closed cells, especially at the surface. While passing the cooling trough, the gas in the voids is shrinking. The resulting underpressure in open cells sucks water from the outside and damping will be poor.
- *Color coding of wires*: Mainly foamed singles for twisted pair cables need to be processed in different colors for later separation. As color pigments have poor tan δ, it is not suitable to colorize the foam layer itself.
- Applying an outer skin layer improves mechanical properties of the insulation, as foam is more soft than solid material.

Because skin layers are optional, foam insulation can be only foam (F), inner skin-foam (SF), foam-outer skin (FS), or inner skin-foam-outer skin (SFS), where most common is SF and SFS. An example of a datacable (twisted pair single) with insulation diameter of 0.9 mm in SFS construction is shown in Figure 5.7.

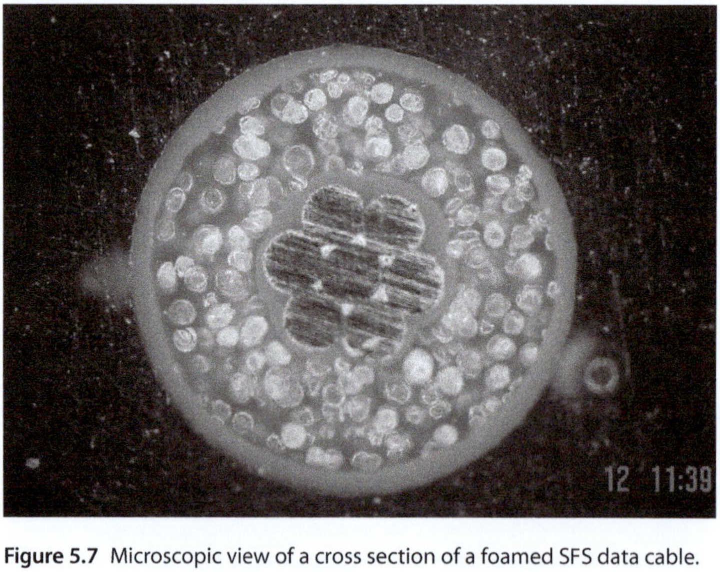

Figure 5.7 Microscopic view of a cross section of a foamed SFS data cable.

5.5.2 Dosing Station

As previously mentioned, SFS is a common setup for physical foaming. The inner skin does not need any additives, so this extruder can be equipped just with a hopper. Foam materials can be provided by polymer suppliers as ready-mix with all necessary components incorporated; for example, many chemical foam grades are available with compounded CBAs. In this case, a simple hopper with loader might be satisfactory. But more variability is given by a dosing station, where mixtures can be blended in place according to the requirements. Dosing stations can be either volumetric or gravimetric. Volumetric is cheaper but less accurate. A typical example is blending a ready-mix in steps of 10% (in volume) with an uncompounded polymer of the same grade to improve attenuation. A gravimetric dosing is somewhat more expensive, but it is much more precise when blending small amounts such as less than 1% CBA masterbatch into a nature polymer. Sometimes, two base polymers are blended and two different additives are used. Therefore, it is not unusual that foam extruders are equipped with fourfold dosing stations. Table 5.5 lists some foam material setups with suitable dosing station requirements.

Powder dosing is not recommended as powder and granules tend to demix in the extruder feeding section, and agglomeration of powder can create unstabilities in the foam process. A typical layout of the first components of an extrusion line for SFS with physical foam is given in Figure 5.8 (top view, not all details are shown).

5.5.3 Gas Injection

Injection of PBA into the barrel is done either by valves or by nozzles. The gas can be supplied by bottles or a central gas supply line.[4] To overcome the melt pressure in the barrel, gas needs to be boosted up to higher pressure than all processing melt pressure values in the injection zone. This and the adjustment of gas flow is done in a separate gas cabinet. Common mass flow devices are based on coriolis effects or differential temperature measurement with a heating element between two sensors. The gas cabinet can provide additional features such as gas accumulation for easy bottle changeover or handling gas at a certain temperature, e.g. with CO_2 to keep gas beyond its critical temperature. Connection to the injector is done by small metal tubes that are suitable for the high gas pressure. Care should be taken to avoid large internal volumes between the gas cabinet and extruder injection point as this affects accuracy of flow as well as latency when changing the flow setpoint.

In the case of nozzles, borehole diameter and applied gas pressure control the amount of injected gas. Nozzles cover only a narrow range of gas amount due to their fixed borehole diameter and have to be exchanged for different ranges. Gas flow is adjusted by pressure drop from gas injection to melt within the barrel.

4 Nitrogen can be supplied as well by generators based on compressed air. Membranes separate air into "low flow" (N_2, Ar) and "fast flow" (O_2, CO_2, H_2O) gases. Care should be taken for proper purity of nitrogen and usually a second stage booster is needed.

Table 5.5 Examples of PE-foam mixtures and required dosing stations.

Product	Material A	Material B	Material C	Material D	Equipment needed
TP cable Small dia	**100% ready-mix** DowDuPont CX 6923 NTCPD				Hopper
Coax cable Medium dia	**90% pure HDPE** Borealis HE1105	**PE with CBA** Borealis MD 1244			Volumetric Twofold dosing
RF cable Large dia	**49% LDPE** DowDuPont CX 1258 NTCPD	**59% HDPE** DowDuPont CX 6944 NTCPD	**1% nucleating** DowDuPont CXA-0012 NT		Gravimetric (or volumetric) Threefold dosing
RF cable Large dia (low loss)	**29.7%LDPE** LyondellBasell Petrothene NA217080	**70% HDPE** LyondellBasell Alathon M5370RF	**0.2% nucleating** Lehmann&Voss Luvobatch PEBA1096	**0.1% stabilizer** BASF Irganox 1010	Gravimetric Fourfold dosing

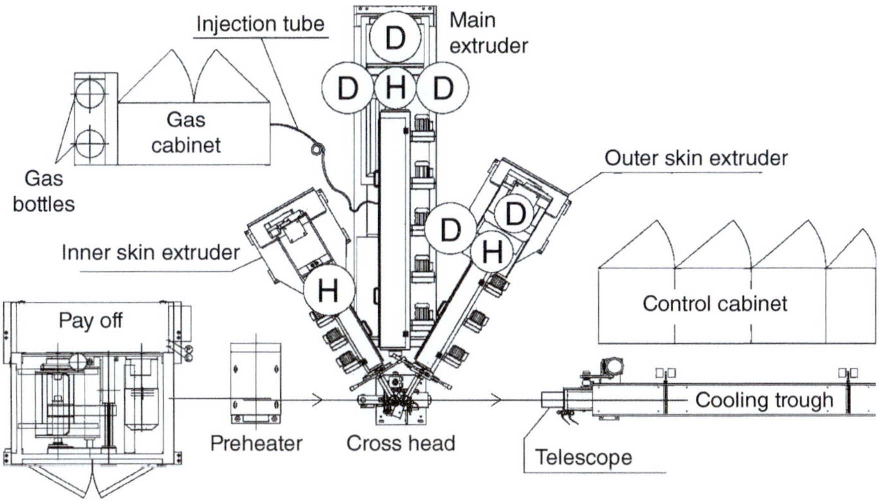

Figure 5.8 Layout of the first part of an extrusion line for physical foam (D, dosing; H, hopper).

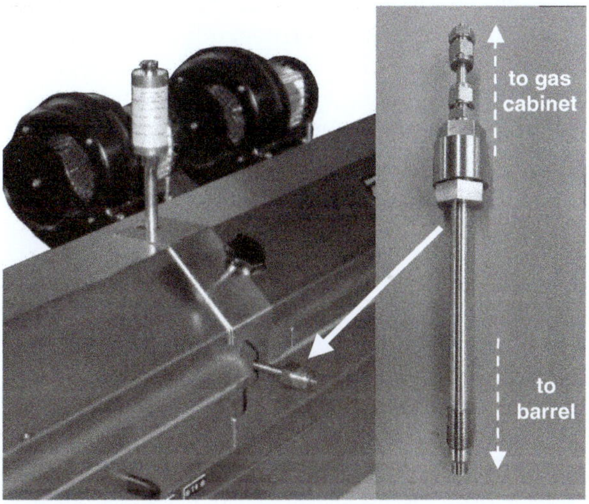

Figure 5.9 Extruder injection point and valve type injection device.

Usage of valves covers the total range of necessary gas amount; additionally they are more independent of applied pressure. The flow is typically measured by mass flow meters in units of normalized volume,[5] which is more meaningful in terms of recipe. An additional advantage of valves is the possibility of closing the injection point against backflow of melt (for example, in case of emergency line stops). Figure 5.9 shows the injection zone of a 65 mm FEP extruder with an injector and a pressure transducer for melt monitoring at the same screw length position.

5 Volume at 0 °C and 1013 mbar, given in flow per minute and indicated by "*n*" (e.g. 1.5 l_n/min)

5.5.4 Melt Transport from Screw to Die

With N_2 or CO_2, polymer solubility is much lower than with (H)CFCs and therefore, the melt can be near the critical saturation. This causes an effect that should not be neglected: even if static system pressure overrides the steam pressure of the blowing agent, local oversaturation and therefore uncontrolled cell growth may happen because of dynamic pressure drops. This is possible in every place where high shear rates reduce local pressure below the critical saturation pressure.

Typical places for high shear rates are all points along the flow channel where the cross-section area has large changes (e.g. braker plate). Flow channel design should be optimized for reduced differential changes in channel cross section. Measurements have shown that flow field is best with maximum angles below 45°. With higher slopes, uncontrolled local bubbles can establish. These bubbles grow and are washed from a critical size into the melt stream and appear as large voids at the die exit.

In high saturated melt processes (= high void applications) pressure loss between screw and die can cause a slightly oversaturation; bubbles grow in the flow channel and destabilize the melt flow resulting in defects of the finished foam structure. Therefore, channel cross section including cross head distributing cartridge should be designed for low pressure drop.

While signal attenuation depends on product material selection and geometry, structural return loss (SRL) is highly affected by machinery equipment. Mostly periodic interferences reduce the signal transfer quality. SRL can be understood as the relationship between the supplied signal energy and the reflected portion. The definition is

$$\mathrm{SRL} = 20 \cdot \log\left(\frac{U_i}{U_r}\right) \tag{5.7}$$

where

U_i magnitude of the incident wave with reference to nominal impedance Z

U_r magnitude of the reflected wave with the cable terminated with Z

Reflection happens at local faults in cable impedance and has its maximum when the disturbance length fits the signal frequency. The corresponding frequency in the used signal band is

$$\nu_R = \frac{c_0}{2\lambda\sqrt{\varepsilon_r}} \tag{5.8}$$

where

ν_R interference frequency

λ disturbance length

For example, foam insulation lines for large cross sections (e.g. 7/8″ cable and bigger) are often equipped with gear pumps between the end of the extruder and the inlet of the cross head. This reduces pressure variations due to bow waves coming from the last flights of the screw. This pressure disturbance happens periodically with screw turn speed and can be found in the frequency band of the cable (Figure 5.10).

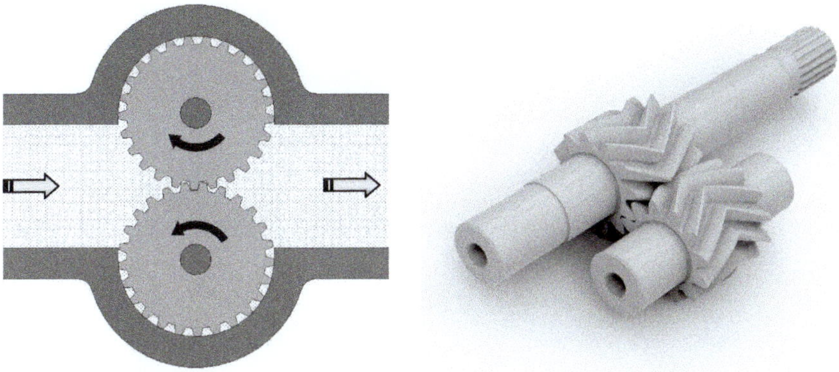

Figure 5.10 (a) Schematic drawing of gear melt pump principle. (b) Herringbone gear pair (Source: WITTE PUMPS & TECHNOLOGY [18]).

A gear pump decouples screw effects from cross head but creates its own chamber pulsation with the disturbance length λ_{GP}:

$$\lambda_{GP} = \frac{v_L}{\frac{\frac{dm_p}{dt}}{\rho V_{ch} n} \cdot Z} \tag{5.9}$$

where

v_L	line speed
dm_p/dt	mass output per time
ρ	melt density
V_{ch}	volume of chamber
n	number of teeth
Z	elements per time/revolution

In most cases, interference caused by λ_{GP} is now out of the cables used signal band and does not matter. (Anyway the pulsation can be reduced using helical or herringbone gearing instead of spur gearing [11].)

5.5.5 Cooling Trough

According to Henry's law, foam starts growing when gas-laden melt exits the cross head and pressure drops. The foam cone in Figure 5.11 has an almost transparent ring at the die as the gas and polymer are still a single phase system. After a few millimeters, cells start to grow and the insulation expands.

A movable entry of cooling trough (the so-called "telescope section") should be installed to define a point of freezing the foam. An uncooled foam would grow until the bubble pressure is almost in equilibrum with outside pressure. As cell membranes would not be stabilized, they tend to break resulting in a collapsing of the foam and a slight reduction in the foam layer diameter. So the latest point of cooling (i.e. stabilizing) is where the foam diameter has its maximum. Earlier cooling prevents cells from reaching the maximum diameter, and the insulation

Figure 5.11 A "foam cone" for a 7/8" cable (23 mm insulation diameter).

diameter is reduced with increasing capacitance. The smaller the wall thickness of the foam layer is, the more sensitive the bubble growth is to this effect. Therefore, large foam cable lines are often equipped with just a manual adjustable telescope, whereas small cable constructions are usually produced with automatic controlled telescopes.

The gas within the voids has a significantly lower heat transfer capability than the solid polymer (e.g. 0.026 W/m K for N_2, 0.4 W/m K for PE); the foam is also a thermal insulation. Similarly, a foam layer needs more time in the water to cool down to the same temperature compared to a solid layer. In Figure 5.12, thermal simulation is plotted for two cables with similar geometry (RG58-50Ohm and RG59-75Ohm), both with a production speed of 200 m/min and a cooling trough length of 50 m. Even the solid cable has around 50% higher polymer output rate; the foam product needs almost double the cooling length to go through the melt temperature all over the cross section.

For large insulation diameter, gravitational and buoyancy effects of the foam cannot be neglected. With the inner conductor held in position by its tension, foam either starts draining when running through air or tends to swim when passing water basins. Therefore alternating immersion and spray cooling sections should be used so that both effects compensate each other.

Another speciality of large cross section cables concerns the inner conductor. For better inner skin control, this layer is preferably applied separately (tandem extrusion). To stabilize the inner skin before entering the foam cross head, mainly with corrugated inner conductor, an additional small cooling trough plus dryer has to be placed in the gap between the inner skin and the main extruder.

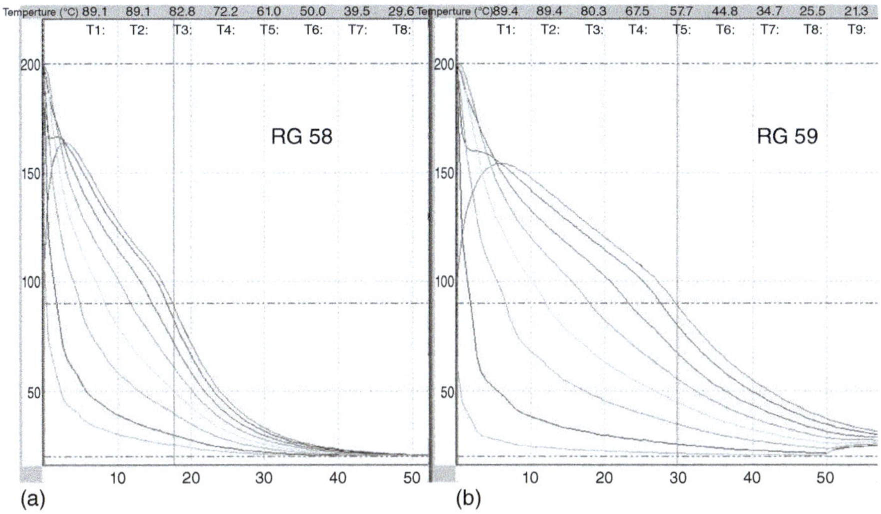

Figure 5.12 Thermal simulation of two cables with similar geometry, (a) solid insulation, (b) foam insulation (*x*-axis is the length of the cooling trough in meters).

5.5.6 Measurement Devices

Compared to solid production, there is no longer a strict relationship between insulation diameter and capacitance, so additional measurement should be installed: a "hot" capacitance device placed after the telescope section gives an early feedback. Even thermal contraction of the polymer in solid state (shrinking) is changing capacitance along the cooling trough, a compare to a second capacitance check ("cold capa") at the end of the water cooling allows a fast control by the deviation of the hot setpoint. Especially for large cross-section cables, a hot diameter control is recommended, as the time required to pass through the cooling trough can be several minutes. Here, a small gap in the water basin after the telecope section is needed to place a camera system. A centricity measurement is also required, as even slight misadjustments of the inner conductor in the cross head are enlarged during the foam expansion.

As mentioned before, inner skin can be applied in tandem extrusion. In this case, an additional skin camera in front of the foam cross head entry is used for monitoring proper layer diameter.

5.6 Processing

Foam extrusion is somewhat more complex than solid extrusion due to additional degrees of freedom that are influencing the process. Especially several slow or delayed reactions in the workflow require some familiarization and patience for parameter adjustments during production.

5.6.1 Extrusion and Die Setup

Once the setup of cable geometry, ε_r and void ratio is established, it is helpful to calculate the unfoamed diameter of the product considering only the polymer output rate using the following equation:

$$d_S = \sqrt{d_{\text{os}}^2 + \text{vr}(d_{\text{is}}^2 - d_f^2)} \tag{5.10}$$

where

d_s	solid size
d_{os}	outer skin diameter
d_{is}	inner skin diameter
d_f	foam layer diameter
vr	void ratio [0–1]

In first order, a foam die size for pressure extrusion should have the minimum diameter d_s. This neglects additional terms such as polymer rheology, polymer swelling, or line speed dependency but is suitable for startup conditions. In case of tube extrusion, standard tip and die calculation[6] can be done based on d_s as well, but DDR should be lower than it would be for solid extrusion. The reason is that growing bubbles are destabilizing the melt cone. The longer the cone is and the higher the DDR is, the more likely cone breaks can happen. For high temperature polymers such as FEP or PFA, the fast insulation surface cooling outside the cross head results in a so called "self-skinning" effect [13], where outside polymer is below the melting point (e.g. some 270 °C for FEP) before bubble growth can establish. In this case, the outer skin can be omitted, but unfoamed region should be considered as skin layer for calculation.

When a new product is initiated, the solid diameter and the correlated solid capacitance should be adjusted first (see "setpoint mass output" curve in Figure 5.3). In a second step, gas can be added to establish the required void ratio. Following Eq. (5.6), the pressure in the barrel should be kept high, while melt temperature with PBAs is normally lowered after the injection point to keep the maximum gas load as high as possible. Figure 5.13 shows some typical temperature profiles for solid production (progressive toward the cross head), chemical foam (gas release adjusted by an almost constant temperature setting after compression), and physical foam (strong heating until injection zone and subsequent cooling).

Dissolved gas acts as a solvent of the gas-laden melt and viscosity of the polymer is reduced. So melt cooling increases additionally the melt strength, an engineering measure of the elongational viscosity defined as the maximum tension that can be applied to the melt without breaking. Similarly, lower temperature stabilizes the membranes between cells during bubble growth. In general, linear polymers such as LLDPE, HDPE, and PP generally have poor melt strength because in extension the linear chains slide by without getting entangled. Branched polymers such as LDPE exhibit higher melt strength and

---

6 Equations for DDR and DRB can be found for example at [12].

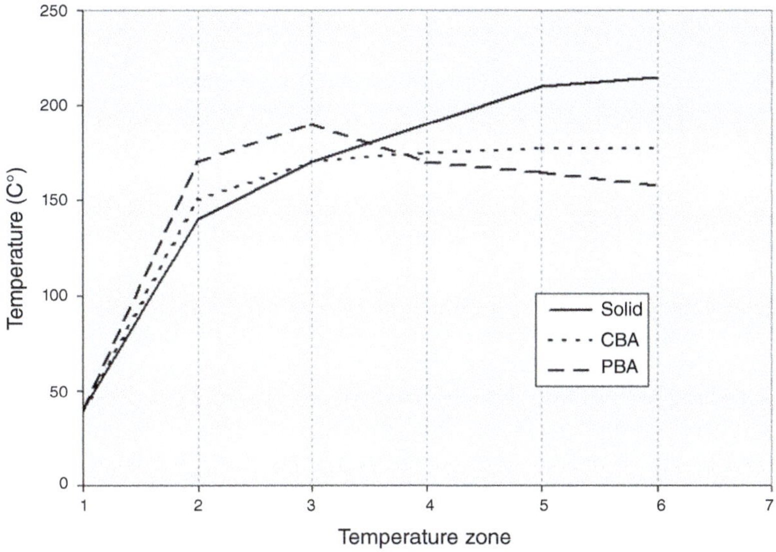

Figure 5.13 Extruder temperature profile examples for solid, chemical, and physical foam. (zone 1 = feeder, zone 2–5 = barrel, zone 6 = first external zone).

elongational viscosity because the long branches get in the way. A blend in the right proportions, e.g. 60/40 of HDPE/LDPE, increases the melt strength compared to pure HDPE and this is very common for foam insulation with large cross section.

Screw rpm (rotation/min) in foam applications is used only in a reduced range. At high turn speeds, shear and friction lead to local overheating and/or dynamic oversaturation due to flow turbulences. On the other hand, too low an rpm may not be sufficient to reach the minimum melt pressure for a certain gas load according to Eq. (5.6). Therefore proper sizing of the extruder is a key factor and an initial approach screw turn speed of approximately 20–60 rpm should cover the production range.

5.6.2 Process Control Modes

Other than in solid production, diameter control by changing the screw speed should be avoided as the working point for gas load in the melt changes as well in this case. Instead of output change of the extruder, diameter control by line speed adjustment is recommended. The following control modes can be implemented into a foam insulation line:

- Diameter control by extruder rpm change, only applicable for solid production.
- Diameter control by line speed change as described, for foam production.
- Capacitance control by an automatic movement of the cooling trough telescope. As mentioned before, shifting the telescope toward the cross head reduces cell growth and capacitance increases, and moving the water entry point away from cross head decreases capacitance. Diameter is affected as

Table 5.6 Examples of gas purity dot notation.

Dot notation	Minimum content in mole fraction
N_2...**2.8**	Nitrogen 99.8%
N_2...**4.0**	Nitrogen 99.990%
Ar...**4.5**	Argon 99.995%
CO_2...**6.0**	Carbon dioxide 99.99990%

well, so this mode is for fine-tuning – if the basic setup is far from final product specification; the telescope would go to the front or rear end position without reaching final production specification.

- Gas flow control by changes of the PBA amount injected into the barrel. This can be used to keep the telescope in the desired position range, but loop time is slow and adjustments are small, as flow changes have to proceed from injection point to cross head before any effect can be monitored.

Additionally, foam insulation is not possible during ramping up/down of line speed in good production as it is often done with solid production. The reason is again that ramping speed requires a synchronous extruder rpm change and the gas load between injection point and cross head is not of the right amount, even when making a synchronous gas flow change. Latest developments to reduce scrap production during ramping up the line are pressure controlled bypasses that keep the extruder working point independent of line speed [14].

5.6.3 PBA Handling

As previously mentioned, gas affects the overall tan δ of the insulation. This concerns mainly water fractions within the used gas. The purity of gases is given as a percentage of mole fraction [15]. The so-called dot notation for industrial gases has widely gained acceptance. This is an abbreviation for indication of purity by means of two digits separated by a dot. The leading digit is the number of "nines" in the percentage of the purity, while the digit following the dot indicates the first decimal deviating from the "nines" (see Table 5.6).

To prevent significant impurities of the gas, the recommended specification should be at least 4.0 (i.e. 99.990% purity).

When using CO_2 from bottles, the gas exists as a liquid inside at room temperature [12]. Depending on the gas injection equipment, particular care must be taken to extract either liquid CO_2 (riser bottles) or gaseous CO_2 (standard bottles). Any unplanned phase change can dramatically affect metering conditions. As a result, pumping, metering, and monitoring are more complex than for other gases such as nitrogen.

5.6.4 Maximum Void and Bubble Size

Different polymer types have different melt strength as well. Therefore, it depends on polymer type, when cell membranes reach their minimum thickness under

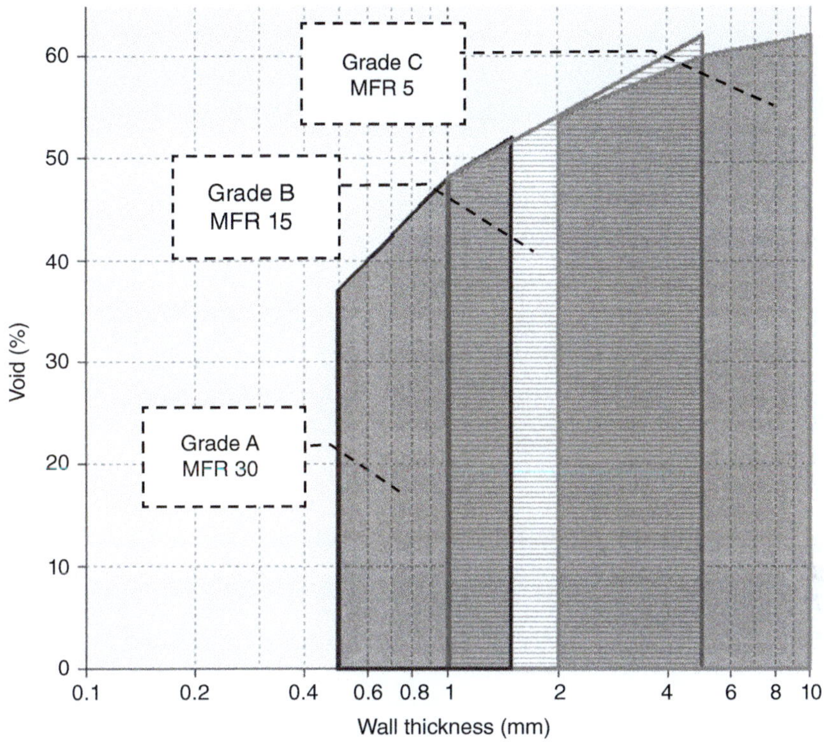

Figure 5.14 Influence of insulation wall thickness and MFR on maximum void.

comparable conditions. As already listed in Table 5.2, polymers such as PS or PE can be processed at higher void ratio than TPU or PFA. In addition, maximum void depends also on the wall thickness of insulation. Because the melt is processed at a relatively low temperature, thin layers lose much faster heat energy to the outside and to the inner conductor so that they freeze quickly and bubble growth stops. For example, PE at a wall thickness of 0.3 mm can reach only about 60% void whereas wall thickness around 2–5 mm can be processed with 75% and products with 10 mm and more allow a maximum void of around 85%.

Especially for fluorinated polymers, melt flow rate (MFR) needs to be selected according to wall thickness as well. Figure 5.14 shows an example of how a combination of FEP grades with different MFR with their maximum void can cover a wide range of product sizes.

There is no strict rule about cell size; the requirement is simply that foam appearance must be homogeneous compared to product cross section and mechanical stability is sufficient for cable applications. Polystyrene, for example, can be processed with high void and very fine cell structure, but mechanical properties are not suitable in most cases. Homogeneity is needed to have no local variations in electrical values. A bubble claiming nearly the complete wall thickness in size looks like a fault in spark testing. So cell sizes not bigger than a third of the wall thickness should be the goal. Figure 5.15 showing the cross sections of a small and a large cable insulation demonstrates this: the small cable

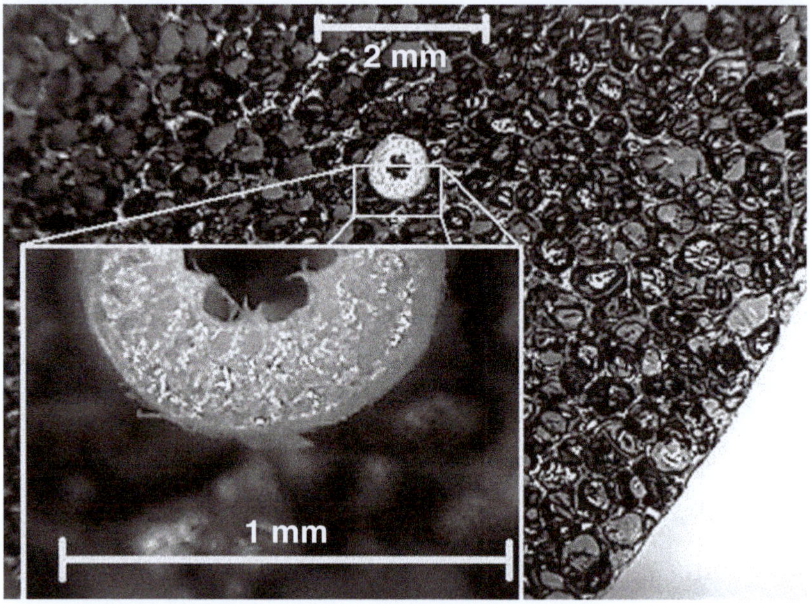

Figure 5.15 Foam structure of a 2-1/4″ (49 mm) SF and 0.75 mm SFS cable insulation.

is placed near the outer rim of the large cable. Both total cross sections look good and have fine dispersed bubbles. But average cell size of the large cable is nearly the same as the total cross section of the small one.

5.6.5 Inline Analysis by FFT

Fourier transformation (FT) is a mathematical algorithm to convert a signal measured versus time into a frequency spectrum of this signal. For example, a sine signal $f(t) = \text{A}\cdot\sin(B\cdot t)$ with a periodical appearance with amplitude A and frequency B is represented after Fourier transformation by one peak at position B in the frequency diagram. Fast Fourier transformation (FFT) is the digital implementation; detailed description can be found for example in [16]. By this means, sensor signals such as melt pressure and insulation diameter can be converted for easier identification of periodical faults in the product. In combination with the already mentioned SRL according to Eq. (5.8), structural defects could be detected during insulation manufacturing and indication for faulty components is possible. Figure 5.16 shows in the upper part a noisy input signal taken on 25 m product length and in the lower part the resulting frequency spectrum with a clear distorsion signal at around 1.04 GHz. With the actual production speed, corresponding evaluation values (right sided) such as fault length or error repetition frequency could be used to indicate and improve faulty components such as pulleys or motor drives.

An important condition for a proper FFT reading is a sensor signal bandwidth that is at least twice as high as the disturbance frequency (known as "Nyquist–Shannon sampling theorem"). The minimum sampling speed for a

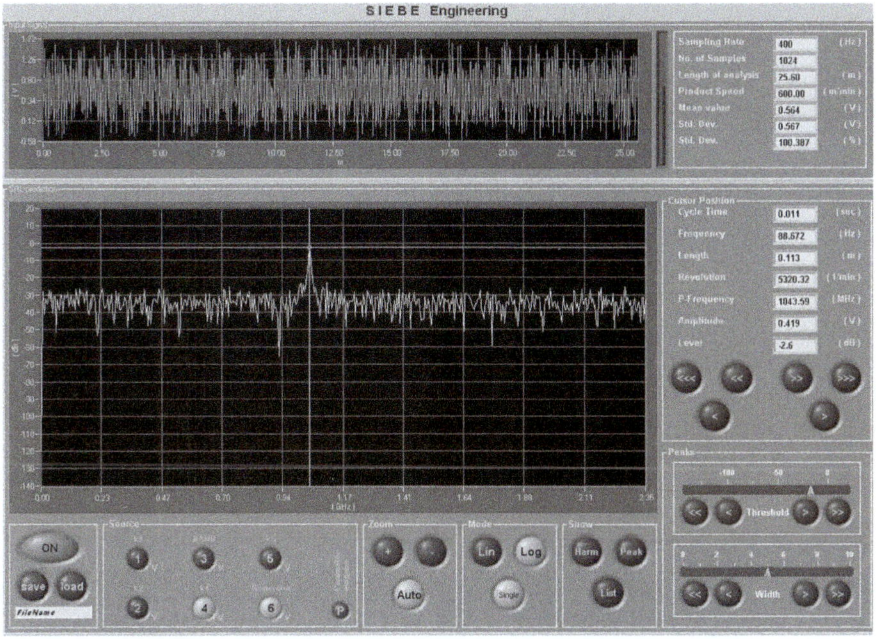

Figure 5.16 Inline FFT evaluation, line speed 600 m/min, sampling rate 400 Hz.

device can be calculated with Eq. (5.11):

$$f_{\text{sampling}} \geq \frac{V_L}{60} \cdot \frac{1}{\lambda} = \frac{V_L}{60} \cdot \frac{2\nu_R\sqrt{\varepsilon_r}}{c_0} \tag{5.11}$$

where

f_{sampling}	sampling frequency/minimum cutoff frequency of sensor signal
V_L	actual line speed in (m/min)

The value of ν_R is here the highest interference frequency of interest for the product of concern. Typical signal sources for FFT are cross head pressure, capacitance (hot and cold), and diameter of insulation.

As shown, processing of foamed cable insulation requires a careful matching of polymer, type and amount of blowing agent and machinery in combination with a good process knowledge for setting up a new product. It is obvious that not every aspect of cable foam applications could be mentioned here. For a deeper insight on foaming, literature such as [10, 17] is recommended, where foam physics and chemistry is treated in detail, but not related to cable production in particular.

Glossary

ADC	azodicarbonic acid diamide or azodicarbonamide or azo(bis)formamide
BA	blowing agent

CATV	community antenna television
CBA	chemical blowing agent
CFC	chlorofluorocarbons
DDR	draw down ratio
DRB	draw ratio balance
ETFE	ethylene tetrafluoroethylene
FEP	fluorinated ethylene propylene
FT	Fourier transformation
FFT	fast Fourier transformation
FS	foam skin
HCFC	hydrochlorofluorocarbons
HDPE	high density polyethylene
HMI	human machine interface
L/D	length to diameter ratio
LDPE	low density polyethylene
LLDPE	linear low density polyethylene
MFR	melt flow ratio
PBA	physical blowing agent
PE	polyethylene
PFA	perfluoro alkoxy alkene copolymer
PP	polypropylene
PS	polystyrene
PVC	polyvinyl chloride
PVDF	polyvinylidene fluoride
RF	radio frequency
RG	radio guide number
SF	skin foam
SFS	skin foam skin
SRL	structural return loss
TP	twisted pair
TPU	thermoplastic polyurethane
UNF	unifies national fine thread

References

1 Jonsson P, Carson S, Davis S et al. Ericsson Mobility Report 2018. https://www.ericsson.com/en/mobility-report/reports/november-2018/mobile-traffic-by-application-category.

2 Cisco global cloud index: forecast and methodology, 2016–2021. White Paper. https://www.cisco.com/c/en/us/solutions/collateral/service-provider/global-cloud-index-gci/white-paper-c11-738085.html (last accessed 2019).

3 Tanner, C. Coaxial cable: it's not just a piece of wire. https://www3.nd.edu/~wzech/CoaxialTransmissionLine.pdf (last accessed 2020).

4 Schäfer, H., Sternin, E., Stannarius, R. et al. (1996). Novel approach to the analysis of broadband dielectric spectra. *Physical Review Letters* 76 (12): 2177–2180.

5 Herter, E. and Lörcher, W. (2003). *Nachrichtentechnik*. Carl Hanser.
6 Terman, F.E. (1943). *Radio Engineer Handbook*. McGraw-Hill.
7 Keysight. (2019). Basics of measuring the dielectric properties of materials. http://literature.cdn.keysight.com/litweb/pdf/5989-2589EN.pdf (last accessed 2019).
8 Elfving, K. (2001). A new grade of expandable microspheres for foaming PP. *Rapra Blowing Agents and Foaming Processes Conference Proceedings*, 13–14 March 2001, Frankfurt, Germany.
9 Scheid, HA. (2002). Thermoplastic foam extrusion with physical blowing agents. *IWCS conference proceedings*, 51st IWCS, November 2002, Charlotte, USA.
10 Reid, R.C., Prausnitz, J.M., Poling, B.E. et al. (2000). *Properties of gases and liquids*. McGraw-Hill.
11 Kremer, H. (2013). Precise and pulsation-free feeding. *Packaging Films 24863*. Vol. 1.
12 Erocarb. (2019). Calculator for tip and die. http://www.erocarb.ch/services-en/ddr-calculator (last accessed 2020).
13 Dupont. (2018). Extrusion foaming of Teflon fluorocarbon resins. http://www2.dupont.com/Cabling_Solutions/en_US/assets/downloads/extrusion_foaming.pdf (last accessed 2019).
14 Siebe. (2019). Bypass control system for extrusion of foamed insulations. *Wire Magazine*. Vol. 1/19.
15 Häring, H.-W. (2008). *Industrial Gases Processing*. Wiley-VCH.
16 Uszczapowski, G. (1985). *Fourier-Transformation/Fourier-Integral*. Demmig.
17 Throne, J.L. (1996). *Thermolastic Foams*. Sherwood Publishers.
18 Wieczorek S. Witte Pumps & Technology. http://www.witte-pumps.com

6

Flame Retardancy of Cables

Günter Beyer

Fire and Polymer, Auenweg 14, Eupen 4700, Belgium

6.1 Introduction

The Center of Fire Statistics (CFS) of International Association of Fire and Rescue Services (CTIF) presented fire statistics for 2017 within its latest report issue 24 from 2019 [1]. Common indicators of fire statistics in all 34 reporting countries of the world in 2017 are reported (Table 6.1).

The distribution of fires by types in 2017 from the report are presented in Table 6.2.

Babrauskas [2] reported that in the United States 16% of the structural fires are electrical fires and the factors that contributed to ignition in the reported structure fires are provided in Table 6.3.

On 11 April 1996, a fire broke out inside the passenger terminal of Düsseldorf Airport in Germany, which killed 17 people [3]. Approximately 1000 firefighters were involved in extinguishing the blaze. The fire was caused by welding work. Droplets of molten metal promoted smoldering combustion inside the polystyrene insulation on the dropped ceiling of the arrival halls below. A total of 11 MT (metric tons) of PS burned as well as numerous PVC cable jackets, releasing highly toxic substances including carbon monoxide, dioxins, hydrogen cyanide, and hydrogen chloride. All victims died from smoke inhalation rather than from heat injuries. Owing to the thick black smoke, nine passengers in the airport lounge of Air France did not find the nearest escape route. Seven people died in two elevators and one passenger died inside a lavatory. One person had initially escaped the fire but died two weeks later due to the consequences of smoke inhalation. Owing to the effects of fire and smoke, two terminals were out of commission for a long time and the total damage from this disaster was estimated at €0.5 billion.

Wires and cables used in enclosed spaces obviously require fire safety properties, which are described in many national and international standards.

There are many areas where flame-retardant (FR) wires and cables are often used:

The Global Cable Industry: Materials, Markets, Products,
First Edition. Edited by Günter Beyer.

Table 6.1 Fire statistics by CFS.

Population	Total number of emergency calls	Total number of fires	Total number of fire deaths	Total number of injuries
1 106 745	48 964 229	3 115 061	16 808	47 948

Table 6.2 Different fire types.

Structure fires	Vehicles	Forests	Grass and brush	Rubbish	Other
37.3%	15.9%	1.6%	20.4%	10.4%	14.4%

Table 6.3 Ignition in structural fires.

Factors contributing to ignition in reported structural fires	Percentage
Unclassified electrical failure or malfunction	7
Unspecified short-circuit arc	4
Short-circuit arc from defective or worn insulation	2
Arc or spark from operating equipment	1
Arc from faulty contact or broken conductor	1
Short-circuit arc from mechanical damage	1
Water caused short-circuit arc	<1
Fluorescent light ballast	<1
Total (all electrical failures or malfunction)	16

Source: Babrauskas [2].

- Building wires for low voltage power and communication networks
- Control cables, for example, for nuclear power station cables or chemical plants
- Industrial power distribution cables
- Portable power cables such as mining cables, welding cables, or elevator cables
- Marine and automotive cables
- Cables for tunnels
- Photovoltaic cables and wind mill cables
- Automotive cables
- Robotic cables and many other applications

The national and international standards for FR wires and cables are very complex and diverse but in general the following groups of tests for FR wires and cables can be distinguished:

(a) Preventing ignition and low flame spread (FS) with low rate of heat release
(b) Reduction of smoke generation and also of toxic/corrosive degradation products
(c) Circuit and functional integrity of cables during a fire.

Often the market requires FR cables with the requirements of (a) and (b); cables in public buildings (security cables) can demand all three requirements (a)–(c).

6.2 Flame Propagation Tests for Wires and Cables

Industry standards and specifications are highly important for the selection and formulation of insulation and jackets materials of cables. There are many international and national cable standards to measure flame propagation by cables in the case of a fire; the most relevant are listed in Table 6.4. Some of these tests measure not only flame spread but also the heat released.

The International Electrotechnical Commission (IEC) as an international standards body prepares and publishes international norms for all electrical, electronic, and related products. Within IEC the technical committee TC 20 and its Working Group 18 (burning characteristics of electrical cables) prepare and maintain standards for test methods relating to burning and characteristics of electrical cables. Underwriters Laboratories Inc. (UL) and the analogous Canadian Standards Association (CSA) are the most important organizations that standardize the wire and cable flame tests in the United States and Canada. In the United States a set of standards and requirements for flame spread and smoke generation of wires and cables installed in office buildings, factories, and residences has been defined by the National Electric Code (NEC); the National Fire Protection Association (NFPA) prepares these codes and publishes also installation guidelines for FR cables. According to the 2008 NEC (NFPA 70) [4], cables can have different flame resistance ratings from the lowest fire resistance level to the highest fire resistance level: UL 1581, VW-1, UL 1685, UL 1666, and NFPA 262 (formerly UL-910 or Steiner tunnel test) as the globally most severe cable fire test. The latter test defines flame propagation and smoke characteristics for a bundle of electrical or optical fiber cables used in the

Table 6.4 Standards to measure flame propagation of cables.

Location	Horizontal single wire	Vertical single wire	Vertical tray, bundled cables	Riser, vertical bundled cables	Plenum, horizontal bundled cables
International		IEC 60332-1-2	IEC 60332-3		
			IEEE 383		
			IEEE 1202		
USA	UL 1581 HB	UL 1581 VW-1	UL 1685	UL 1666	NFPA 262 (UL 910)
Canada		CSA FT-1	CSA FT-4		CSA FT-6
UK		EN 60332-1-2	BS 4066-3		
Japan	JIS C-3005	JIS C-3005			
Europe		IEC 60332-1-2	EN 50399		

horizontal air-handling spaces (plenums) of buildings. The energy input by the flame for this test is 88 kW under ventilated conditions. The UL 1666 riser test simulates the flame propagation of installation of cables in a building riser shaft with a 154 kW flame attack. The cables must both pass flame spread and heat generation, which is measured via the temperature rise. UL 1685 and IEC 60332-3 measure the performances of cables installed on a vertical ladder by a 20 kW flame energy input. The classification of cables in Europe under the CPR (Construction Products Regulation) by flame spread, peak heat release, fire growth rate (FIGRA), and other parameters from EN 50399 will be discussed in Chapter 7 within in this book.

6.3 Smoke, Corrosivity, and Toxicity Tests for Wires and Cables

Cables to be used in enclosed spaces with a low level of smoke generation are necessary to improve fire safety because reduced visibility in a real fire situation makes it more difficult to escape from the fire, thus increasing the danger to human life by toxic gases and heat. The standards IEC 61034-1, IEC 61034-2, NES 711, and BS 7622-1 define smoke generation tests. IEC 61034 is also known as 27 m^3 cube test; burning alcohol as the fire source attacks horizontal mounted cables and the reduction of the light transmission in a 3 m × 3 m × 3 m chamber is measured.

The corrosivity of combustion gases from cable materials can damage electronic devices as well as communication equipment. To measure the corrosivity of combustion gases acidity tests such as IEC 60754-1 are used. This test is conducted on cable insulations or cable jackets instead of the finished cables. One gram of the sample is heated to 800 °C for 20 minutes and the amount of halogen acid gas (except hydrofluoric acid) is measured by titration with an upper allowed limit of 0.5 wt%. IEC 60754-2 and BS 6425-2 determine the degree of acidity of gases evolved during the combustion of materials taken from electric cables by measuring the pH and electrical conductivity. The material is heated up to 935 °C for 30 minutes and the combustion gases are dissolved in water. To pass the test, pH has to be above 4.3 and the electrical conductivity has to be less than 10 μS/mm.

Toxicity is also a topic for the global cable industry. NES 713 as a Naval Engineering Standard involves a close 0.7 m^3 enclosure in which a small sample of cable material such as insulation or jacket is burned by means of an air balanced Bunsen burner. The combustion gases are sampled and quantitatively analyzed in terms of critical components such as carbon monoxide (CO), carbon dioxide (CO_2), sulfur dioxide (SO_2), hydrogen chloride (HCl), hydrogen cyanide (HCN), hydrogen fluoride (HF), and others. The toxicity index is a summation in numerical form of the toxicity factor of individual gases as they are produced during the process of total combustion of the cable material under controlled conditions. Table 6.5 shows typical combustion gases and also the concentration of gas in parts per million considered fatal to man for a 30 minutes exposure.

Table 6.5 Toxicity of combustion gases.

Combustion gas component	Concentration (ppm) of gas component considered fatal to man for 30 minutes exposure
Carbon dioxide (CO_2)	100.000
Carbon monoxide (CO)	4.000
Hydrogen sulfide (H_2S)	750
Ammonia (NH_3)	750
Formaldehyde (HCHO)	500
Hydrogen chloride (HCl)	500
Acrylonitrile (CH_2CHCN)	400
Sulphur dioxide (SO_2)	400
Nitrogen oxides	250
Phenol (C_6H_5OH)	250
Hydrogen cyanide (HCN)	150
Hydrogen bromide (HBr)	150
Hydrogen fluoride (HF)	100
Phosgene ($COCl_2$)	25

6.4 Circuit Integrity and Functional Integrity for Security Cables

To ensure the continuous functionality of fire safety equipment such as sprinklers, water pumps, and emergency signal wires it is necessary to maintain the functionality of cables under fire conditions. Standards such as IEC 60331, BS 6387, EN 50200, UL 2196, or DIN 4102-12 test security cables with reference to this requirement.

Circuit integrity (insulation integrity) refers to fire tests for cables only. This is denoted by FE 180 in some European countries. The underlying standard is IEC 60331, which was originally carried out alone for flame attacks to a cable under rated voltage for a period of 180 minutes at 750 °C. To better simulate real fire conditions such as mechanical stresses to cables by collapsing walls and the presence of water, the testing conditions have been regularly modified over the years by changing the duration, increasing the temperature of the flames, and by adding mechanical stresses and water spray. Another standard, BS 6387, is still used in many countries and the security cables are categorized by a letter symbol or series of letter symbols according to the categories they meet. Table 6.6 provides the test conditions for the different categories.

DIN 4102-12 as an example describes the test for the functionality of the so-called cable sets, which consists of a group of power cables,

Table 6.6 Categories and test conditions for BS 6387.

Test conditions for BS 6387	Category
Resistance to fire alone	
650 °C for 3 h	A
750 °C for 3 h	B
950 °C for 3 h	C
950 °C for 20 min (short duration)	S
Resistance to fire with water	
15 min at 650 °C, then 15 min at 650 °C with water	W
Resistance to fire with mechanical shock	
15 min at 650 °C, then 15 min at 650 °C with mechanical shock	X
15 min at 750 °C, then 15 min at 750 °C with mechanical shock	Y
15 min at 950 °C, then 15 min at 950 °C with mechanical shock	Z

Table 6.7 Categories for DIN 4102 part 12.

Category ranking of security cable	Cable system function maintenance in the case of fire (minutes)
E 30	30
E 60	60
E 90	90

telecommunication cables, etc. These cables are fixed to support systems such as ladders, clay trays, or hanging items. Cables are powered by voltages up to 400 and 110 V AC for telecommunication. Functional integrity is tested by short-circuit of insulations or discontinuity of wires in a fire chamber test station with minimum dimensions 2 m × 3 m × 2.5 m (width/length/height). The effects of thermal expansion and mechanical load during a fire are also taken into the account. The temperature must follow the standard fire curve; for cables tested for 30, 60, or 90 minutes the temperature within the chamber goes up to 830, 930, or 1000 °C. There are three categories of function maintenance as given in Table 6.7.

The duration of the cable functionality is determined not only by design and selection of used cable materials but also often by the construction and selection of the support system and its deformation at high temperature. To bring these cables to the German market, testing and qualification can only be done by independent test institutes. Functionality integrity and circuit integrity are not comparable to their results.

6.5 Laboratory Tests for the Flammability of Wire and Cable Materials

To investigate new materials and composites to be used in wire and cable applications it is necessary for small scale laboratory tests to obtain information on flammability, smoke generation, and heat release. The LOI (limited oxygen index) test is based on ISO 4589 and ASTM D 2863. It measures the minimum percentage of oxygen in nitrogen that will support candle-like burning of a specimen for at least three minutes. It is still widely used to measure flammability of polymers as well as cable compounds. In merchantable cable compound datasheets and customer specifications the LOI is also reported as a measure of the materials' flammability. Since it is a simple, low-cost, and reliable test, it is sometimes used for material specifications and quality control of compounds for the cable industry.

The cone calorimeter was developed in the early 1980s at the NBS (National Bureau of Standards), now the NIST (National Institute of Standards and Technology). Currently, it is the most commonly used bench-scale calorimeter and is used by universities and industry. The apparatus and test procedure are standardized in the United States by ASTM E 1354 and NFPA 271 and internationally by ISO 5660 parts 1 and 2. A square specimen of 100 mm × 100 mm is exposed to the radiant flux of an electric heater. The heat fluxes that are used regularly by the cable industry are 35 or 50 kW/m^2.

The following are the material properties directly measured by cone calorimeter:

- Heat release rate
- Time to ignition
- Mass loss rate
- Smoke release rate
- Effective heat of combustion
- Rates of release of combustion gases such as carbon monoxide

Because char formation (see later in this chapter) and its character are important to improve the flame retardancy of compounds, a large sample size as requested for cone calorimeter tests is helpful to rate the char properly after combustion.

The Fire Performance Index (FPI) is a parameter frequently used to characterize the flame retardancy of materials. It is an empirical variable calculated as the ratio of the time to ignition and the peak rate of heat release by cone calorimeter measurements. The FPI is important because it correlates to the "time to flashover" indicating the time available to escape. Therefore the FPI may be considered as a good individual indicator of overall fire hazards. The higher the FPI value, the better is the flame retardancy of a material.

6.6 Polymers for Flame-Retardant Wires and Cables

PVC and polyolefins are globally the most commonly used polymers for flame-retardant wire and cable applications. Fluoropolymers, silicones, flame-retardant thermoplastic polyurethanes (TPU), and other polymers such as chlorosulfonated polyethylene (CSM) or chlorinated polyethylene (CPE) have niche applications within the wire and cable market. The book chapter by AMI describes the polymers used by the global cable industry in more detail.

- PVC

PVC was introduced to the cable and wire industry over 80 years ago because of its dielectric strength, volume resistivity, mechanical strength, water resistance, UV stability, good processability, and inherent flame retardancy. Moreover, its cost-effectiveness makes PVC the material of choice.

- Polyolefins

The next important polymers for flame-retardant applications after PVC are polyolefins and their copolymers. In contrast to PVC, polyolefins or their copolymers are not intrinsically flame retardant and therefore they need the addition of flame retardants. There is a strong global trend toward flame-retardant polyolefin-based cables with flame retardants that are not based on halogens.

- Fluoropolymers

Fluoropolymers are also a class of flame-retardant polymers for the cable industry. They demonstrate high permanent service temperatures, excellent dielectric performances, and chemical resistance; they are inherently flame retardant. The most used fluoropolymer within the cable industry is FEP, a copolymer of tetrafluoroethylene and hexafluoropropylene, which is melt-processable. The most important application for FEP is to serve as insulation material in plenum cables. Polyvinylidene fluoride (PVDF) is a homopolymer and the second important fluoropolymer in the cable industry.

- Silicone polymers

The cable industry utilizes silicone elastomers if high temperature resistance and high flexibility are needed. Silicones formulated with flame retardants show excellent char formation and low heat release values. By proper formulation of the silicones and the cable design, the char acts as a mechanically strong ceramic and can form an electrically insulating layer useful for circuit integrity applications for standards such as UL-2196 or DIN 4102 part 12.

6.7 Flame Retardants for Flame-Retardant Wires and Cables

The global cable industry consumes about 20% of all flame retardants [5, 6] and the flame retardants used principally to produce flame-retardant cables are listed in Table 6.8.

Table 6.8 Flame retardants used for flame-retardant cables.

Flame retardant	Polymer
Aluminum hydroxide (ATH)	PVC and polyolefins
Magnesium hydroxide (MDH)	PVC and polyolefins
Antimony trioxide (ATO)	PVC
Zinc borate	PVC and polyolefins
Organoclay for nanocomposites	Polyolefins
Brominated plasticizers	PVC
Chlorinated paraffins	PVC
Phosphate plasticizers	PVC
Smoke suppressants	PVC

Sources: Crozier [5]; Ceresana [6].

6.8 Flame-Retardant PVC

In 1997, a four-volume book described the knowledge at that time about the flame retardants for PVC [7]. Weil and Levchik [8] reported on the actual knowledge in 2005 and Wypych published a compendium with detailed flame-retardant PVC formulations in 2009 [9]. In 2011 Levchik et al. [10] resumed also the knowledge of smoke generation and degradation products by PVC. An actual paper by Hirschler [11] in 2017 describes in detail the flame retardancy properties and also the toxicity of smoke gases of plasticized PVC compounds.

6.8.1 Flame Retardants for PVC and Flame-Retardant PVC Cable Formulations

6.8.1.1 Phthalate-Based Plasticizers and Other Plasticizers

The flame-retardant properties of plasticized PVC are mainly controlled by the type and amount of the plasticizer. According to the applications of PVC, different plasticizers are used. Typical plasticizers are DIDP (diisodecylphthalate), DINP (diisononylphthalate), and also DOP (dioctylphthalate). Trimellitates are used as plasticizers in PVC applications for higher temperature applications; plasticizers from the chemical classes of adipates or sebacates are used in PVC formulations with low temperature flexibility. In principle, all these plasticizers are low molecular weight organic liquids with high volatility and combustibility in case of fire (see Figure 6.1).

Flame-retardant properties of PVC formulations can be improved easily with the usage of antimony trioxide and halogen-containing plasticizers.

6.8.1.2 Antimony Trioxide

Antimony trioxide (Sb_2O_3) is a very effective flame-retardant additive in the presence of halogen-containing polymers such as PVC. Often it is added to PVC in the range of 3–7 phr (parts per hundred resin) (Table 6.9). The mechanism of action is based on the formation of volatile antimony trichloride, which acts as

Figure 6.1 Flame propagation of cables during a cable fire test.

Table 6.9 PVC formulations for cable jacket; quantities for PVC, fillers, etc. by weight.

PVC	100	100
Plasticizer DINP	55	55
Calcium carbonate	10	10
ATH	50	40
Antimony trioxide	—	5
Thermal stabilizer	4	4
Lubricant PE-wax	0.5	0.5
LOI (limited oxygen index, % O_2)	27	31
Tensile strength (MPa)	17.5	16
Elongation at break (%)	220	250

a very effective flame retardant in the gas phase. Primarily it is exported from China and due to health issues of antimony trioxide dust, it is offered mostly as a master batch up to 90% active ingredient content or as a dust-free powder.

6.8.1.3 Brominated Phthalate Plasticizers

The most important material of flame-retardant plasticizers is di-2-ethylhexyl tetrabromophthalate. For instance, it is known as Uniplex FRP-45 offered by Lanxess and it is a very effective flame retardant due to its high bromine content of 45% with additionally high thermal stability and relative low smoke formation. Often, it is used in combination with antimony trioxide.

6.8.1.4 Chlorinated Paraffins

This class of flame retardants are often used as secondary plasticizers and serve as sources of additional halogen content. Today the cable industry is using chlorinated paraffins, e.g. by Dover, with higher molecular weights/longer chains that allow the reduction of overall costs of flame-retardant PVC formulations. It is reported that a formulation with 50 phr of DOP and 8 phr of antimony trioxide can be substituted for the same flame-retardant performance by a formulation of 40 phr DOP, 12 phr chlorinated paraffins, and 4 phr of antimony trioxide. To avoid migration of plasticizers resulting in exudation of cables the content of chlorinated paraffins must be as low as possible and is restricted by the plasticizers DIDP or DOP. Using DOP, for instance, the content of chlorinated paraffins should not be higher than 20 phr. Due to higher Cl content, one also has to watch out for the thermal stability of the PVC formulation containing chlorinated paraffins.

6.8.1.5 Aluminum Hydroxide (ATH) and Magnesium Hydroxide (MDH)

Both metal hydroxide fillers are used as flame retardants in plasticized PVC. The flame retardancy effect is based on the endothermic decomposition of the aluminum hydroxide (ATH) or magnesium hydroxide (MDH) within the solid phase and the release of water into the gas phase, thus quenching radical reactions. Within the global cable industry both fillers find great applications. Table 6.9 shows typical formulations for cable jackets with ATH or ATH/antimony trioxide as flame retardants.

The challenge for such formulations is to use enough metal hydroxide for sufficient flame retardancy and to achieve also good rheological and mechanical properties. Table 6.10 shows a typical formulation for PVC-based cable insulations used in the United States according to UL requirements [12].

For cable jackets and insulations fine precipitated metal hydroxides are used because of the optimized particle size distributions for enhanced mechanical

Table 6.10 PVC formulation for a cable insulation according to UL-requirement; quantities for PVC, fillers, etc. by weight.

PVC	100
Plasticizer DINP	40
MDH	30
Calcium carbonate	20
Antimony trioxide	3
Thermal stabilizer	3
Lubricant PE-wax	0.5
LOI (limited oxygen index, % O_2)	31
Tensile strength (MPa)	18
Elongation at break (%)	139

Source: Based on Marietta [12].

properties. Nabaltec and Huber offer their ATH products Nabaltec 40 CD and Martinal 104 LE or Martinal 104 LEO.

To reduce the smoke density for PVC cable compounds regularly MDH is used. By its basic character it reduces simultaneously also the acidity of the PVC degradation gases during the fire. Often in literature MDH is described as the more effective flame retardant due to its more favorable enthalpy of decomposition (enthalpy: −298 kJ/mol for ATH and −380 kJ/mol for MDH). But there are also technical papers by Huber, which reported contrary results [13]. Maybe the improved flame retardancy is also linked to the better charring by MDH-based formulations. Moreover, a synergistic flame effect is also reported for the combination of ATH and MDH.

6.8.1.6 Zinc Borate

This additive is used in PVC cable formulations because in combination with antimony trioxide it has a strong flame retardancy effect through enhanced char formation and simultaneously such formulations also show reduced smoke density. Firebrake ZB by Rio Tinto/U.S. Borax is the optimum product; its formula is $2\ ZnO{\cdot}3\ B_2O_3{\cdot}3.5\ H_2O$. Best flame-retardant properties can be obtained if the ratio of zinc borate/antimony trioxide is between 3/1 and 1/1 [14, 15]. Sometimes, a reduction in the thermal stability of PVC by zinc borate is observed; this is related to the zinc moiety of zinc borate rendering negative thermal stability but is not as detrimental as zinc oxide. This negative effect can be compensated by enhancement of the thermal PVC stabilizer in the formulation. The clear improvement of the flame retardancy by zinc borate is often due to improved char formation during the degradation of PVC; this char acts as a good thermal barrier against the external heat flux in the case of a fire. The synergy in flammability improvement and smoke reduction between antimony trioxide and zinc borate is particularly effective in the presence of ATH and MDH.

6.8.1.7 Phosphate Plasticizers

These plasticizers are used in flame-retardant PVC formulations if a high demand for flame retardancy is required. Often alkyldiphenyl phosphates are used, which are commercialized by Valtris Specialty Chemicals or ICL-IP under the trade names Santicizer or Phosflex. The cable industry uses these flame retardants if cables are tested by the US norm UL 910, the Steiner tunnel test, which is the globally most severe flame propagation test. A typical PVC plenum cable formulation [16] is shown in Table 6.11.

6.8.1.8 Smoke Suppressants

Plasticized PVC cables often demonstrate a high smoke density in case of fire. To reduce the smoke density effectively different smoke suppressants are used. Usually molybdenum or zinc compounds are used, e.g. additives from Huber under the trade name of Kemgard. The leading smoke suppressant for PVC applications is ammonium octamolybdate. Studies have shown that molybdates work in the solid phase and they reduce smoke by catalyzing the formation of *trans*-olefin structures from *cis*-olefin structures of the dehydrochlorinated PVC residue; this transfer reduces smoke generation, which is started by cyclization reactions of

Table 6.11 PVC formulation for a plenum cable jacket; quantities for PVC, fillers, etc. by weight.

PVC	100
Santicizer 2418 (alkyl aryl fire-retardant plasticizer)	30
Bis(2-ethylhexyl) tetrabromophthalate (FR-plasticizer)	20
ATH	30
Ammonium octamolybdate	30
Zinc molybdate	10
Thermal stabilizer	7
Lubricants PE-wax/stearic acid	1

Table 6.12 PVC formulation for a cable jacket with reduced smoke emission; quantities for PVC, fillers, etc. by weight.

PVC	100
Pentaerythriol ester (plasticizer)	31
Bis(2-ethylhexyl) tetrabromophthalate (FR-plasticizer)	10
Alkyl diphenyl phosphate (FR-plasticizer)	5
ATH	60
Antimony trioxide	5
Zinc borate	2
Calcium carbonate	10
Kaolin	5
Zinc-Molybdate complex (smoke suppressant)	10
Thermal stabilizer	7
Lubricant PE-wax	0.25

cis-polyene segments. Table 6.12 [17] shows a typical PVC formulation for low smoke density.

6.8.1.9 Nanocomposites

A new flame-retardant system, called nanocomposites, is the topic of many international conferences on flame retardancy of polymers over the last 15 years [18]. A so-called organoclay (a layered silicate such as montmorillonite, which has been modified by a quaternary ammonium compound; e.g. from Minerals Technologies Inc.) is dispersed at the nanometer level by classical polymer compounding machines in the polymer matrix. The extremely fine dispersion changes the degradation mechanism of the polymer and a very good char formation is generated, which acts as a thermal barrier against the external heat from the fire to the underlying polymer. Unfortunately, to date, this concept is not successful for PVC because the thermal stability of PVC is collapsed by the strong basic character of the nitrogen-containing modifier within the organoclay [19]. But excellent improvements of flame retardancy by nanocomposites are reported for halogen-free flame-retardant (HFFR) compounds.

6.9 Flame-Retardant Polyolefins

Beside PVC, which is still the most used polymer within the global cable industry, an interesting market, especially in Europe and the Far East, has been established. This market is based on HFFR cables, which are also named as low smoke zero halogen (LS0H) cables. This market and the flame retardants used will be described. Also, the new European classification for flame-retardant cables by the European Construction Product Regulation (CPR) will be summarized. More details on the CPR and its impact on the cable industry are presented in Chapter 7 within this book.

It should be noted that although HFFR cables have taken more and more market share, the use of bromine-based flame retardants in polyolefin cables still occupies a major market share of the global flame-retardant polyolefin cable market.

As indicated by Applied Market Informations Ltd. [20] the segmentation of the global cable market of compounds for cable insulations and jackets today is given in Table 6.13.

The following segmentation for the compounds as reported by Applied Market Informations Ltd. [20], only looking at the European market, is given in Table 6.14.

Table 6.13 Global share of different compounds for cable insulations and jackets.

Compounds	Portion (by percentage)
PVC	60
PE- and cross-linked PE (mostly bromine-containing)	26
HFFR compounds	6
Other compounds (cross-linked elastomers, TPU, …)	8

Table 6.14 Share of different compounds for cable insulations and jackets in Europe.

Compounds	Portion (by percentage)
PVC compounds	43
PE- and cross-linked PE	32
HFFR compounds	16
Other compounds (cross-linked elastomers, TPU, …)	9

Table 6.15 Global geographical locations of cable fabrication.

Continent	Portion (by percentage)
Far East including China	86.2
Europe	4.5
North America	3.4
South America	2.8
Mideast and Africa	3.1

The higher market portion of HFFR cables in Europe is based on market requirements and the necessity to meet norms. Beside a low smoke density according to the norm IEC 61034, there is often the request for low acidity of fire effluents and the chemical proof of absence of halogens according to the norm IEC 60754.

Besides, approximately 350 European cable companies are using about 3% of the total European polymer market. The top 10 of the European cable companies serve 60% of the market in Europe. In the following years the European cable market will expand by 3.3% per year. Globally, more than 8000 cable companies exist and Table 6.15 lists the geographical allocations.

6.9.1 Flame Retardants for Polyolefins and HFFR Cable Formulations

6.9.1.1 Aluminum Hydroxide (ATH) and Magnesium Hydroxide (MDH)

Both metal hydroxides are used intensively as flame retardants in HFFR cables. ATH is more widely used because MDH is much more expensive. The flame-retardant effect is based on the endothermic degradation of the metal hydroxide by the release of water. Table 6.16 shows exemplarily a typical ATH-based HFFR compound for cable jackets [21].

The challenge of metal hydroxide based HFFR formulations is related to the content of the flame retardants within the formulation: the content must be as high as possible to achieve a sufficient flame-retardant level and on the other hand the level should be as low as possible to have good mechanical and rheological properties. For both cables' insulation and jacket almost exclusively fine precipitated metal hydroxides are used due to their optimal particle size distributions. Nabaltec or Huber (formerly Martinswerke) offer with their products Apyral 40 CD or Martinal 104 LE and Martinal 104 LEO optimized products. A polymer blend by LLDPE and EVA is often used, because EVA demonstrates a high acceptance of ATH or MDH, and LLDPE as a blend component contributes to high mechanical properties. To compatibilize the flame retardant with the polymer matrix silanes (such as aminosilanes coupling agent) or maleic anhydride, grafted polymers such as polyethylene (PE-g-MSA) are often used.

Table 6.16 Aluminum hydroxide-based HFFR cable jacket formulation.

Component	Portion (by weight percentage)
Polymer LLDPE	10
Polymer EVA	24.8
Compatibilizer (PE-g-MSA)	5
ATH	60
Thermal and UV stabilization system	0.2
Properties	
Tensile strength	>12 MPa
Elongation at break	>200%
LOI (limited oxygen index) in % oxygen	>32

Source: Based on PINFA [21].

6.9.1.2 Zinc Borate and Polysiloxanes

These additives are used as synergists in HFFR cable formulations. The clear improvements in flame-retardant properties of zinc borate are related to improved char formation by a ceramification reaction between alumina oxide from the ATH degradation and zinc borate. This improved char reduces the release of combustible gases from the polymer degradation while also forming a protective layer that shields the underlying material from the heat of the external heat flux. The best product is Firebrake ZB by Rio Tinto/U.S. Borax with the following composition:

$$2\,ZnO \cdot 3\,B_2O_3 \cdot 3.5H_2O$$

Substitution of 5% of ATH or MDH by zinc borate demonstrates a significant improvement in flame-retardant properties. Shen [22] reported also about additional synergistic effects of polysiloxanes in HFFR compounds. Compared to an EVA/MDH formulation with a weight ratio of 35/65 EVA/MDH the combination of 55/5/5 weight ratio of ATH/zinc borate/polysiloxane demonstrates an optimal improved char and therefore the best flame-retardant properties. A good choice of polysiloxane is Genioplast Pellet S from Wacker, which can be handled easily as granulates. Table 6.17 demonstrates a HFFR cable compound that can be used as insulation and jacket [23].

6.9.1.3 Nanocomposites

A new flame-retardant system for HFFR cable compounds has been reported a few years ago [24] and is based on nanotechnology. An organoclay (e.g. by Minerals Technologies Inc.) based on a layered silicate such as montmorillonite and organically modified by a quaternary ammonium compound is dispersed in the polymer matrix at the nanometer level resulting in a nanocomposite. This very fine filler dispersion changes the degradation mechanism of the polymer

Table 6.17 Aluminum hydroxide-based HFFR cable insulation and jacket formulation.

Component	Portion (by weight percentage)
EVA (Escorene UL 00328 by Exxon Mobile)	21.3
LLDPE (BPD 3642 by Ineos)	9.9
Maleic anhydride grafted EVA (Fusabond C 250 by Dow Dupont)	2.75
Maleic anhydride grafted LLDPE (Fusabond MB-226 D by Dow Dupont)	1.25
Fine precipitated ATH	63.5
Thermal stabilizer (Irganox 1010 by BASF)	0.2
Thermal stabilizer (Irgafos 168 by BASF)	0.1
Poysiloxane (Genioplast Pellet S by Wacker)	1.0

and results in strong char formation in case of fire. Therefore, the heat release measured by cone calorimeter (ISO 5660) can be reduced dramatically by this technology as confirmed in many scientific publications. The organoclays for HFFR applications are combined with ATH or MDH and act as very effective synergists. Therefore, nanotechnology enables compounds to pass fire tests with high requirements. This concept was pioneered by the author and awarded in 2003 and 2004 by IWCS (International Wire & Cable Symposium, Inc.).

For industrial applications it is important to ensure an optimal thermodynamic compatibility between the organoclay and the polymer matrix by polar interactions. Therefore, it is important that the polymer is polar to enable the dispersion of the polar filler organoclay to the nanometer level. Thus, EVA is a perfect polymer. For the nonpolar polymer LLDPE, it is important to increase the polarity; this can be done by adding maleic anhydride grafted LLDPE or LDPE. Using a proper polymer the nanocomposite formation is then achieved easily by existing extruders or internal mixers. Paul and Robeson [25] reviewed in their paper additional possibilities for the synthesis of nanocomposites.

6.9.1.4 Ceramifiable Compounds

It is a logical further development for HFFR cables to look for polymer compounds that can be transferred easily by heat to a ceramic as a mechanically stable char [26]. Today, this is possible and some companies such as Wacker or Elkem are offering ceramifiable compounds to the cable industry. Most commonly, these compounds are based on cross-linked silicone-based rubbers with additives for ceramification reactions. But also analog compounds based on cross-linked EPDM are described. Ceramifiable compounds are used as insulation in security cables with electrical functionality in case of fire; see part 6.4 of this chapter.

Table 6.18 Euroclasses of cables.

Euroclass	Test procedure	Classification criteria	Additional requirements
A_{ca}	EN ISO 1716	PCS $\leq$ 2.0 MJ/kg	
$B1_{ca}$	EN 50399 (30 kW burner) and EN 60332-1-2	FS $\leq$ 1.75 m; $THR_{1200s} \leq$ 10 MJ; PHRR $\leq$ 20 kW; FIGRA $\leq$ 120 W/s; dripping of burning droplets/particles $H \leq$ 425 mm	Smoke production and acidity
$B2_{ca}$	EN 50399 (20.5 kW burner) and EN 60332-1-2	FS $\leq$ 1.50 m; $THR_{1200s} \leq$ 15 MJ; PHRR $\leq$ 30 kW; FIGRA $\leq$ 150 W/s; dripping of burning droplets/particles $H \leq$ 425 mm	Smoke production and acidity
C_{ca}	EN 50399 (20.5 kW burner) and EN 60332-1-2	FS $\leq$ 2.00 m; $THR_{1200s} \leq$ 30 MJ; PHRR $\leq$ 60 kW; FIGRA $\leq$ 300 W/s; dripping of burning droplets/particles $H \leq$ 425 mm	Smoke production and acidity
D_{ca}	EN 50399 (20.5 kW burner) and EN 60332-1-2	$THR_{1200s} \leq$ 70 MJ; PHRR $\leq$ 400 kW; FIGRA $\leq$ 1300 W/s; dripping of burning droplets/particles $H \leq$ 425 mm	Smoke production and acidity
E_{ca}	EN 60332-1-2	$H \leq$ 425 mm	
F_{ca}	Does not fulfill requirement of class E_{ca}	$H >$ 425 mm	

6.10 CPR (Construction Products Regulation)

Since July 2017 the CPR regulates the requirements for construction products such as cables for the European market. Four important properties are to be measured by EN 50399:

- Flame spread (FS)
- Peak heat release rate (PHRR)
- Total heat released rate (THRR)
- Fire growth rate

The underlying norm EN 13501-6:2019-05 led to seven euroclasses and additionally further requirements such as smoke production or acidity of smoke (Table 6.18).

For public buildings in Europe now CPR rated cables very often must be halogen free with low smoke emission. For the important classes B2 and C, the fire test EN 50399 is much more severe compared to the well-known international cable bundle test IEC 60323-3. Therefore, it is a general direction of flame-retardant cable developments to improve clearly the char formation and its mechanical stability in case of fire.

References

1 Brushlinsky, N.N., Ahrens, M., Sokolov, S.V., Wagner, P. (2019). World of Fire Statistics. Report 24. www.ctif.org.

2 Babrauskas, V. (2008). Research on electrical fires: the state of the art fire safety science. In: *Proceedings of the Ninth International Symposium*, 3–18. Karlsruhe/Germany: International Association of Fire Safety Science.

3 National Fire Protection Association (1996). *Fire Investigation Summary.* Düsseldorf, Germany: NFPA.

4 National Fire Protection Association (2008). *NFPA 70: National Electrical Code.* NFPA.

5 Crozier, J. (2018). Trends in FR requirements and demand, as seen by the European PIN FR industry. *Presentation at the AMI Conference Fire Resistance in Plastics* (10–12 December 2018), Cologne, Germany.

6 Ceresana (2019). *6th Flame Retardants Market Report*. Constance, Germany: Ceresana.

7 Nass, L.I. and Heiberger, C.A. (1997). *Encyclopedia of PVC*, 2e. New York, NY: Marcel Dekker.

8 Levchik, S.V. and Weil, E.D. (2005). *Polymers for Advanced Technologies* 16 (10): 707–716.

9 Wypych, G. (2009). *PVC Formulary.* Toronto, ON: Chemtech Publishing.

10 Levchik, S.V., Hirschler, M., and Weil, E. (2011). *Practical Guide to Smoke and Combustion Products from Burning Polymers – Generation, Assessment and Control.* Shawbury, UK: iSmithers.

11 Hirschler, M. (2017). *Fire and Materials* 41 (8): 993–1006.

12 Marietta, M.. (1999). Technical Paper: MAGSHIELD in PVC.

13 Chen, T. (2006). *Proceedings of the SPE Conference – Vinyltec* (16–18 October 2006), Atlanta, GA, USA.

14 Shen, K. 2003. Firebrake zinc borates as multifunctional flame retardants in wire and cables. *52nd International Wire and Cable Symposium (IWCS)*, Philadephia, PA, USA (17–20 November).

15 Al-Mosawi, A.I., Al-Zubadi, A.A., and Al-Maamori, M.H. (2014). *Journal of Engineering and Materials Sciences* 1 (1): 1–4.

16 Kroushl, P.W.. (1993). Polyvinyl chloride based plenum cable. US Patent 5 227 417, filed 24 January 1992 and issued 13 July 1993.

17 Linsky, L.A., Andries, J.C., Quellette, D., Buono, J. A., and Tao, T. (1999). Flame retardant composition. US Patent 5 886 072, filed 7 August 1995 and issued 23 March 1999.

18 Beyer, G. (2005). *Journal of Fire Sciences* 23 (1): 75–87.

19 Beyer, G. (2007). *Journal of Fire Sciences* 25 (1): 65–78.
20 AMI. (2018). Prospects and developments for the wire and cable industry. International Conference Cables, AMI, Cologne/Germany (6–8 March 2018).
21 PINFA. *Innovative and Sustainable Flame Retardants in Building and Construction*, Technical Documentation. PINFA.
22 Shen, K. (2004). Boron-based flame retardants in non-halogenated polymers. In: *Non-Halogenated Flame Retardant Handbook* (eds. A.B. Morgan and C.A. Wilkie), 201–235. Scrivener Publishing/Wiley.
23 Schombourg, J (+), private communication.
24 Beyer, G. (2001). *Fire and Materials* 25 (5): 193–197.
25 Paul, D.R. and Robeson, L.M. (2008). *Polymer* 49 (15): 3187–3204.
26 Beyer, G. (2017). Developments of low smoke zero-halogen compounds for cable applications within the last two decades – a review. *International Conference Polymers in Cables*, Pittsburgh, PA, USA (20–21 June).

7

CPR Testing of Cables

Franck Poutch

Crepim, Parc De La Porte Nord, Rue Christophe Colomb, Bruay-la-Buissière 62700, France

7.1 Introduction

The fire properties of cables used in European Construction are of great importance from a fire safety perspective. The path to new fire tests system for the cable industry has taken time, as shown below:

- Definition of a new set of fire testing methods, via an exploratory work financed by the European commission known as the Fire Performance of Electric Cables (FIPEC) program (1996–1999) [1],
- Setting up of a new set of rules related to extended application, via the CE MArking of Cables (CEMAC) II round robin program (2006–2010) [2],
- Integration of these outputs to the standardization work of CENELEC TC 20, in connection with CENELEC TC 127, in link with the Mandate M443 [3],
- Publication of a new classification system EN 13501-6 in 2014 [4], for power and communication cables, which has been included in the classification system under the European Construction Product Regulation (CPR),
- Publication by CENELEC TC 20 [5] of a product standard for cables in September 2014, EN 50575 [6], relayed by the Official Journal of the European Union (OJEU) in mid-2015, associated with a two years transitional period,
- From this date, all cable manufacturers who intend to sell cables for use in permanent installations in buildings in Europe must obtain CE-marking for their products.

This process has been triggered by the necessity to adopt an adequate fire safety classification for cables used in construction under the CPR framework [7] (Table 7.1).

A large part of the workload has been achieved for other construction products by using the EN 13823 (Single Burning item) for linear product and wall/ceiling covering solutions, and for floor covering by using EN ISO 9239-2 (flooring radiant panel) (Figures 7.1 and 7.2).

The challenge was to adapt or develop fire tests for cables in line with the new fire parameters such as Heat Release Rate (HRR) and Smoke Production Rate (SPR), measured in dynamic conditions in over-ventilated scenario.

The Global Cable Industry: Materials, Markets, Products,
First Edition. Edited by Günter Beyer.

Table 7.1 Euroclasses versus fire contribution level (based on Room corner test ISO 9705, 100 kW heat source) [8].

	Level of fire contribution	Flash over	Time to flash over
A1 class	No Contribution	No	
A2 class	Low	No	
B class	Medium	No	
C class	Significant	Yes	>10 min
D class	High level	Yes	>2 min
E class	Very high level	Yes	<2 min

Figure 7.1 EN 13823 test [9].

We propose to go ahead in the process featured above to explain clearly the key drivers linked to cable fire safety.

7.2 FIPEC Program

The FIPEC [11, 12] project is a research project funded by DG XII of the European Commission and co-financed by several european cable manufacturers, materials suppliers, cable users, and governmental research bodies.

The FIPEC program was initiated by the European Commission and goal of the program consists in developing of new measurement techniques for assessing the fire performances of electric cables.

Figure 7.2 EN ISO 9239-2 flooring radiant panel test [10].

Finally, guidance documents have been drafted and made available to the national standardization group and CEN/CENELEC.

This project has been a large program involving several partners:

- IC (Interscience Communication Ltd., UK) for project management
- SP (Sweden) for financial management
- ISSeP (Belgium)
- CESI (Italy)

The FIPEC project has led to the development of different levels of testing, ranging from small-scale, cone calorimeter test procedures developed for cables and materials, to full-scale-test procedure based on the IEC 60332-3 [2], but utilizing additionally HRR and SPR measurements, and a real-scale test.

This program consists of 18 work packages, each of which is being undertaken by several laboratories. These are shown in Table 7.2.

7.2.1 FIPEC Approach

This program has been set up in the framework of Construction Product Directive in order to develop an adequate set of fire tests for cables. Indeed, the current set of tests designed for construction product was not adapted to cable geometry. The idea was to assess the same type of parameters with fire test featuring realistic fire scenario for cables.

The experiments have been carried out at four scales ranging from small material samples to real-scale cable installations. The real-scale scenario test was considered as a reference scenario:

1) Real-scale scenario tests carried out on model electric cable installations,
2) Full-scale standard tests carried out on cable trays (based on IEC 60332-1-3) [13],
3) Small-scale tests on cables carried out in a cone calorimeter,
4) Small-scale tests on materials carried out in a cone calorimeter,

Table 7.2 Work packages of the FIPEC program.

WP	Title
WP 1	Review of European cable installations and planning of real-scale scenarios test series
WP 2	Investigate the effects of variables on a full-scale test
WP 3	Develop a small-scale cable test that can determine the essential parameters
WP 4	Develop a small-scale material test
WP 5	Construct cables
WP 6	Conduct real-scale fire tests
WP 7	Conduct full-scale standard test
WP 8	Conduct small-scale cable tests
WP 9	Conduct small-scale tests on materials (sheaths and insulations)
WP 10	Validate full-scale standard tests by correlation to real-scale tests in the European scenarios defined in WP 1
WP 11	Correlate the small-scale test results to full-scale standard test results
WP 12	Develop bases for correlation of results of small-scale tests on materials to small-scale test results on cable used in WP 11 to predict full-scale test data
WP 13	Develop mathematical model for the prediction of heat release rate and flame propagation of burning cables in real fires and full-scale tests from the results of small-scale tests on cable specimens
WP 14	Develop a measurement system proposal based on the existing measurement system and the information obtained
WP 15	Prepare guidelines for use in the production of standards
WP 16	Data management
WP 17	Program management
WP 18	Final report

7.2.1.1 Real-Scale Scenario

Regarding the real-scale scenario, the review identified several vertically and horizontally orientated cable installations and several test configurations were selected which were representative of these end use conditions.

Four different configurations have been identified: open, semi-closed, closed with ventilation, and closed without ventilation. Table 7.3 illustrates the selected test configurations that can represent scenarios in each of the four studied installation scenarios.

Table 7.3 Overview of horizontal and vertical real-scale test setups.

Open	Semi-closed	Closed without ventilation	Closed with ventilation
Power plant	Power plant	Power plant	Tunnels
Vehicles		Vehicles	
Tunnels	Tunnel	Tunnels	
Occupancies		Occupancies	Occupancies

Source: Based on Van Hees et al. [14].

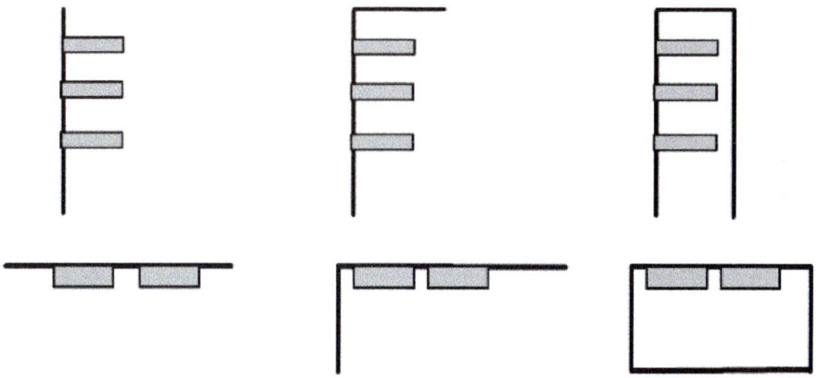

In total 22 unique tests were performed, and the work program finally set up featured the following:

- Four heat source programs (40, 100, 200, and 300 kW combinations),
- Four different categories of cables,
- Eight different setups (horizontal and vertical),
- Three different ventilation levels.

7.2.1.2 Cable Selection

Regarding cable selection, the selection was based on four end use applications [14]:

- Power plants
- Vehicles (trains, ships, and aircraft)
- Tunnels
- Occupancies (e.g. control rooms, underfloor voids, ceiling voids, and riser shafts) (Table 7.4)

The FIPEC cable test program also included all the different kinds of polymeric matrix (11 systems from polyvinyl chloride materials to cross-linked polyethylene materials).

Table 7.4 Cable selection for power plant applications.

Type of cable	Construction type	Approximate number of cables present	Main orientation
Medium voltage	$3 \times 150\,mm^2$	10	H + V
Low voltage	$7 \times 1.5\,mm^2$	50	H + V
Data cables	12 pairs $0.5\,mm^2$	100	H + V
Wires	$1.5\,mm^2$	1000–2000	H + V
Optical	Multiple diameter 1 cm	10	H + V

Source: Based on Van Hees et al. [14].

Table 7.5 Horizontal and vertical setup real-scale fire test [15].

Horizontal testing setup	Vertical testing setup
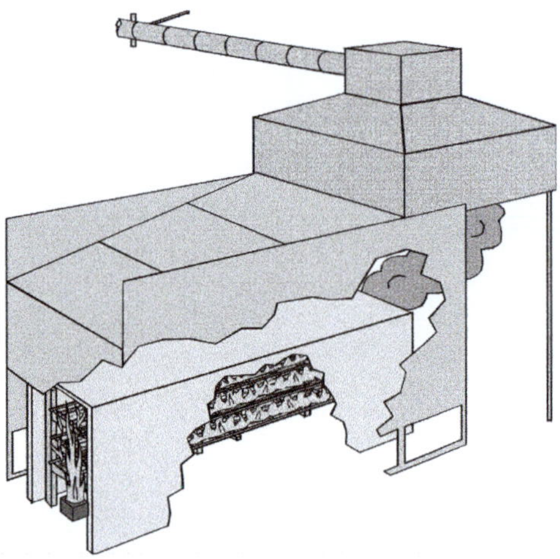	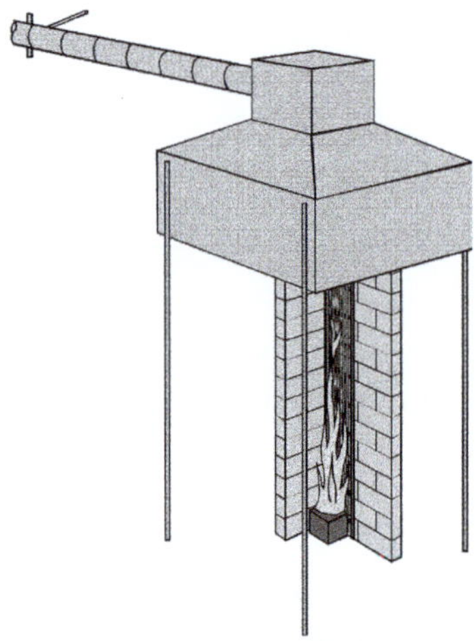
Closed configuration	Semi-closed configuration (corner)
40–100–300 kW sand box burner	40–100–300 kW sand box burner
Three trays with 4 m of cables	One ladder with 4 m of cables
Mounting with spacing	Mounting with spacing

7.2.1.3 Real-Scale Fire Tests

The real-scale fire tests led to the establishment of one horizontal and one vertical configuration for testing real-scale database cables (Table 7.5). Both test configurations utilized a stepwise heat source program from 40 to 300 kW.

7.2.1.4 Full-Scale Fire Test

The full-scale fire test, was based on the former IEC 60332-1-3 test and has been updated with online measurement of heat and smoke releases and ignitability. Oxygen consumption calorimetry was used for measuring HRR, and SPR was measured using a dynamical method with photometric system.

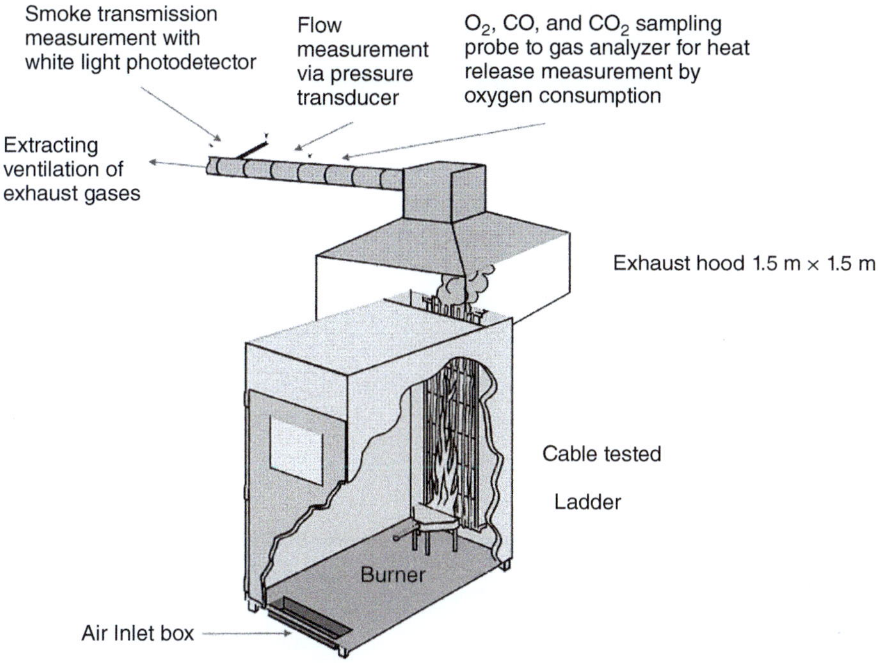

Figure 7.3 Modified former IEC 60332-1-3 large-scale test. Source: Based on Van Hees et al. [14].

Please note that as for the other construction product (EN 13023 SBI [15]), the HRR measurements were performed using oxygen consumption techniques (Figure 7.3).

Two different test protocols (fire scenario) have been used for the database tests. These are both based on the IEC 60332-1-3:

Scenario 1:

- This scenario corresponded to a 20.5 kW burner and differs from the IEC 60332-1-3 method in two major ways regarding air flow rate and cable fitting:
  - Air flow in the test chamber is increased from 5000 to 8000 l/min in order to improve the response of the heat release measurement and for increasing the HRR measuring range.
  - All cables with a diameter greater than 5 mm are mounted individually with a spacing of one cable diameter between cables.
  - Cables with a diameter less than 5 mm are mounted in bundles (non-twisted) of 1 cm diameter with a 1 cm distance between bundles.

Scenario 2:

- This second test procedure uses the IEC 60332-1-3 burner with an increased burner output equal to 30 kW. In this case, a non-combustible backing board of calcium silicate was mounted on the rear of the cable ladder.

7.2.1.5 Capability Study

The modified cable test has been confirmed by a complete capability study in order to assess the impact of test variables on cable performance.

Some critical points of these validations are featured below.

First and most importantantly, repeatability and reproducibility were assessed by testing a large number of cables. The reproducibility of HRR was less than 5% – differences between peak HRR and their producibility for smoke porduction rate (SPR) were in the same magnitude. This results qualified the test method as being able to deliver results with a high level of confidence (Figure 7.4).

Furthermore, the results were comparable with those obtained with the traditional IEC 60332-1-3 series cable test.

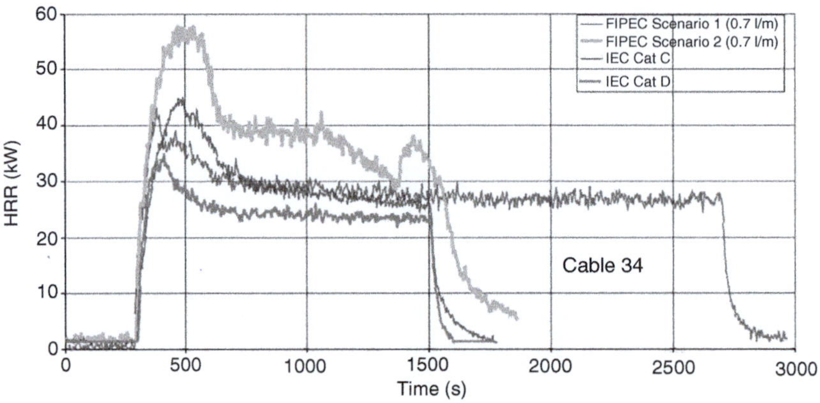

Figure 7.4 Comparison with traditional IEC 60332-1-3 series. Source: Based on Van Hees et al. [14].

A large array of other parameters has been screened to assess the impact on test result deviation:

- Loading, bundling, grouping, and spacing of cables on the ladder,
- Burner output,
- Presence of thermal boundaries,
- Burner position, e.g. angle between burner and cables,
- Level of ventilation through the chamber,
- Influence of layers, etc.

The conclusions, fully detailed in the FIPEC Report, showed that (Figure 7.5):

- Spacing is more severe than non-spacing for larger diameters,
- Spaced bundles are more severe for smaller cables,
- Increasing the loading by more layers does not make the test more severe (Figure 7.6),
- Thermal boundary together with burner increase has the most effect in increasing severity,
- Angle of the burner does not change significantly the results,
- **Mounting is a dominating factor in the flame spread of cables.**

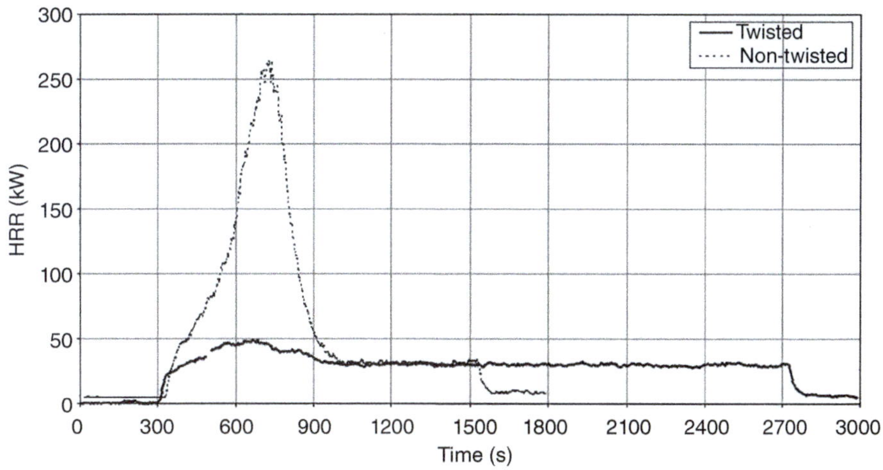

Figure 7.5 Influence of twisted and non-twisted bundles. Source: Based on Van Hees et al. [14].

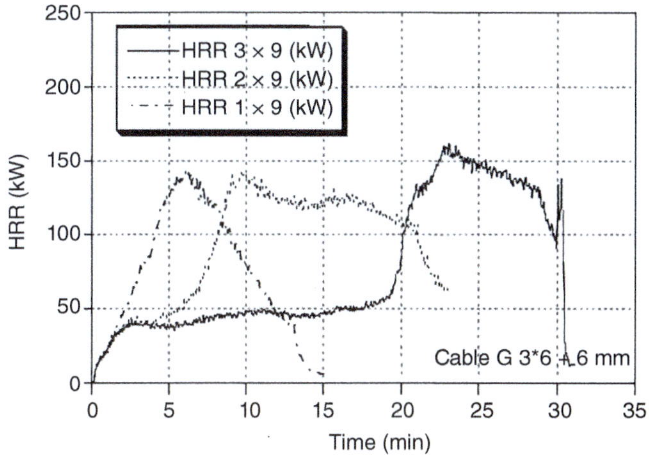

Figure 7.6 Influence of layers. Source: Based on Van Hees et al. [14].

This confirmed that the variance value linked to the test specimen confection was responsible for 70% of the result dispersion, while systematic error linked to the fire test technology accounted only for 30%.

7.2.1.6 Correlation Between Real- and Large-Scale Test

Correlation studies have been finalized between IEC 60332-1-3 Scenarios 1 and 2 and vertical and horizontal large-scale tests.

The parameters retained for correlation between fire models encompassed the flame spread, ignition, peak of HRR (kW/m^2), Fire Growth Rate Index (FIGRA, w/s), Total Heat Release (THR, kJ), peak of SPR (m^2/s), SPR, Smoke Growth Rate Index (SMOGRA, m^2/s^2), and Total Smoke Production (TSP, m^2) (Annex 7.A).

The analysis is detailed in the FIPEC report and can be summarized as follows:

- The linear correlation between **IEC 60332-1-3 Scenario 1** and vertical real-scale tests is relevant for both heat release and smoke parameters. IEC 60332-1-3 Scenario 1 and horizontal real-scale tests also have a very relevant linear correlation, which confirms the strong link between the IEC 60332-1-3 tests and real-scale scenarios.
- For **IEC 60332-1-3 Scenario 2** and vertical real-scale tests is relevant for smoke parameters but less relevant for heat release parameters. This can be explained by the fact that the cables tested in the IEC 60332-1-3 Scenario 2 were mostly high-performance cables that did not burn much in Scenario 1.

The results of the full-scale modified IEC 60332-1-3 test has been valorized for printing out a classification for the ranking of the fire development and smoke development rates of the cable (Table 7.6), which has fueled the commission decision of 27 October 2006 (OJEU 4 November 2006) as regards to the classification of the reaction to fire performance of construction products [16].

As a conclusion, FIPEC project also formalized all the standard proposals related to the modified IEC 60332-1-3 series, which became the EN 50399 cable test. It has been considered as the essential technical brick toward fire classification of cables at the European level.

7.3 Construction Product Regulation (CPR) Framework

The CPR [7] has been set up to facilitate the free circulation of construction products in the EU, removing trade barriers eventually created by technical specification. This is achieved by providing a unique base of technical standards offering uniform assessment methods for the performance of construction products throughout the European Economic Area.

Construction product regulation is applicable to all construction products put in a permanent manner in a building. For cables, those applicable are intended to be used for the supply of electricity and communications permanently installed in buildings and other civil engineering works. CPR became mandatory for cables since 1 July 2017.

The process toward harmonized standards for cables has been a long road and is summarized in Table 7.7.

CPR applied to cables is organized around two series of technical standards.

Reference standards:

- Harmonized standard hEN 50575, which details the content of the essential requirement and features (in the Annex Z) the Essential Requirements, the Assessment and Verification of Constancy of Performance (AVCP), and the labeling description [6].
- CLC/TS 50576 [17] for extended applications, completed by some Guidance from the Group of Notified Bodies for the Construction Products Directive 89/106/EEC (SH02).

Table 7.6 FIPEC proposal classes of reaction to fire performance for cables.

Class	Test method(s)	Classification criteria	Additional classification
A_{ca}	EN ISO 1716	$PCS \leq 2.0$ MJ/kg[a)] $PCS \leq 2.0$ MJ/kg[b)]	
$B1_{ca}$	$FIPEC_{20}$ Scenario 2[f)]	$FS \leq 1.75$ m $THR_{1200s} \leq 10$ MJ Peak $HRR \leq 20$ kW $FIGRA \leq 120$ W/s	Smoke production[c),g)] and flaming droplets/particles[d)] and acidity[e)]
	EN 50265-2-1	$H \leq 425$ mm	
$B2_{ca}$	$FIPEC_{20}$ Scenario 1[f)]	$FS \leq 1.5$ m $THR_{1200s} \leq 15$ MJ Peak $HRR \leq 30$ kW $FIGRA \leq 150$ W/s	Smoke production[c),h)] and flaming droplets/particles[d)] and acidity[e)]
	EN 50265-2-1	$H \leq 425$ mm	
C_{ca}	$FIPEC_{20}$ Scenario 1[f)]	$FS \leq 2.0$ m $THR_{1200s} \leq 30$ MJ Peak $HRR \leq 60$ kW $FIGRA \leq 300$ W/s	Smoke production[c),h)] and flaming droplets/particles[d)] and acidity[e)]
	EN 50265-2-1	$H \leq 425$ mm	
D_{ca}	$FIPEC_{20}$ Scenario 1[f)]	$THR_{1200s} \leq 70$ MJ Peak $HRR \leq 400$ kW $FIGRA \leq 1300$ W/s	Smoke production[c),h)] and flaming droplets/particles[d)] and acidity[e)]
	EN 50265-2-1	$H \leq 425$ mm	
E_{ca}	EN 50265-2-1	$H \leq 425$ mm	
F_{ca}	No performance determined		

Symbols used: PCS, gross calorific potential; FS, flame spread (damaged length); THR, total heat release; HRR, heat release rate; FIGRA, fire growth rate; TSP, total smoke production; SPR, smoke production rate; H, flame spread.

a) For the product, excluding metallic materials.
b) For any external component (i.e. sheath) of the product.
c) $s1 = TSP_{1200} \leq 50\ m^2$ and peak $SPR \leq 0.25\ m^2/s$
s1a = s1 and transmittance in accordance with EN 50268-2 ≥ 80%
s1b = s1 and transmittance in accordance with EN 50268-2 ≥ 60% < 80%
$s2 = TSP_{1200} \leq 300\ m^2$ and peak $SPR \leq 1.5\ m^2/s$
s3 = not s1 or s2.
d) For $FIPEC_{20}$ scenarios 1 and 2: d0 = no flaming droplets/particles within 1200 s; d1 = no flaming droplets/particles persisting longer than 10 s within 1200 s; d2 = not d0 or d1.
e) EN 50267-2-3: a1 = conductivity < 2.5 μS/mm and pH > 4.3; a2 = conductivity < 10 μS/mm and pH > 4.3; a3 = not a1 or a2. No declaration = no performance determined.
f) Air flow into chamber shall be set to 8000 ± 800 l/min
$FIPEC_{20}$ Scenario 1 = prEN 50399 with mounting and fixing according to Annex 2
$FIPEC_{20}$ Scenario 2 = prEN 50399 with mounting and fixing according to Annex 2.
g) The smoke class declared for class $B1_{ca}$ cables must originate from the $FIPEC_{20}$ Scenario 2 test.
h) The smoke class declared for class $B2_{ca}$, C_{ca}, D_{ca} cables must originate from the $FIPEC_{20}$ Scenario 1 test.

Source: Based on Van Hees et al. [14].

Table 7.7 Milestones overview.

OJEU (Official Journal of the European Union) 11 February 89	Council Directive 89/106/EEC of 21 December 1988 on the approximation of laws, regulations, and administrative provisions of the Member States relating to construction products.
OJEU 4 November 2006	2006/751/EC: Commission Decision of 27 October 2006 amending Decision.
	2000/147/EC implementing Council Directive 89/106/EEC as regards the classification of the reaction-to-fire performance of construction products.
	This decision pointed out a distinct fire classification for cables, and setup instruction for cable fitting and mounting – all data fueled by the FIPEC program.
OJEU 4t April 2011	Regulation No. 305/2011 (Construction Products Regulation or CPR) of the European Parliament and of the European Council is a regulation of 9 March 2011 that lays down harmonized conditions for the marketing of construction products and replaces Construction Products Directive (89/106/EEC).
	Application deadline 1 July 2013.
18 May 2009	M/443 Mandate from Commission to CEN and CENELEC concerning the execution of standardization work for harmonized standards on power, control, and communication cables related to supply of electricity, communications, and fire detection and alarm.
10 June 2016	***Harmonized standard EN 50575***, amended in 2016-03-25, related to power, control, and communication cables. Cables for general applications in construction works subject to reaction to fire requirements.
	Mandatory for cables since 1 July 2017.

The reference standards use technical standards, including those related to fire test methods:

- *EN 13501-6* [4], which details the fire classification and refers to fire test methods below:
  - o *EN ISO 1716* [18]: Determination of the gross calorific value
  - o *EN 60332-1-2* [19] (*ex EN 50265-2-1*): Test for flame spread on vertically mounted single cable
  - o *EN 50399* (*ex IEC 60332-3*): Test for flame spread on vertically mounted bunched wires or cables, issued from FIPEC program
  - o *EN 61034-1 and 2* [20]: Smoke opacity measurement in a 27 m^3 smoke box on horizontally mounted bunched wires or cable
  - o *EN 60754-2* [21]: Test on gases evolved during combustion of materials from cables in the tubular furnace. Determination of acidity (by pH measurement) and conductivity.

7.3.1 EN 13 501-6

The EN 13501-6 [4] turns over the interpretation of test results into Euroclasses applicable to cable, and completes the array, especially the EN 13 501-1 [22] dealing with reaction to fire with linings, floorings, and linear system (pipe insulation, etc.).

Table 7.8 EN 13501-6 overview.

Class	Level of fire contribution	Test required		
		EN ISO 1716	EN 60332-1-2	EN 50399
A_{ca}	No contribution	X		
$B1_{ca}$	Low		X	X
$B2_{ca}$	Medium		X	X
C_{ca}	Significant		X	X
D_{ca}	High level		X	X
E_{ca}	Very high level			X
F_{ca}	No performance determined			

The classification is based on heat release and flame spread, smoke production, burning droplets, and acidity.

EN 13501-6 describes different levels of reaction to fire considering the heat release, smoke opacity, and smoke corrosivity.

Regarding heat release rate and flame spread classes, seven classes of cables are defined: A_{ca}, $B1_{ca}$, $B2_{ca}$, C_{ca}, D_{ca}, E_{ca}, and F_{ca}. These classes combines the test results issued form EN ISO 1716 [18] (Class A_{ca}), EN 60332-1-2, EN 50399, and EN 60332-1-2 [19] (Classes $B1_{ca}$, $B2_{ca}$, C_{ca}, D_{ca}, E_{ca}).

The performances of the different heat release and flame spread classes can approximately be described as in Table 7.8.

Regarding dripping of flaming droplets, three classes are considered, d1, d2, and d3, and are issued from the results of the EN 50399 test.

Regarding rate of smoke production [23]:

- Three classes are defined, s1, s2, s3, and are issued for the results of the EN 50399 test. The measurement is made in dynamic conditions and under over-ventilated fire sceneric,
- Two complementary smoke classes are considered – s1a and s1b. These complementary classes are linked to the test results from EN 61034-2 test, corresponding to cumulative values measured in an under-ventilating fire scenario (Table 7.9).

Regarding smoke corrosivity, three classes are defined: a1, a2, a3, and are issued from the results of the EN 60754-2 test.

As indicated, EN 50399 becomes a central test in Euroclasses classification because it combines heat release measurement (classes $B1_{ca}$, $B2_{ca}$, C_{ca}, D_{ca}), smoke transmission measurement (s1, s2, s3), and flaming droplet assessment (d0, d1, d2).

Table 7.9 Smoke classes for cables.

Smoke classes	s1	s1a	s1b	s2	s3
EN 50399	$TSP_{1200} \leq 50\ m^2$ Peak SPR $\leq 0.25\ m^2/s$	$TSP_{1200} \leq 50\ m^2$ Peak SPR $\leq 0.25\ m^2/s$	$TSP_{1200} \leq 50\ m^2$ Peak SPR $\leq 0.25\ m^2/s$	$TSP_{1200} \leq 300\ m^2$ Peak SPR $\leq 1.5\ m^2/s$	Nor s1 or s2
EN 61034-2		Transmittance in accordance with EN 61034-2 $\geq 80\%$	Transmittance in accordance with EN 61034-2 $\geq 60\%$ and $< 80\%$		

The classification system from EN 13501-6 is summarized below (Table 7.10):

Table 7.10 Reaction to fire classes for electric cables.

Class	Test method(s)	Classification criteria	Additional classification
A_{ca}	EN ISO 1716	PCS ≤ 2.0 MJ/kg[a)]	
$B1_{ca}$	EN 50399 (30 kW flame source)	FS ≤ 1.75 m THR_{1200s} ≤ 10 MJ Peak HRR ≤ 20 kW FIGRA ≤ 120 W/s	Smoke production[b),e)] and flaming droplets/particles[c)] and acidity[d)]
	EN 60332-1-2	H ≤ 425 mm	
$B2_{ca}$	EN 50399 (20.5 kW flame source)	FS ≤ 1.5 m THR_{1200s} ≤ 15 MJ Peak HRR ≤ 30 kW FIGRA ≤ 150 W/s	Smoke production[b),e)] and flaming droplets/particles[c)] and acidity[d)]
	EN 60332-1-2	H ≤ 425 mm	
C_{ca}	EN 50399 (20.5 kW flame source)	FS ≤ 2.0 m THR_{1200s} ≤ 30 MJ Peak HRR ≤ 60 kW FIGRA ≤ 300 W/s	Smoke production[b),f)] and flaming droplets/particles[c)] and acidity[d)]
	EN 60332-1-2	H ≤ 425 mm	
D_{ca}	EN 50399 (20.5 kW flame source)	THR_{1200s} ≤ 70 MJ Peak HRR ≤ 400 kW FIGRA ≤ 1300 W/s	Smoke production[b),f)] and flaming droplets/particles[c)] and acidity[d)]
	EN 60332-1-2	H ≤ 425 mm	
E_{ca}	EN 60332-1-2	H ≤ 425 mm	
F_{ca}	No performance determined		

Symbols used: PCS, gross calorific potential; FS, flame spread (damaged length); THR, total heat release; HRR, heat release rate; FIGRA, fire growth rate; TSP, total smoke production; SPR, smoke production rate; H, flame spread.

a) For the product as a whole, excluding metallic materials, and for any external component (i.e. sheath) of the product.

b) s1 = TSP_{1200} ≤ 50 m² and peak SPR ≤ 0.25 m²/s
s1a = s1 and transmittance in accordance with EN 61034-2 ≥ 80%
s1b = s1 and transmittance in accordance with EN 61034-2 ≥ 60% < 80%
s2 = TSP_{1200} ≤ 300 m² and peak SPR ≤ 1.5 m²/s
s3 = not s1 or s2

c) d0 = no flaming droplets/particles within 1200 s; d1 = no flaming droplets/particles persisting longer than 10 s within 1200 s; d2 = not d0 or d1.

d) EN 60754-2: a1 = conductivity < 2.5 μS/mm and pH > 4.3; a2 = conductivity < 10 μS/mm and pH > 4.3; a3 = not a1 or a2. No declaration = no performance determined.

e) The smoke class declared for class $B1_{ca}$ cables must originate from the test according to EN 50399 (20.5 kW flame source).

f) The smoke class declared for class $B2_{ca}$, C_{ca}, D_{ca} cables must originate from the test according to EN 50399 (30 kW flame source).

7.3.2 EN ISO 1716

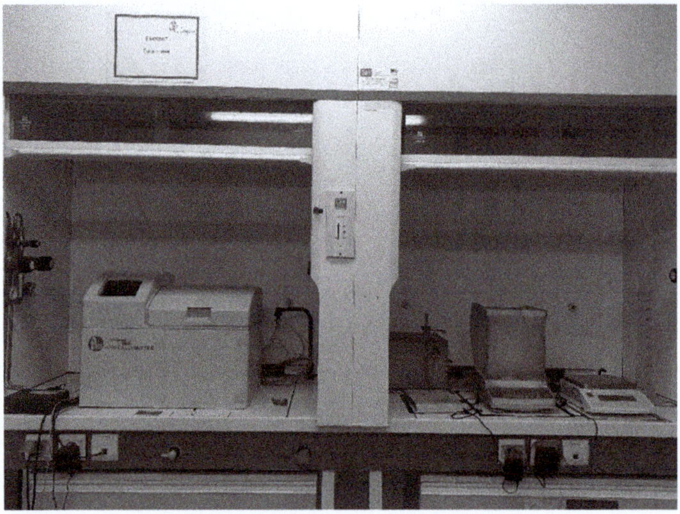

Figure 7.7 ISO 1716 apparatus. Source: Courtesy of CREPIM.

This standard characterizes the potential maximum value of heat release when a product is completely burned in an over-ventilated fire. The calorific potential of a material is measured in a bomb calorimeter.

This test is used to assess the polymeric part of electric cables and each individual layer needs to be assessed. Before testing, the matrix must be finely grounded with a cryogenic grinder.

The sample is placed in a calorimetric bomb under high oxygen pressure and under pure oxygen atmosphere.

The test is used for *Class* A_{ca} level, which corresponds to the highest performance products that practically cannot burn, i.e. ceramic products (Figure 7.7).

7.3.3 EN 60332-1-2

This test is used to assess the vertical flame propagation of a single insulated wire or cable with a 1 kW pre-mixed flame.

The cable specimen (600 ± 25 mm length) is attacked by a Bunsen burner 175 mm flame with 40 mm inner blue cone.

The flame shall be applied continuously for a time period linked to the diameter. At the end of the specified test duration, the burner shall be removed, and the flame of the burner extinguished (Table 7.11).

The wire or cable shall pass the test if the distance between the lower edge of the top support and the onset of charring is greater than 50 mm (Figure 7.8).

Table 7.11 Time for flame application.

Overall diameter of test piece(mm)	Flame application time(s)
$D < 25$	60 ± 2
$25 < D \leq 50$	120 ± 2
$50 < D \leq 75$	240 ± 2
$D > 75$	480 ± 2

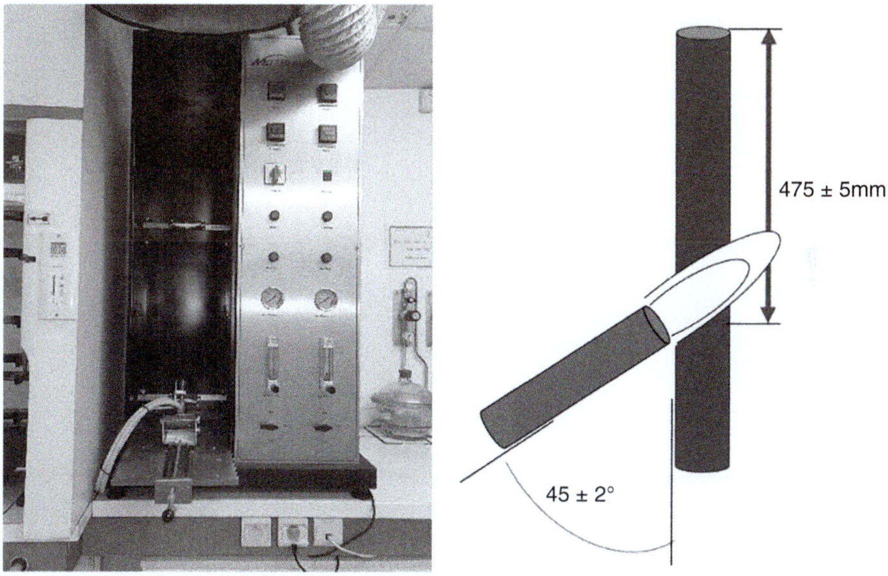

Figure 7.8 EN 60332-1-2 principle. Source: Courtesy of CREPIM.

The onset of char/changing state area is determined by assessing the surface state change. The area where the surface changes from a resilient to a brittle ("crumble effect") surface indicates the onset of charring. Any trace of soot is to be ignored. Softening or any deformation of the nonmetallic is also to be ignored.

The Critical point in this test is the burner output calibration, which must be carried out according to EN 60695-11-2 [1]. The burner output is controlled with the system pictured in Figure 7.9 [1].

After flame stabilization and verification of the burner alignment, the time measured to pass from 100 to 700 °C shall correspond to 46 ± 6 s. The calibration must be repeated three times per set, and must be done regularly to secure the burner capability.

This test is not adequate for low section cables and optical cables, which must be tested according to EN 60332-1-3 [24]. In this case, the dripping of flaming particles is assessed with a filter paper put below the burner (cellulosic based, 80 ± 15 g/m^2, harsh yield $\leq 0.1\%$), stabilized before test at 23 ± 2 °C, and $50\% \pm 5\%$ relative humidity.

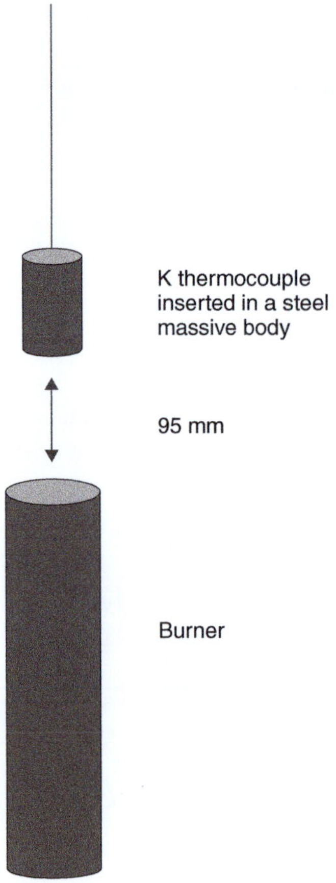

Figure 7.9 Burner calibration according to EN 60695-11-2.

7.3.4 EN 50399

This test is an evolution of the former IEC 60332-1-3 [13] test and has been adapted during FIPEC program to characterize the reaction to fire of a cable in the framework of the CPR. The test measures the vertical spread of flame on cables positioned vertically, with an ignition source positioned below (Figure 7.10).

Some measuring techniques have been added, including heat release and smoke production measurements. Compared with existing test methods described in the former IEC 60332-3, they enable a more comprehensive assessment system, which is both more precise and sensitive, and enables a wider range of fire performance levels.

The test chamber is 1 m wide, 2 m deep, and 4 m high. The test rack is placed vertically into the chamber with the cables facing the burner.

Table 7.12 reviews the key differences between both test methods. EN 50399 has been fitted to correspond to the CPR requirement regarding Euroclass (Figure 7.11).

The mounting phase is critical because it clearly influences a large part of the test results and controls the upstream convective flow of heat. The mounting procedure is described below in Table 7.13 and the number of cables is linked

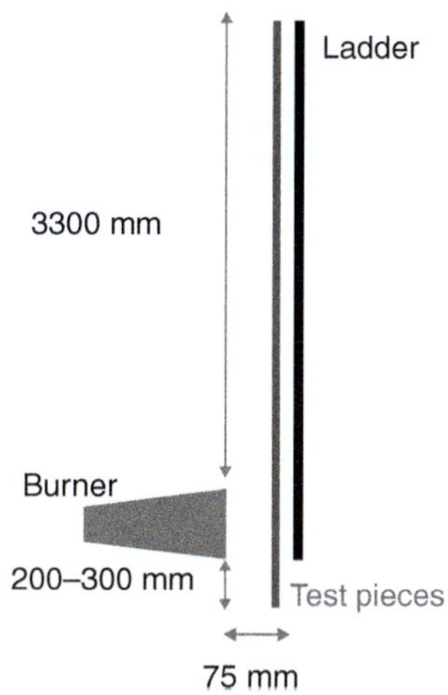

Figure 7.10 EN 50399 principle.

to their diameter. Please note that cable mounting implies a space between test pieces.

The parallelism of the cables must be controlled during mounting and on the final test specimen before testing. Calibrated spacers use is recommended to check out the global geometry of the specimen and the space between cables (Figure 7.12).

The conditioning of cable at room temperature before mounting gets them mechanically relaxed and easier to fit – less curbing. It is recommended to set the cables under mechanical tension to get a parallel straight line of cables.

Regarding HRR measurement, the apparatus is regularly calibrated with reference gases (propane at 20.5, 30, and 50 kW burner outputs) and methanol, to cross-check the calibration data. The difference between correction factors linked to each of the gases shall be less than 10%.

Regarding SPR measurement, the apparatus is regularly calibrated with 99% purity heptane tested in a tray, and the TSP divided by mass of heptane-ratio shall be within the range $110 \pm 25\,m^2/1000\,g$.

At last, linearity, sensibility, and drift of oxygen measurement devices and photometric system are controlled before each tests series.

7.3.5 EN ISO 61034-1 (Apparatus) and -2 (Test Procedure)

Historically, this is a key test for the cable industry to assess the smoke release. The cable is placed above a tray containing standardized fuel (ethanol 90%/methanol 4%/water 6%) in a $27\,m^3$ ($3\,m \times 3\,m \times 3\,m$) chamber. This chamber is equipped with a halogen photometric system, with a 3 m path length.

The smoke released by the cable stressed by the fire below is measured in terms of light attenuation, and the result delivered is the transmittance (%). The chamber is equipped with a $10–15\,m^3/min$ ventilator to homogenize the smoke density (Figure 7.14).

The number of cables tested depends on the cable diameter and the specimens must be bundled at 300 mm in front of each end. The test pieces are positioned

Table 7.12 EN 50399 comparison versus former CEI 60332-1-3.

	EN 50399	CEI 60332-1-3
Flow rate	8000 ± 400 l/min Less than 10% of variation during the test	5000 ± 500 l/min 10% of variation during the test
Test duration	20 min	20 or 40 min depending on the cable category
Sample length	3.5 m, vertical	3.5 m vertical
Sample conditioning	16 h at 20 ± 10 °C	16 h at 20 ± 10 °C
Sample details	Non-jointed cable mounted on the front of the standard ladder Number of cables depending of the cable's diameter	Jointed (cables ≤ 35 mm^2) and non-jointed (cables > 35 mm^2) mounted on the front of the standard ladder Number of cables depending on the volume of nonmetallic material per meter of ladder Category A: 7 l/m Category B: 3.5 l/m Category C: 1.5 l/m Category D: 0.5 l/m This value is calculated considering the density of polymeric material, determined according to the EN 60811-13
Burner	95% min propane, gas burner One burner used at two scenarios 20.5 kW for Euroclass $B2_{ca}$, D_{ca} 30 kW for Euroclass $B1_{ca}$, cable associated with calcium silicate board (11 mm thickness, 870 ± 50 kg/m^3)	Gas burner One or two burners depending of the cable section (more or less than 35 mm^2)
Parameters registered	**Heat release rate** based on oxygen consumption technique THR 1200s MJ (total heat release) HRR peak kW (heat release rate – peak) FIGRA w/s (Fire Index of Growth rate) **Smoke production rate** based on white or laser light photometric measurement TSP 1200 m^2 (total smoke production) SPR peak m^2/s (smoke production rate – peak) **Flaming droplets** and particles persisting longer than 10 s	Vertical spread of flame
Criteria	Euroclasses calculation details in the 206/751/CE document of the Official Journal of the European Communities	Maximum height of the charred portion does not exceed 2.5 m above the bottom edge of the burner

(a) (b)

Figure 7.11 EN 50399 apparatus. (a) During and (b) after test. Source: Courtesy of Nexans.

horizontally and centered above the tray, and the distance between the bottom exposed face of the sample and the bottom of the tray is set at 150 ± 5 mm (Table 7.14).

The test is considered as over if there is no decrease in light transmittance for 5 minutes after the fire source has extinguished, or when the test duration reaches 40 minutes.

The linearity of the photometric system must be verified regularly, and the tension applied on the halogen light shall be stabilized at 12 V for all the test duration, to get a stable nominal luminous flux from 2000 to 3000 lm.

The response of the chamber must be checked regularly with two reference mixtures of alcohol/toluene at two different contents of toluene (4% and 10%). The acceptance criteria are based on TSP: from 8 to 1.2 m^2 for 4% toluene mix and 0.18 to 0.26 m^2 for 10% toluene mix.

Compared with the EN 50399 test method, this test corresponds to an under-ventilated fire, and the transmission measurement corresponds to a cumulative value (Table 7.15).

This test is used to fuel the s1a (final transmittance $\geq$ 80%) and s1b (final transmittance $\geq$ 60%) categories of the Euroclass.

7.3.6 EN 60754-2

This standard details corresponds to the determination of the degree of acidity of gases, which consists in two parameters: pH value and conductivity.

Table 7.13 Mounting of the test sample.

	Number of 3.5 m long test piece per test	Spacing between test pieces	Mounting details
Cables with a diameter greater than or equal to 20.0 mm	Test piece = cable $N = \text{int}\left(\frac{300+20}{\text{dc}+20}\right)$	20 mm spacing between cables	The test sample shall be mounted on the front of the standard ladder
Cables with a diameter greater than 5.0 mm but less than 20.0 mm	Test piece = cable $N = \text{int}\left(\frac{300+\text{dc}}{2\text{dc}}\right)$	One cable diameter spacing between cables	The first test piece or bundle of test pieces shall be positioned in the center of the ladder
Cables with a diameter less than or equal to 5.0 mm	15 test pieces Each test pieces corresponding **to a 10 mm diameter bundle** Each bundle containing n cables $n = \text{int}\left(\frac{100}{d^2c}\right)$ For bundles, apply a metal wire around the bundle at each rung position	10 mm spacing between non-twisted bundles	Further test pieces shall be added on either side so that the whole array of test pieces is centered on the ladder Each test piece or bundle of test pieces shall be attached individually to each rung of the ladder by means of a metal wire using crossed wire method of fixing shown in Figure 7.13 For cables up to and including 50 mm diameter, wire between 0.5 mm, and up to and including 1.0 mm in diameter shall be used For cables above 50 mm diameter, wire between 1.0 and 1.5 mm in diameter shall be used The lower part of each test piece of test pieces shall extend between 200 and 300 mm under the lower edge of the burner face.

dc, measured diameter of the cable (in mm and rounded to the nearest mm); int, the integer part of the result (the value rounded down).

Figure 7.12 Test pieces fitted on the ladder. Source: Courtesy of Nexans.

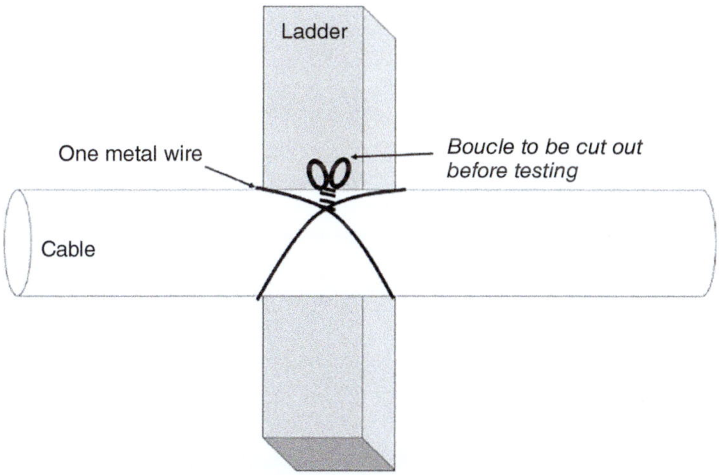

Figure 7.13 Crossed wire method for fixing cable.

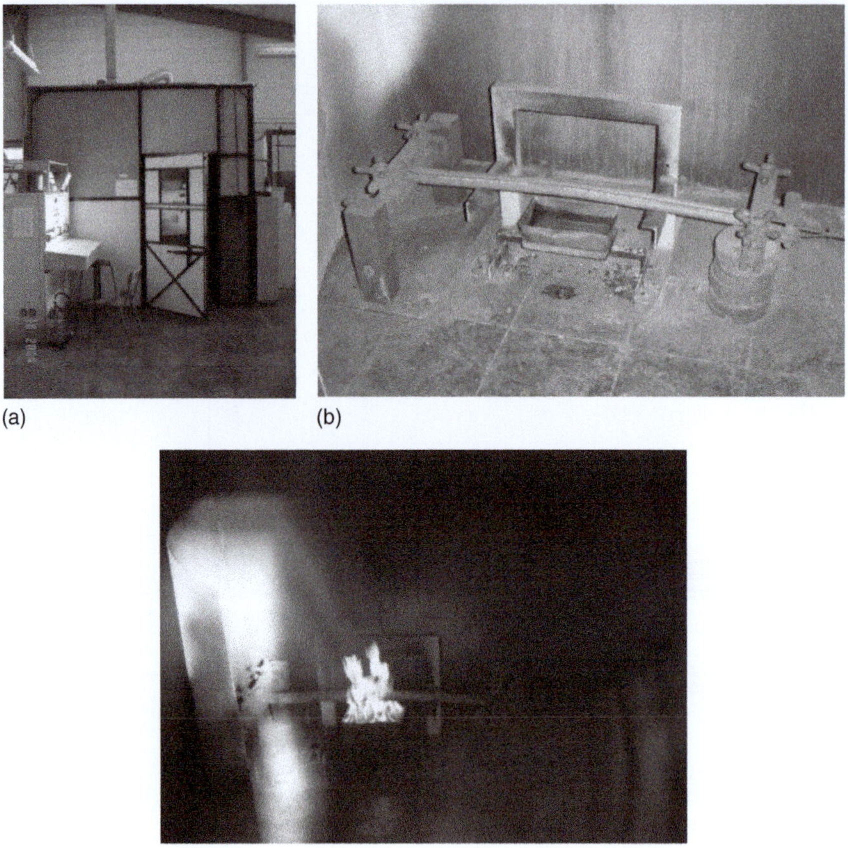

(a) (b) (c)

Figure 7.14 EN ISO 61034-1 apparatus. (a) Apparatus overview, (b) Test specimen. (c) Fire test. Source: Courtesy of Nexans.

This standard provides a method for determining the acidity (by pH measurement) and conductivity of an aqueous solution of gases evolved during the combustion of materials.

The material under test shall be heated in a stream of dry air and generally an ambient air sucked system is used. The flow rate of air introduced into the quartz tube has to be adjusted according to the actual internal cross-sectional area of the tube, such that the speed of air flowing across the sample is approximately 20 ml/mm^2/h, i.e. around 40 l/h for a 50 mm internal diameter quartz tube. The heating system is adjusted such that the temperature at the designated position for the boat is not less than 935 °C and not more than 965 °C (Figure 7.15).

The combustion procedure is done for 30 minutes in the furnace, under air flow conditions.

Three test specimens for the general method, or two for the simplified method, each consisting of (1000 ± 5) mg of the material to be tested, are be prepared.

Table 7.14 Number of cables per test piece.

Length of cable = 1 m Diameter of the cable, mm	Number of cables per test piece
$D < 40$	1
$20 < D \leq 40$	2
$10 < D \leq 20$	3
$5 < D \leq 10$	$N_1 = 54/D$ mm N_1, number of cable sections
$1 < D \leq 5$	$N_2 = 45/3D$ mm N_2, number of bundles confectioned from seven test pieces of cable sections twisted together

Table 7.15 Fire scenario comparison.

EN 50399	EN ISO 61034-1 and 2
Over-ventilated fire model	Under-ventilated fire model
Dynamic measurement	Cumulative measurement
20.5, and 30 kW/m^2 fire model	Less than 20 kW/m^2 fire model

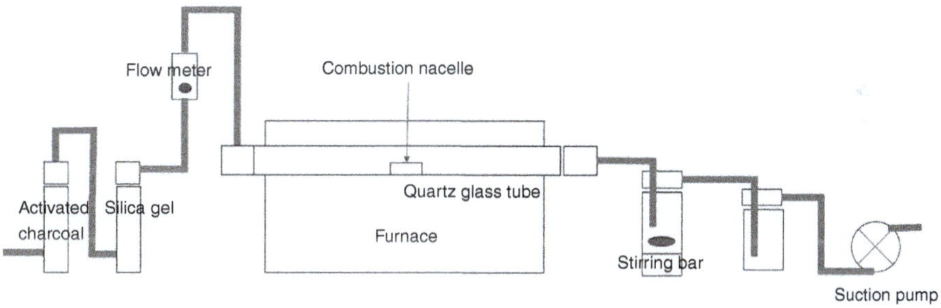

Figure 7.15 EN 60 754-2 Method 3: ambient suck air system by means of suction pump.

The evolved gases shall be trapped by bubbling through wash bottles filled with distilled or demineralized water. The acidity of the resulting solution is assessed by determination of its pH value. The conductivity of the solution is also determined using a conductometer electrode apparatus.

Each test specimen shall be taken from a sample representative of the final material, which means that each individual constitutive nonmetallic part of the cable must be tested. Each test specimen is subdivided into several smaller pieces and the final values of pH and conductivity are pondered by the weight contribution of each subpart.

This test is used for determining the categories a1, a2, and a3 of the Euroclasses:

- a1 = conductivity < 2.5 μS/mm and pH > 4.3;
- a2 = conductivity < 10 μS/mm and pH > 4.3;
- a3 = not a1 or a2.

This test does not determine whether a material has zero halogen or not.

References

1 EN 60332-1-2. *Tests on electric and optical fibre cables under fire conditions – Part 1-2: Test for vertical flame propagation for a single insulated wire or cable – procedure for 1 kW pre-mixed flame.* https://www.boutique.afnor.org.

2 Sundström, B., Johansson, P., Försth, M. et al. (2010). *CEMAC – CE Marking of Cable.* SP Report 2010:27. SP Technical Institute of Sweden, Interscience Communications, UK, ISSEP, Belgium, LSF, Italy, VDE, Germany, Europacable. Boras. ISBN: 978-91-86319-65-6, ISSN 0284-5172.

3 M/443 Mandate from Commission to CEN and CENELEC concerning the execution of standardization work for harmonized standards on power, control and communication cables related to supply of electricity, communications and fire detection and alarm, 18 May 2019.

4 EN 13501-6:2014. *Fire classification of construction products and building elements – Part 6: Classification using data from reaction to fire tests on electric cables.* https://www.boutique.afnor.org.

5 European Committee for Electrotechnical Standardization. Technical Committee 20

6 EN 50575. *Power, control and communication cables – cables for general applications in construction works subject to reaction to fire requirements.* https://www.boutique.afnor.org.

7 Construction Products Regulation (CPR) (2011). Regulation No. 305/2011 of the European Parliament and of the European Council is a regulation of 9 March 2011 that lays down harmonized conditions for the marketing of construction products and replaces Construction Products Directive (89/106/EEC). *Official Journal of the European Union*, L88 54.

8 ISO 9705-1:2016. *Reaction to fire tests – room corner test for wall and ceiling lining products – Part 1: Test method for a small room configuration.* https://www.boutique.afnor.org.

9 EN 13823+A1:2015. *Reaction to fire tests for building products – building products excluding floorings exposed to the thermal attack by a single burning item.* https://www.boutique.afnor.org.

10 EN ISO 9239-2:2002. *Reaction to fire tests for floorings – Part 2: Determination of flame spread at a heat flux level of 25 kW/m*^{2}. https://www.boutique.afnor.org.

11 A review of the FIPEC project. Van Hees, P., Vercellotti, U., Breulet, H., and Grayson, S. (1999). *Fire and Materials Conference 1999*, San Antonio, TX, USA.

12 An overview of the real scale tests in the FIPEC project (Fire Performance of Electrical Cables), Interflam 1999, Interflam '99. Proceedings of the eight international Interflam conference. Vol. 1, Interscience Communications, 1999, 379–391.

13 CEI 60332-3:2000+ A1:2008. (1992). *Test on electric cable under fire conditions. Test on bunched wires or cables. This standard has been replaced in 2000 by a series of international standards: IEC 60332-3-10 Apparatus, -21 Cat A F/R, -22 Cat A, -23 Cat B, -24 Cat C, -25 Cat D.* https://www.boutique.afnor.org.

14 Assessing the fire performance of electric cables. Van Hees, P., Axelsson, J., Vercellotti, U., Breulet, H., Green, A., and Grayson, S. (2000). *FR 2000 conference*, London, UK.

15 Grayson, S.J., Van Hees, P., Green, A.M. et al. (2000). *The Fipec Report: Fire Performance of Electric Cables.* Interscience Communications Ltd, UK, SP, Sweden, Fire Testing Technology Ltd, UK, ISSeP, Belgium, and CESI, Italy, Interscience Communication. ISBN: 0-953-2312-5-9.

16 2006/751/EC (2006). Commission Decision of 27 October 2006 amending Decision 2000/147/EC implementing Council Directive 89/106/EEC as regards the classification of the reaction-to-fire performance of construction products. *Official Journal of the European Union*, L305 49: 8.

17 CLC/TS 50576. *Electric cables – extended application of test results.* https://www.boutique.afnor.org.

18 EN ISO 1716:2018. *Reaction to fire tests for products. Determination of the gross heat of combustion (calorific value).* https://www.boutique.afnor.org.

19 EN 50399. *Common test methods for cables under fire conditions – heat release and smoke production measurement on cables during flame spread test – test apparatus, procedures, results.* https://www.boutique.afnor.org.

20 EN 61034-1 and 2. *Measurement of smoke density of cables burning under defined conditions.* https://www.boutique.afnor.org.

21 EN 60754-2. *Test on gases evolved during combustion of materials from cables. Determination of acidity (by pH measurement) and conductivity.* https://www.boutique.afnor.org.

22 EN 13501-1. *Fire classification of construction products and building elements. Classification using data from reaction to fire tests.* https://www.boutique.afnor.org.

23 Gyppaz, F. *Smoke and Safety in Case of Fire.* Nexans Brochure of September 13. Lyon, France: Nexans Research Center.

24 EN 60332-1-3. *Tests on electric and optical fiber cables under fire conditions – test for vertical flame propagation for a single insulated wire or cable – procedure for determination of flaming droplets/particles.* https://www.boutique.afnor.org.

25 Babrauskas, V. (1993). Ten years of heat release research with the cone calorimeter. In: *Heat Release and Fire Hazard*, vol. I (ed. Y. Hasemi), III-1–III-8. Tsukuba, Japan: Building Research Institute.

7.A Measuring Heat Release Rate (HRR) by Oxygen Consumption Technique, and Smoke Density

7.A.1 Measure of the Heat Release Rate (HRR) by Oxygen Consumption Technique

The HRR from a fire is the most important single parameter for fire characterization. The HRR is a measure of fire intensity or fire powerfulness. It is, however, only during the last 25 years that this parameter could be measured in a fire situation.

Hugget explained the principle of the measurement of the HRR. He showed that HRR from a fire involving conventional organic fuel or material is 13.1 kJ/g of oxygen consumed, with an accuracy of ±5% or better [25].

Incomplete combustion or a variation in the nature of the fuel only has a minor effect on the results.

7.A.1.1 Burning of Methane

The combustion of methane can be described as follows:

$$CH_4 + 2O_2 \rightarrow CO_2 + 2H_2O \tag{7.A.1}$$

The net heat of combustion Δ Hc, is 50 kJ/g of methane burnt (reaction considered at 25 °C water evolves to vapor).

The molecular masses of the molecules are

- $M(CH_4) = 16\,g/mol$
- $M(O_2) = 32\,g/mol$

Thus, $E = 50 \times 16/64 = 12.5\,kJ/g\ O_2$ for this reaction.

E is the heat released per gram of oxygen consumed by burning methane and can be written as

$$E = \frac{\Delta Hc}{r_0} \tag{7.A.2}$$

where r_0 is the oxygen mass to fuel mass ratio, m_{O_2}/m_{fuel}.

The value of E from a fire involving conventional organic fuel or material is 13.1 kJ/g of oxygen consumed, with an accuracy of ±5% or better (Table 7.A.1).

This fact is the underlying principle for oxygen consumption calorimetry and allows the measurements of HRR from fires.

As shown in Figure 7.A.1, there is a perfect symmetry between oxygen consumption during fire due to carbon oxidation and HRR evolution.

The HRR ($\dot{q}$) can be expressed as

$$\dot{q} = E(m^0_{O_2} - \dot{m}_{O_2}) \tag{7.A.3}$$

where $\dot{q}$ is the HRR (kJ/s); $m^0_{O_2}$ is the mass flow of oxygen contents in the incoming air; and $\dot{m}_{O_2}$ is the mass flow of oxygen contents in the exhaust combustion gas.

In the oxygen analyzer, the percentage or molar fraction of oxygen in the gas.

Table 7.A.1 *E* for typical synthetic polymers.

Fuel	*E* (kJ/g O_2)
Polyethylene	12.65
Polypropylene	12.66
Polyisobutylene	12.77
Polybutadiene	13.14
Polystyrene	12.97
Polyvinylchloride	12.84
Polyvinylidene chloride	13.61
Polyvinylidene fluoride	13.32
Polymethylmethacrylate	12.98
Polyacrylonitrile	13.61
Polyethylene terephthalate	13.21
Polycarbonate	13.12
Cellulose triacetate	13.23
Nylon-6,6	12.67
Isobutene polysulfone	12.59
Average	13.02

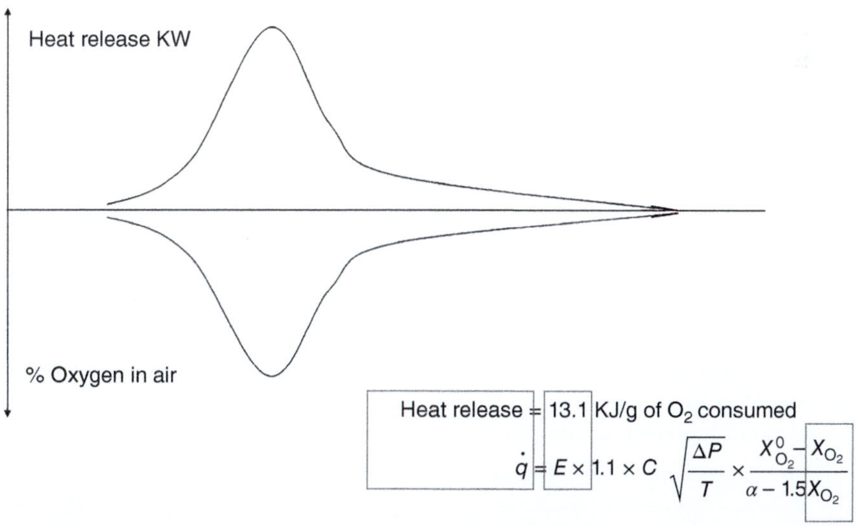

Figure 7.A.1 Heat release measurement by oxygen consumption.

Thus,

$$X_{O_2} = \frac{\frac{\dot{m}_{O_2}}{M_{O_2}}}{\frac{O_2}{M_{O_2}} + \frac{\dot{m}_{N_2}}{M_{N_2}}} \tag{7.A.4}$$

Consequently,

$$m^0_{O_2} - \dot{m}_{O_2} = \dot{m}_{N_2} \times \frac{M_{O_2}}{M_{N_2}} \times \frac{X^0_{O_2} - X_{O_2}}{(1 - X_{O_2})\cdot(1 - X^0_{O_2})} \tag{7.A.5}$$

where $X^0_{O_2}$ is the fraction of oxygen in the incoming air; X_{O_2} is the fraction of oxygen contents in the exhaust combustion gas; $\dot{m}_{N_2}$ is the mass flow of nitrogen (kg/s); M_{O_2} is the molecular weight of oxygen (≈32 g/mol); and M_{N_2} is the molecular weight of nitrogen (≈28 g/mol).

We must consider the following relation:

$$\frac{\dot{m}_a}{M_a} = \frac{\dot{m}_{N_2}}{M_{N_2}} + \frac{m^0_{O_2}}{M_{O_2}} \tag{7.A.6}$$

This leads to

$$\dot{q} = \dot{m}_a \times E \times \frac{M_{O_2}}{M_a} \times \frac{(X^0_{O_2} - X_{O_2})}{(1 - X_{O_2})} \times (1 - X^0_{H_2O}) \tag{7.A.7}$$

where $X^0_{H_2O}$ is the fraction of water in the incoming air.

The missing parameter is now the exhaust flow rate $\dot{m}_e$. Owing to the dilution linked to combustion reaction and stoichiometry evolution, the incoming flow rate of air $\dot{m}_e$ is not equal to the exhaust flow rate $\dot{m}_a$.

7.A.1.2 Determination of the Mass Flow Rate ($\dot{m}_a$)

Considering the burning of methane under nitrogen:

$$CH_4 + 2O_2 + 7.5 \cdot 2N_2 \rightarrow CO_2 + 2H_2O + 7.5 \cdot 2N_2 \tag{7.A.8}$$

Incoming gas molecules: 9.52; outgoing gas molecules: 10.52.

The incoming volume of gas that has completely reacted has expanded by 10.52/9.52 = 1.105. The expansion factor is called (α). In the test setup, only a fraction of the incoming air is fully depleted of its oxygen.

$$\dot{m}_e = \dot{m}_a - \Phi\dot{m}_a + \alpha\Phi\dot{m}_a \tag{7.A.9}$$

where $\dot{m}_e$ is the rate of mass flow of outgoing gas of combustion; $\dot{m}_a$ is the rate of mass flow of exhaust air.

The oxygen depletion factor (Φ) is

$$\Phi = \frac{m^0_{O_2} - \dot{m}_{O_2}}{m^0_{O_2}} = \frac{X^0_{O_2} - X_{O_2}}{(1 - X_{O_2})X^0_{O_2}} \tag{7.A.10}$$

The waste pipe of exhaust fumes is equipped with a venturi tube, two thermocouples, and pressure pick-ups. This device determines the mass flow ($\dot{m}_e$) of exhaust combustion gas applying the Bernouilli law:

$$\dot{m}_e = C \cdot \sqrt{\frac{\Delta p}{T}} \tag{7.A.11}$$

where ΔP is the difference in pressure (Pa); C is the commissioning factor – constant of calibration (SI); and T is the temperature of gases in the device (K).

Substituting $\dot{m}_e$ for $\dot{m}_a$ in the HRR equation, we get

$$\dot{q} = E \times 1.1 \times C\sqrt{\frac{\Delta P}{T}} \times \frac{X^0_{O_2} - X_{O_2}}{\alpha - 1.5X_{O_2}} \tag{7.A.12}$$

where C is a proportionality constant. $X^0_{O_2}$ and X_{O_2} are oxygen analyzer readings of incoming and exhaust gas respectively.

To limit the deviation linked to fast flashing materials and to smooth the response, HRR data are generally calculated over a 30 s running average.

This equation is available in all standards dealing with oxygen consumption calorimetry such as cone calorimeter standard – ISO 5660-1 [2], EN 50399, and EN 13823.

As explained in the EN 50399 Annex E, the C value (commissioning factor) results from the aggregation of different parameters. The first one is the factor k_c linked to the flow profile in the pipe-averaging of the velocity profile in the section of the exhaust duct, which is modified by correction factors issued for propane calibration (20.5, 30, and 50 kW) and methanol calibration.

EN 50399 show up in Table 7.A.2 an example of commissioning factor calculation.

The deviation between the factor from flow profile k_c and the final value of the commissioning factor C must be less than 10%.

Table 7.A.2 Example of commissioning factor calculation.

Type of calibration	Correction factor	Average	C factor
k_c factor from flow profile			0.90
Propane 20.5 kW	1.03	1.013	
Propane 30 kW	1.025		
Propane 50 kW	0.985		
Methanol (4 l)	1.06	1.06	
Final correction factor		1.037	
Commissioning C			0.93

7.A.1.3 Smoke Opacity

The measurement is associated with the oxygen consumption technique because it can easily be set up on the same flow rate. The measurement is realized under dynamic and flow through methods.

Transmittance is the ratio of transmitted light intensity through smoke to the incident light intensity, under specified conditions. It is dimensionless and is usually expressed as a percentage.

The quantity that is being measured is the extinction coefficient according to the Beer–Lambert law:

$$I = I_0 e^{-kL} \tag{7.A.13}$$

where I is the intensity of a light beam after passing through smoke; I_0 is the initial intensity of the light beam; k is the extinction coefficient; and L is path length through the environment with an extinction coefficient k.

Because of the nature of the flow-through system, the extinction coefficient will have to be related to the volume flow rate.

7.A.1.3.1 Smoke Opacity Measurement Applied to EN 50399

Smoke production rate calculation is based on the following formula:

$$\text{SPR} = k \times Q_v \tag{7.A.14}$$

where k (m^{-1}), Q_v (m^3/s), SPR, is the smoke production rate (m^2/s).

Total smoke production calculation is based on the following formula:

$$\text{TSP} = \text{SPR} \times t \tag{7.A.15}$$

where t is the time (s) and TSP is the total smoke production (m^2).

According to EN 50399, the apparatus is regularly calibrated with 99% purity heptane tested in a tray, and the ratio TSP divided by mass of heptane shall be within the range $110 \pm 25\,m^2/1000\,g$.

7.A.1.3.2 Smoke Opacity Measurement Applied to EN 61034-1 and 2

The optical density is defined by the equation:

$$D = \ln 10 \frac{I_0}{I} \tag{7.A.16}$$

where D is the measured optical density.

Optical density D is given for a specific path length:

$$D_0 = \frac{1}{L} \ln 10 \frac{I_0}{I} \tag{7.A.17}$$

where D_0 is the optical density corresponding to the path length L, and L is the path length (m).

7.A.1.4 Calculation of FIGRA and SMOGRA Index

Heat release rate is a key indicator to assess fire powerfulness: the higher the peak of HRR, the higher the heat emitted in the environment, and so the spread of flame.

However, this approach points out some limit linked to the time at which the HRR peak occurs. As shown in Figure 7.A.2 and 7.A.3, both systems tested under building fire scenario are characterized by an equivalent value of HRR peak; the only difference lies in the time of occurrence.

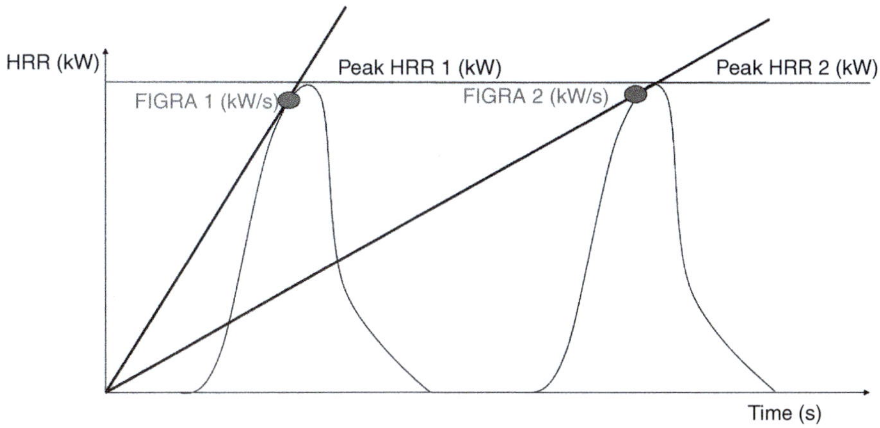

Figure 7.A.2 Heat release rate peak occurrence of two systems.

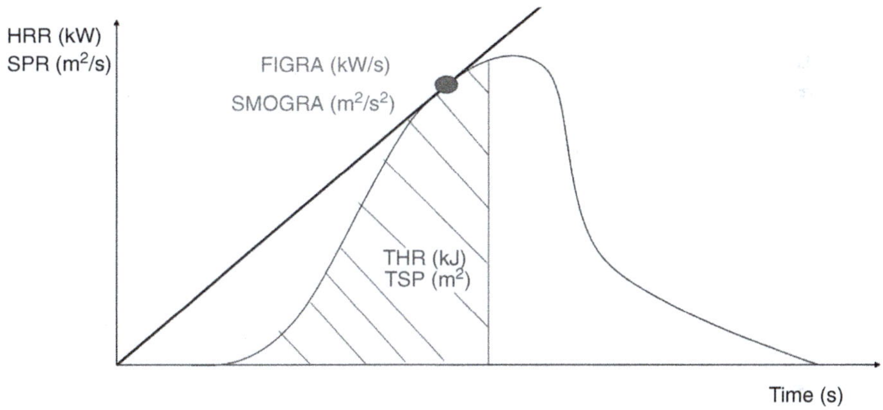

Figure 7.A.3 FIGRA and SMOGRA calculation.

The product corresponding to FIGRA value has a far lower impact of flame spread because the HRR peak occurs later.

This demonstrates that HRR peak is not considered alone as a reliable parameter to assess the real fire risk of a material/system, because it does not integrate the occurrence time.

The indices FIGRA and SMOGRA have been introduced to combine HRR peak and occurrence time.

Mathematically, FIGRA and SMOGRA indices are obtained by operating on HRR and SPR values with a second derivative, which corresponds to an acceleration. FIGRA corresponds to the maximum value of the function that divides HRR by elapsed test time.

FIGRA and SMOGRA represent finally the maximum rate of acceleration for HRR and SPR (Figure 7.A.3).

8

Crosslinking Technologies

Ron Goethals

Smaron bv, Zuidergracht 12, Soest 3763 LV, The Netherlands

Impression of the "electron-beam at work." Source: Courtesy of IBA, Louvain-la-Neuve, Belgium.

8.1 Introduction

The author of this chapter, with a background in chemistry, has over 35 years of experience in industries, where crosslinking plays an important role in the total process of manufacturing various products in these industries.

Without crosslinking the world would be different. Not only paints, light weight shoes, and foams would be hard to find, but also products such as PEX-pipes, heat-shrinkable tubing, XLPE-insulated wires, and high temperature resistant halogen free cables would not be available and therefore would make the world less safe.

The Global Cable Industry: Materials, Markets, Products,
First Edition. Edited by Günter Beyer.

Many technical articles, chapters and books about crosslinking, curing, and vulcanizing have been written and published in the past 50 years. Some of these articles have been listed in References.

The main reason for creating this chapter is a result of almost four decades of traveling, pioneering, selling, and working in the cable industry. During this period it was clearly noticed that there is lack of general and practical understanding of the e-beam crosslinking process, including misunderstandings and insufficient knowledge about this clean and environmentally friendly method.

Even today, many people in the cable industry still consider the e-beam crosslinking process as difficult to understand, scientific and expensive. On top of that there is another reason, which contributes to the reluctancy of people to invest in e-beam crosslinking technology. That reason is known as the word "radiation."

To get a better understanding for an e-beam production line and the e-beam crosslinking process, the silane and peroxide crosslinking process will be reviewed in this chapter as well. In a practical approach the reader will be informed about the versatile, but unfortunately (*still*) under-evaluated method of e-beam crosslinking technology.

8.2 Crosslinking, Curing, Vulcanizing

Crosslinking, curing, and vulcanizing are three terms that are used in the cable industry for the process of forming bonds, to join two polymer chains together.

Although in polymer chemistry the aforementioned three terms are used for specific, but different, "polymer compositions" (*curing for thermoset materials, vulcanizing for rubber-like materials, and crosslinking for thermoplastic polymers*), all three terms may be used in this chapter for the process of joining two polymers together.

A simplified way of the understanding of what a "crosslink" is, is visualized in Figure 8.1a (fishing net), Figure 8.1b (a building construction), and Figure 8.2 (a three-dimensional network of crosslinked polymers).

When polymer chains are crosslinked, the properties of the polymer matrix will change. The changes in (for instance, mechanical) properties depend strongly on the crosslink density.

Low crosslink densities decrease the viscosities of polymer melts.

Many leading polymer producers are able to create a range of MFI's for the same polymer by varying the crosslinking degree of the polymer. In some cases the variation of MFI is even achieved by post-treating the pelletized polymers under an e-beam (Figure 8.3).

Intermediate or medium crosslink densities transform polymers into materials that have rubber (also known as elastomeric) properties. Elastomer-based products have the ability to regain its original shape at room temperature, when a load/weight is removed from the material.

High crosslink densities can cause materials to become rigid, glassy, or brittle.

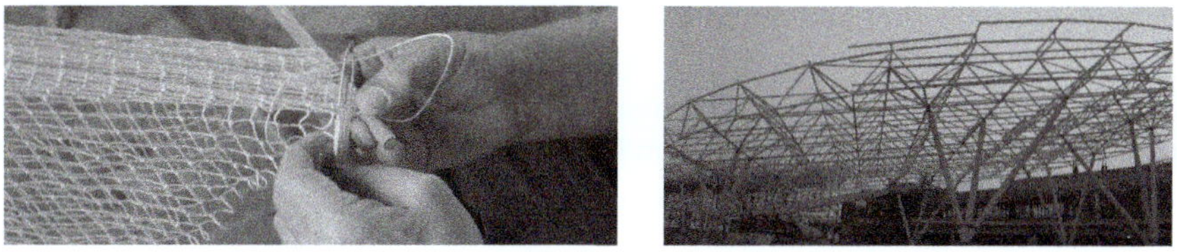

(a) (b)

Figure 8.1 A simplified visualization of "crosslinks" (=joints). It is all about joining individual "parts" to build a three-dimensional structure.

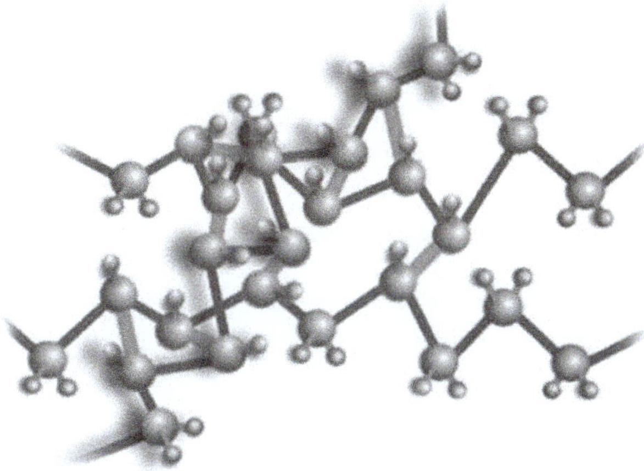

Figure 8.2 Crosslinked polymer chains.

Figure 8.3 MFI modification by e-beam treating. Source: Courtesy of Ron Goethals.

8.3 Crosslinking Processes

Since many years the following three crosslinking methods are mainly used in the cable industry:

- *Peroxide crosslinking*, also known as Continuous Vulcanization (CV)
- *Silane crosslinking*, also known as moisture-crosslinking.
- *e-beam crosslinking* electron-beam (e-beam), also known as e-beam radiation, beta-radiation, or β-radiation.

Cable compounds need to be adapted to the crosslinking process, which is used in the production process of the cable manufacturer. These adaptions or modifications need to be applied to ensure that the final properties of the insulation and or jacket in the cable construction will meet the applicable requirements and national or international specifications.

For all three crosslink processes, the composition of the cable compound (*for both insulation and jacketing applications*) may typically contain around 5–15 of the following ingredients:

- 1–4 polymers and/or elastomers (*the base fundament to meet temperature and chemical resistance*)
- 1–2 compatibilizers and/or coupling agents (*to improve the additive–polymer compatibility*)
- 1–4 antioxidants and stabilizers (*to improve heat aging and UV resistance and to reduce degradation*)
- 1–3 flame retardants (*halogenated, non-halogenated, or combinations*), *to reduce burning behaviour*.
- 0–2 lubricants or plasticizers (to improve processing, flexibility, cold impact, etc.)
- 0–3 fillers (*to modify hardness, stiffness, etc., but also to reduce material costs*)
- 0–3 accelerators, retarders, and co-agents (*to modify the efficiency and speed of crosslinking*)

The peroxide and the silane crosslinking process, which are both considered as "chemical crosslinking processes," can be described as follows:

8.4 The Silane-Crosslinking Process

The silane crosslinking process can be split in three different methods or processes, all used in the cable industry.

8.4.1 The Sioplas® Process

This process was developed in 1968 by Midland Silicones. The essential step in this process is the grafting of the vinyl-silane onto the polymer chains, by using low amounts of peroxide. This step is usually carried out during the compounding process. A mixture of silane and peroxide is added to the polymer melt by means of very accurate dosing. The result is a silane grafted polymer or compound, which can be extruded on conventional wire and cable extruders in a next step (two-step process). A catalyst masterbatch will be added during extrusion to increase the crosslinking speed. The C—Si—O—Si—C bond is not as strong as a C—C bond (dominating in the peroxide and e-beam crosslinking process), as the average bond energy at 0 °C for a C—C bond is 346 kJ/mol, whereas the C—Si bond (the weakest part in the C—Si—O—Si—C bond) has a bond energy of 318 kJ/mol.

Figures 8.4 and 8.5 show the chemical reaction process of the Sioplas method, whereas Figure 8.6 gives the overview of all the steps involved to produce a finished and crosslinked product.

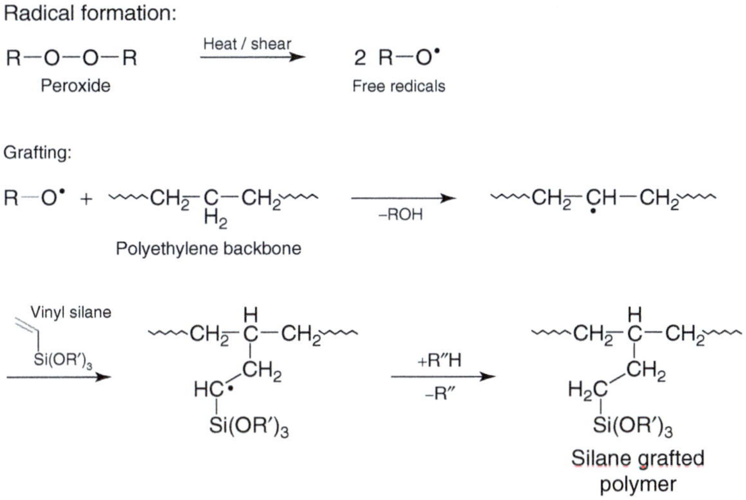

Figure 8.4 Formation of a silane grafted polymer.

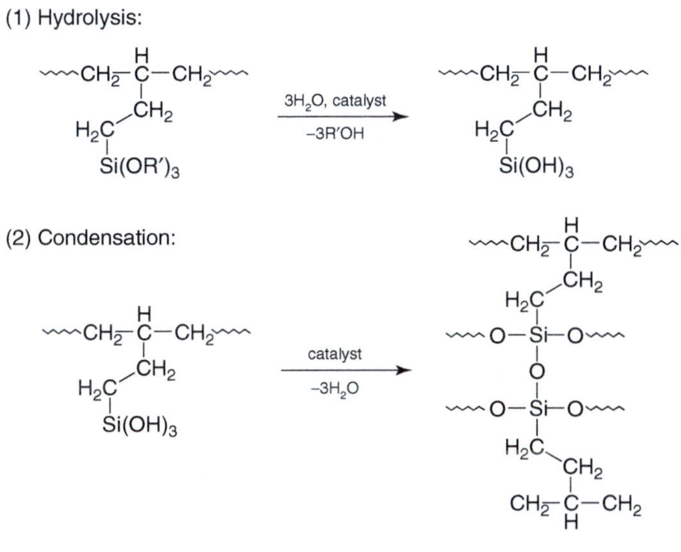

Figure 8.5 Formation of a bond between two polymers.

8.4.2 The Monosil® Process

This process was developed by BICC and Maillefer in 1974. In this process the grafting, the addition of the catalyst, and the extrusion of the wire or cable is made in one and the same process (*one-step process*). This process is far more challenging as it requires special extrusion equipment. The screw design and the technology for feeding the (*reactive*) additives are crucial for this process.

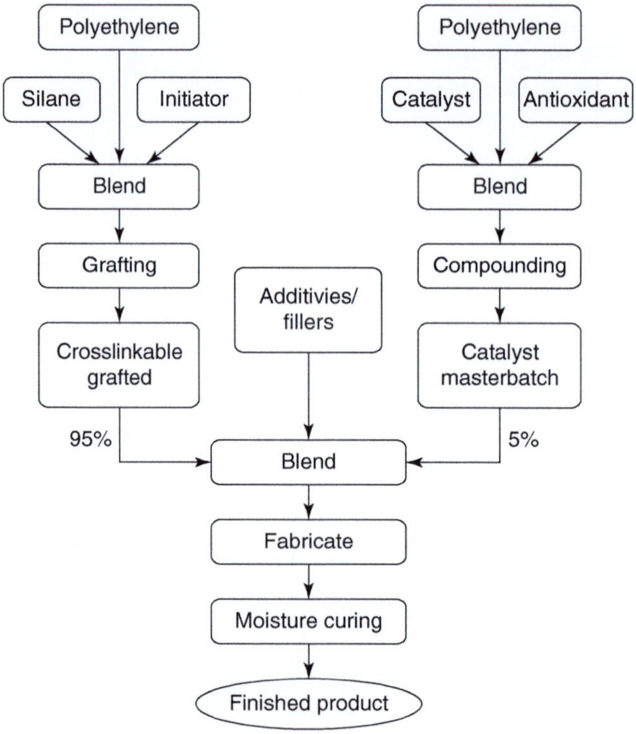

Figure 8.6 Overview of the steps, which are involved to produce a finished and crosslinked product, when using the Sioplas process.

8.4.3 Silane Copolymers

By copolymerizing ethylene (or other monomers) and, for instance, vinyltrimethoxysilane, polymers can be obtained in which the reactive silane groups are incorporated within the backbone of the polymer. Generally speaking such silane copolymers show a lower reactivity to crosslinking, compared with grafted grades produced by the Monosil or Sioplas process.

However, this lower reactivity has a positive effect on the shelf life of silane copolymer based compounds and it also reduces the sensitivity for premature crosslinking (*scorch*).

Extrusion of these products can be carried out on standard extrusion equipment, in a way similar to the Sioplas process. The addition of a catalyst masterbatch during extrusion will increase the speed of crosslinking.

8.4.4 Reactivity of the Silane Crosslinking Process

In both the Monosil and the Sioplas crosslinking system, following is key for the success of this process:

1) The crosslinking reaction needs to be in balance with the residence time in the extruder, the crosslink conditions after extrusion, and the wall thickness of the insulation or jacket.

2) The polymer or polymers used will play an important role in the final degree of crosslinking.

The base of silane crosslinking is the combination of a vinyl-silane and a peroxide. The reactivity of the silane crosslink system can be adapted by adjusting the following:

- Type and amount of the vinyl-silane or mixtures of vinyl-silane
- Type and amount of the peroxide

The properties of the polymer to be grafted (like MFI, density, and crystallinity), the targeted MFI of the grafted polymer, the processing temperature during grafting, the presence of additives, and the presence of fillers all play an important role in the overall reactivity of the system.

A catalyst (masterbatch) that is added during extrusion of the cable will increase the crosslinking speed, but will not affect the final crosslink degree (gel content) of the fully crosslinked system.

The humidity and temperature, which are needed for the crosslinking process, have a significant influence on the speed of crosslinking.

It is not uncommon that gel contents of less than 10% will be measured on the insulation or jacket of the cable, after an exposure of (*only*) six hours in ambient conditions. Especially in geographical areas where the humidity is low to very low, full crosslinking (*) may take more time than anticipated.

The same product may show a gel content of well above 60%, provided that the product is exposed to six hours at humidity of 100% at 80 °C (*sauna*).

Some companies offer specially designed catalysts (masterbatches), which shorten the time of crosslinking in ambient conditions. These are the so called "ambient curing catalysts."

Another important, but limiting, factor for the speed of the silane crosslinking process is the diffusion of moisture. In practice it comes down that doubling the wall thickness will result in squaring the crosslink time.

8.4.5 Advantages and Disadvantages of Silane Crosslinking

Advantages:

- Low (*in case of steam room*) or no initial investment needed for the crosslinking process, as ambient curing or curing in a humidity and temperature controlled room (*steam room*) will do the job.
- Rather low power consumption (*when using a steam room*) or no power consumption (*when ambient curing is used*) is needed for this crosslinking process.
- It is a reasonably easy understandable technology, which is in use since the end of the 1970s.

Disadvantages:

- It is a scorch sensitive process, which may result in pre-mature crosslinking, as shown in Figure 8.7, if:
  - The moisture content of the materials, added to the extruder is not well controlled.

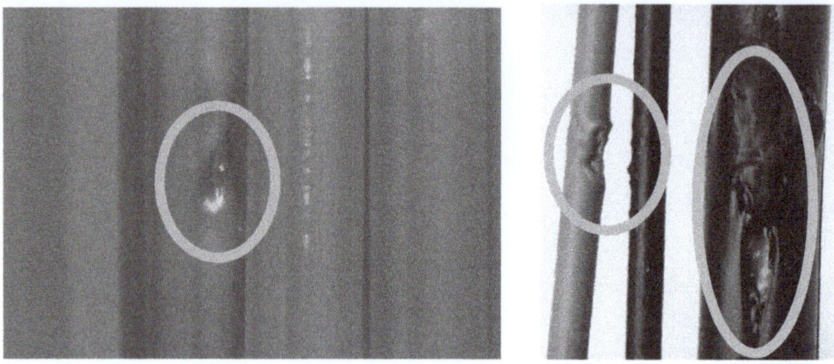

(a) (b)

Figure 8.7 Examples of premature crosslinked polymers during the extrusion process of insulated wires, using the Sioplas process.

  - The screw, head, and tooling design show spots where material is not or hardly flowing.
- In practice the silane crosslinking process is (only) effective for rather thin wall wire and cable constructions, as increasing the wall thickness by a factor 2 will result in squaring the time for crosslinking.
- Moderate costs for the investment in an extrusion line for the Monosil crosslinking process.
- This process needs a rather long "curing time" (*hours to weeks*) to achieve complete crosslinking.
- Non-predictable crosslinking time, when using ambient conditions.
- A rather limited range of polymers and elastomers (*mainly limited to polyolefins, some PE-copolymers, EPR, and EPDM*) can be used for this crosslinking process.
- The limited selection of polymers limits compound designers to develop certain compounds in a silane crosslinkable version.
- The shelf life of Sioplas and silane-copolymer based compounds is limited.
- Skilled operators needed.
- Increased waste due to the fact that extruders needs to be cleaned out with thermoplastic material, before it can be stopped (*or shall have a bypass*) for the changing of colors or dimensions.
- During longer storage, the MFI of the polymers and compounds tend to decrease, resulting in inconsistent process parameters.
- Undesired test results of some properties, if material is tested before full crosslinking has been achieved.

8.5 The Peroxide Crosslinking Process (CV Curing)

During the peroxide vulcanization (Figure 8.8), the peroxide in the cable compound decomposes to produce highly reactive free radicals, which as a result may react with the polymer matrix in the compound to form crosslinks. One of the

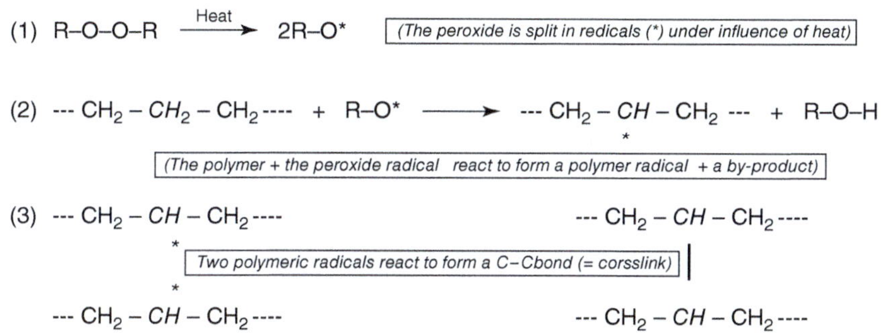

Figure 8.8 Peroxide crosslinking reaction.

strongest bonds in a polymer matrix is the carbon–carbon, also known as C—C bond. During the peroxide vulcanization this bond is the dominating bond. These C—C bonds are stronger compared to the C—S—C bonds, which are the dominating bonds in rubbers using the sulfur curing process to get vulcanization. The average bond energy at 0 °C for a C—C bond is 346 kJ/mol and for a C—S bond this is 272 kJ/mol. The sulfur curing process is hardly used in the cable industry.

8.5.1 Polymer Selection for Peroxide Crosslinking

Peroxide crosslinkable (CV-curable) compounds are temperature sensitive, as the peroxide (*whether added to the compound during the compounding process, or directly injected in the melt during the extrusion process*) starts to react, once the mass-temperature in the extruder reaches the activation temperature of the peroxide.

For that reason peroxide crosslinkable compounds shall always be extruded at temperatures, which are well below the activating temperature of the peroxide.

Most of the commonly used peroxides in the wire and cable industry demonstrate safe processing temperatures in the range of 125–135 °C. Higher temperature profiles or so called "hot spots" in the extruder may result in premature crosslinking, which may show up as undesirable gels.

In other words, the safe processing temperature of the peroxides limits the selection of the wide range of available polymers on the market. Those polymers, having DSC melting points above the safe processing temperature of the peroxide are basically excluded from the CV-vulcanization process.

The limited selection of polymers therefore also limits a compound designer to develop a suitable compound. Generally speaking, CV curing is mainly used for compounds, which are based on elastomers (*rubbers*) and/or elastomer blends.

One of the elastomers, which are extensively used in the wire and cable industry and which are often crosslinked by peroxides in the CV vulcanization process, is EPDM. EPDM shows extensively excellent moisture and weather resistance, in combination with very good electrical properties. Also ethylene–propylene (EP) copolymer is used in high volumes in the cable industry.

Compounds, based on these elastomers are used in power transmission cables, portable power cables, control cables, mining cables, flexible cords, automotive ignition wires, and appliance wires. These elastomers are used for both insulation and jacketing compounds.

Another elastomer, which is used in the CV vulcanization process in the cable industry, is ethylene vinyl acetate (EVA), a saturated copolymer of ethylene and vinyl acetate (VA). The saturation of this copolymer provides outstanding ozone, weather, and temperature resistance. By varying the VA content in the EVA, the properties of compounds based on EVA can be adapted. EVA is actually also known as EVM, depending on the VA content. In general EVA copolymers with VA contents less than 40% are thermoplastics and are known as EVA. Those copolymers having a VA content between 40% and 80% are elastomers and are known under the name EVM. It is known that the higher the VA content the better the heat, oil, and solvent resistance.

However, due to the polarity of the VA group, compounds based on EVA or EVM are far less popular for wire and cable insulation, compared with EPR and EPDM.

Chlorinated polyethylene (CPE) is the third elastomer, which is used in high volumes in the CV curing process. The chlorination of HDPE, to form CPE, turns the PE in a highly saturated polymer, showing excellent ozone, weather, heat, and oil resistance, which are all very desirable at various applications in the wire and cable industry. CPE-based compounds are typically used as jackets for flexible power transmission, portable power, control, and mining cables.

8.5.2 Advantages and Disadvantages of Continuous Vulcanization

Advantages:

- This process is reliable and well proven for over 60 years.
- The resulting carbon–carbon bond (C—C bond) is stable, resulting in good heat-aging properties.
- It is a reasonably easy understandable technology.
- The crosslinking process is completed after take-up on the cable drum.
- Thick wall cables can be crosslinked by CV vulcanization, as heat can be transferred through wall thicknesses of 10 mm and more.

Disadvantages:

- (Very) high investment costs.
- It is a scorch sensitive process, which may easily result in pre-mature crosslinking.
- This technology requires a temperature profile within a very tight operating window.
- Continuous vulcanization curing is a rather slow process, as the heat needs to be transferred through the entire wall thickness.
- This process needs a relatively long "curing time" (*of minutes*) to achieve complete crosslinking.
- The extrusion line-speed is low, as it is fully depending on the speed of the vulcanization process.

- Although a wider range of polymers can be crosslinked by this method, compared with silane crosslinking, still a limited range of polymers and elastomers is available, compared with e-beam crosslinking.
- The shelf life of CV-curable compounds is limited.
- Peroxides may give undesired (side) reactions with some of the materials, which are part of the compound composition.
- Many elastomer-based CV-curable compounds tend to be sticky and therefore are not available in pellet-form (*granules*), but only in strips or slabs.
- To reduce stickiness of many of the CV-curable compounds talcum is applied. The use of talcum may result in a dusty and unhealthy environment during handling.
- Skilled operators are needed to understand and control the extrusion and vulcanizing process.
- The CV-curing process is less favorable for smaller size wires and cables, as the stress on these small size products exceeds the mechanical strength, due to the weight of the product in the long CV tube.
- Due to high temperature and high pressure, the shape integrity of the cable construction needs attention.
- High pressure steam may cause injuries.
- Increased waste due to the fact that extruders needs to be cleaned out with thermoplastic material, before it can be stopped (*or shall have a bypass*) for the changing of colors or dimensions.
- The power consumption of a CV line is rather high.
- Continuous vulcanization lines need quite some space (see Figure 8.9).

8.6 e-Beam Crosslinking

Not many people are aware that many products, which they touch, see, or use in their daily lives have a background to e-beam technology.

Before explaining the more "scientific side" of e-beam crosslinking, the reader should be aware of the principle of an e-beam.

In the rest of Section 8.6, the words or names "e-beam," "electron-beam" or "e-Beam" may be used for two different products:

1) The *machine* that is able to produce a beam (ray) of electrons
2) The *beam of electrons*, produced in the machine

The context in which these names will be used will make it clear if it concerns "the machine" or the "beam of electrons."

8.6.1 Awareness of e-Beams in Our Daily Life

Probably one of the most well-known e-beams, which we all know and which were installed in many living rooms, bed rooms, hotel rooms, restaurants, and sports bars during several decades, is the TV (*monitor*), which was popular till the early 2000s, as shown in Figure 8.10.

Figure 8.9 The heating tubes in CV lines may have lengths up to and over 100 m and therefore CV-lines need quite some space.

Figure 8.10 TV monitors, e-beams that were used all over the globe.

Besides TV monitors (*and PC monitors*), some other widely used applications of e-beams are:

- *Electron microscopes*, which use a beam of accelerated electrons as a source of illumination.

- *Electron-beam welding* (EBW), which is a fusion welding process in which a beam of high-velocity electrons is applied to two materials to be joined.
- *Surface treating*, which is typically based on polymerization processes. Nowadays many paints, varnishes, and inks are applied under environment friendly, solvent-free conditions and get their final properties by e-beam treatment.
- *Electron lithography*, which uses e-beams with diameters ranging from 2 nm up to hundreds of nanometers. Electron lithograph is, for instance, also used to produce holograms.
- *Sterilization* of medical tools and materials.
- *Food processing*: For instance, for inactivation of salmonella; replacing microbial pathogens, insecticides, and chemical preservatives in food; shelf-life extension of products such as herbs and spices; anti-sprouting for products such as potatoes, onions, and garlic.
- *Wastewater treatment*: An additive-free process that uses the short lived reactive species formed during the radiolysis of water for efficient decomposition of pollutants therein.
- *Curing of composite materials*: Applications for projects in the aerospace and defense industry.
- *Modification of material surfaces* for biological, medical, and chemical applications.
- *Coloring or color altering of gemstones*, such as topaz, quartz, tourmaline, aquamarine, emerald, and even diamond.
- *Cargo inspection*, for the detection of contraband, fissile and radioactive materials, explosives, and drugs.

8.6.2 The Principle of an e-Beam

The principle of an e-beam is simple.

Free electrons are produced by an electron emitter (*such as an electron gun or even a lamp* [*light bulb*]) in a high vacuum environment/area, to avoid or minimize collision with other molecules. These free electrons can be manipulated by electric and magnetic fields to form a fine e-beam.

The "power" of the e-beam is depending on the number of electrons (*measured in micro- or milliampere*) and the speed of the electrons. The speed of the electrons is depending on the voltage difference (*accelerating voltage*), which is applied between the electron emitter (cathode) and an anode. As electrons have a negative charge, they will flow to the anode area.

Due to the "accelerating voltage," e-beams are also known as "e-beam accelerators."

Usually all e-beams have one thing in common: they are created to hit a target. When an e-beam hits a solid matter heat or kinetic energy is generated. In Figure 8.11 the comparison of the principle of a 25 kV e-beam in an old TV and a 500–5000 kV e-beam in a high voltage accelerator is visualized.

The amount of energy in a rather small volume can be precisely controlled and can be used in applications, such as those listed in Section 8.6.1.

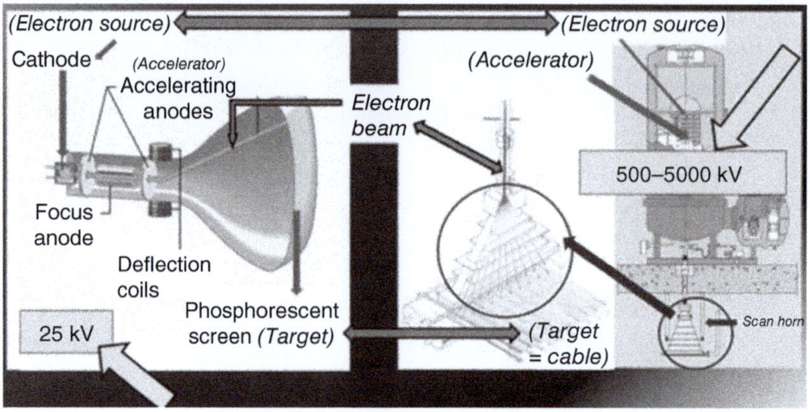

Figure 8.11 Principle of an e-beam in an (old) TV monitor (left) and an e-beam for crosslinking of cables.

8.6.3 Creating a High Voltage e-Beam

Starting at the end of the nineteenth century until mid of the twentieth century, many physicists were working on creating high voltages for scientific research. The Dutch–American Robert J. Van de Graaff, physicist and lecturer at both Princeton University and Massachusetts Institute of Technology, was the inventor of the Van de Graaff generator, a device that in the 1930s was able to create voltages of well over 5 million volt.

During World War II, Robert J. Van de Graaff was one of the three founders of the US High Voltage Radiographic Project and (*after World War II*) was the co-initiator of High Voltage Engineering Corporation (HVEC) with Dennis M. Robinson and John Trump, the paternal uncle of Donald Trump, 45th president of the United States.

The Van de Graaff accelerator has been the heart of many e-beams, which use the high voltages of several million volt, to accelerate free electrons into a beam of high velocity electrons.

8.6.4 Scan Horn and Beam Window

The accelerated stream of electrons is directed within the high vacuum accelerator tube, toward the beam transmission window. This window is a thin (*typically 25 μm or thinner*) titanium foil. Its main function is to maintain the vacuum within the acceleration chamber itself. Being extremely thin, this foil does not absorb much of the e-beam energy (*less than 2%*), and it permits most of the accelerated electrons to travel through the foil and impinge/hit the target.

As the window still absorbs some energy, and it does become hot during operation of the e-beam. Usually air cooling is used to keep the titanium foil (*beam window*) at ambient temperature. In the course of use, the e-beam can imprint a heat pattern on the window. To avoid an implosion, which could carry materials and unwanted dust into the acceleration chamber, it is good practice to change the titanium foil according to a routine preventative maintenance program.

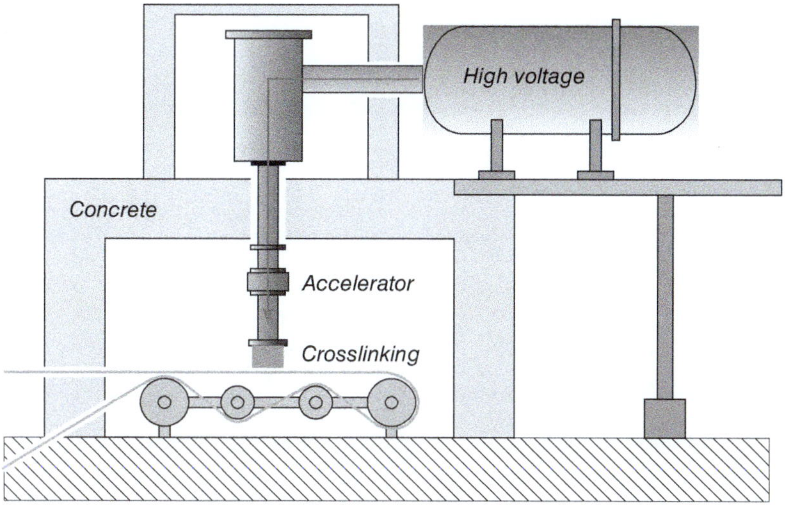

Figure 8.12 Schematic drawing of an e-beam facility, including the high voltage generator, the accelerator tube (in which the electrons are accelerated to the desired energy [voltage]), the cable in the under-beam-handling-system (four rolls) and the shielding provisions (concrete bunker).

The titanium foil, some rubber gasket O-rings, and accelerator filament, the initial source of the electrons, are the only three components of an e-beam accelerator, which require routine replacement. As an example the service use for a filament under normal production use is around 10 000 hours (=*well over one year*). Like every machine, repeated on and off switching will shorten the life of such a filament. A titanium foil typically needs to be replaced twice per year. Should an e-beam slowly lose vacuum, this is an indication of a pinhole in the foil.

Nowadays internal diagnostics are part of the e-beam systems to provide customers critical information to further improve the system's availability by displaying parameter trending and expected lifetime of critical components or subsystems.

The typical diameter of the beam of electrons is around 10–15 mm. This beam of electrons needs to be evenly "distributed" (*divided or scanned*) over the "production area." This is the area where the wires and cables get hit by the e-beam (see Figures 8.12 and 8.13). The "evenly distribution" of the beam of electrons is realized by the implementation of a high speed scanning system, which follows a pattern as a function of a magnetic field and shows a similarity to the old style TV monitors.

8.6.5 Under-Beam Handling System (UBHS)

This system guides the insulated wires and cables under the e-beam.

A few different under-beam handling systems (UBHSs) are available for optimal guiding of the wires and cables under the e-beam, to ensure the most uniform dose of electrons (Figures 8.14–8.16).

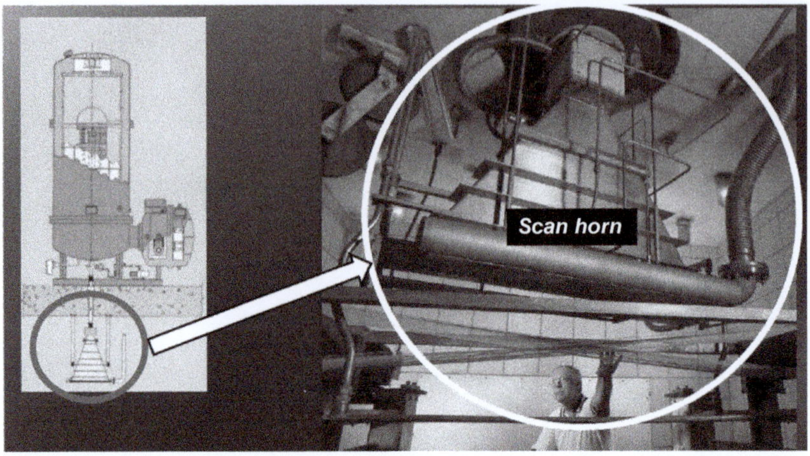

Figure 8.13 The scan horn: under the e-beam (left) and positioned above a cable which is ready to be treated. The triangle-shaped scan horn shows the degree of the deflection of the beam of electrons. Source: Courtesy of IBA(B).

Figure 8.14 Example of an UBHS. Courtesy of vom Hagen und Funke GmbH, Germany.

Each UBHS has its individual advantages with regard to outer diameter (OD), wall thickness, cable construction, batch size, etc.

For large size batches a UBHS, which can hold some 100–150 m of wire or cable, will not be very efficient for cable drums, which only contain 200 or 250 m of cable, but are suitable for cable drums that can contain quantities of up to 10 000 m or more.

Figure 8.15 Larger wires and cables usually require a larger bending radius and can be processed with a four-drum fixture.

Figure 8.16 Four-drum UBHS. Source: Courtesy of vom Hagen & Funke GmbH, Germany.

Small wires will twist slightly between successive passes on the UBHS when getting treated by the e-beam. On top of that the electrons of the e-beam will be diverged in the scan horn. That will ensure that also the sides of the wire or cable will get hit by the electrons.

The outer sheaths on very large cables can be treated by rotating the cables as they pass through the e-beam along the scanning direction. With this method, both the pay-off and take-up reels are rotated outside of the shielded treatment room. The electron energy needs to be just enough to penetrate the radial thickness of the sheath. In this way, a nearly uniform dose distribution around the cable will result. The maximum crosslinking speed for such "rotating drum system" is limited.

8.6.6 Control System

To ensure consistent and evenly crosslinked wires and cables, the process parameters of the e-beam system will need to be correctly set and monitored.

A software system, which is "product" driven, will link all essential parts and parameters of this system. Detailed reports for each meter of crosslinked wire or cable can be produced, to ensure that crosslinking was within specification and to ensure traceability for QC purposes.

8.6.7 Safety

8.6.7.1 Radiation and Radioactivity

The two aforementioned words "radiation" and "radioactivity" are probably the most "sensitive words" of e-beam crosslinking technology.

For that reason an understandable and clear explanation is needed to avoid misunderstandings and decisions, which are based on emotions.

8.6.7.2 Bremsstrahlung or X-rays

The high-speed electrons of the beam are created to hit the wire or cable, but may also pass straight through the insulation or jacket and hit other parts, such as the UBHS.

In any case the speed of the electrons will be slowed down dramatically, when hitting a product. This (*enormous*) deceleration produces radiation, which is known as "Bremsstrahlung".

Bremsstrahlung is a German word, which can be split in: "Brems" (*bremsen* = *braking*); "strahlung" means radiation.

A simple way of understanding and visualizing "Bremsstrahlung" is to compare this "Bremsstrahlung" with the effect of a car, when it needs to stop within two or three seconds, while driving at a speed of 200 km/h. The result is the brake disks of the car will turn red instantaneously because of the heat, which is produced during the braking process.

Heat is also a "radiation" process! See Figures 8.17 and 8.18.

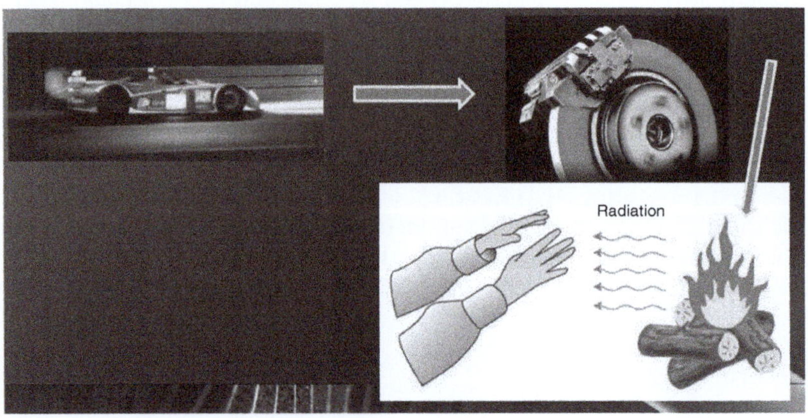

Figure 8.17 Heat, produced by deceleration of a car.

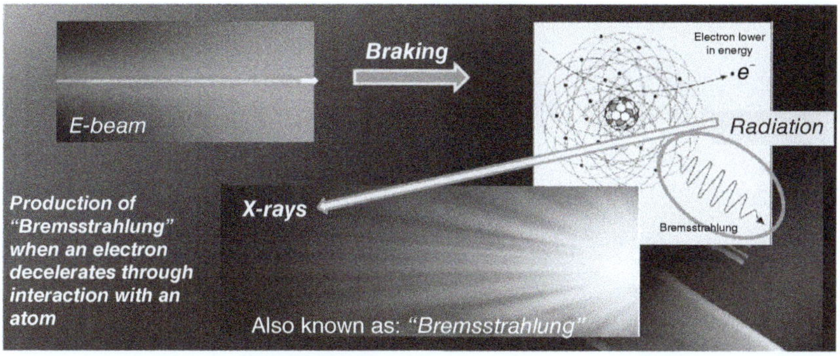

Figure 8.18 X-rays, produced by deceleration of high speed electrons.

8.6.7.3 Shielding and Radioactivity

8.6.7.3.1 Safety Systems The "Bremsstrahlung" creates a spectrum of X-rays, which is only generated as long as the e-beam is working. To avoid workers or other people to be exposed to these X-rays, the manufacturers of e-beams have to encase the critical areas of their e-beams with a shield.

Low-energy e-beam accelerators, typically up to 1.0 or 1.5 MeV, encase these critical areas in lead. Typically thicknesses of several centimeter of lead are sufficient to stop any X-rays generated by these e-beams. Such lead casing is used around the entire accelerator chamber and around the UBHS. Arranging the angle and length of openings to transport the wires and cables to and through the e-beam system enables the e-beam manufacturer to eliminate any possible X-ray emission, coming through these openings.

e-beam systems are equipped with X-ray detection devices, which will read the background level and will be interlocked with the system to shut it down instantly should an increased level of background X-rays be detected.

It should be noted that lead shielding is practical up to somewhere around 1.0–1.5 MeV. At higher voltages separate concrete or steel bunkers must be constructed in order to eliminate any X-ray emissions, as shown in Figure 8.19. The "low energy" e-beam systems are often called "self-shielded," as shown in Figure 8.20.

When electrons (*from the e-beam*) are absorbed in a wire or cable, heat is generated. For that reason usually also care is taken to see that the area immediately under or opposite of the beam exit is cooled. In most situations, a water-cooled metal plate is positioned directly below the beam.

Due to the nature of X-rays, there is no residual radiation when the e-beam is switched off and **products are not made radioactive**. Also the shielding bunker does never become radioactive.

8.6.7.3.2 Desired Production of X-rays Although many people would not immediately recognize the advantages of X-rays or Bremsstrahlung, there are various applications, where the only goal of the beam of electrons is creating X-rays. Some of these applications are:

Figure 8.19 A concrete bunker functions as protective shield. Source: Courtesy of IBA, Belgium.

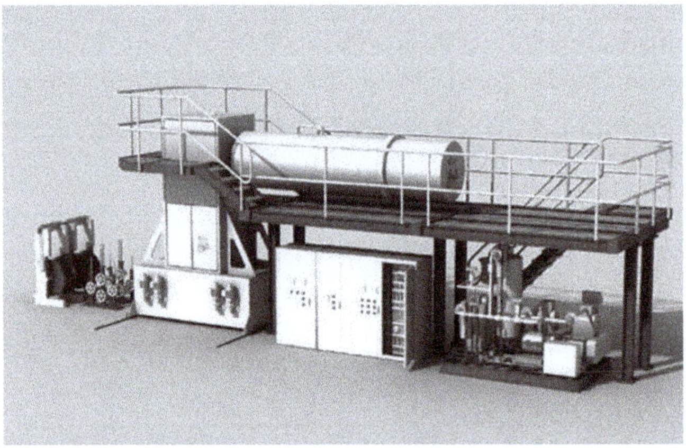

Figure 8.20 A 0.8 MeV "self-shielded" e-beam for e-beams with voltages >1.0 or 1.5 MeV. Source: Courtesy of IBA, Belgium.

* In medicine for diagnosis or cure.
* For dental inspection of your mouth.
* To detect flaws (mistakes), cracks, and air bubbles in finished goods, such as parts of airplanes, but also to inspect the quality of the welding of metal pipes in the oil industry.
* For cargo inspection of containers in a harbor.

8.6.7.4 Other Safety Systems

All e-beams are provided with multiple interlock systems to ensure the safety for the operators. These safety systems will be activated as soon as the access doors are open(ed) or anyone is still inside the bunker. This will prevent the e-beam system from working.

8.6.7.5 Ozone

Inside high-energy e-beam facilities ozone can be generated by exposing normal oxygen gas to energetic electrons. Ozone is toxic and governmental regulations have been established to protect people from excessive exposures to this gas. Each e-beam facility is provided with a system for ozone removal.

8.6.7.6 IAEA

e-beams fall under the Safety Standards of IAEA (International Atomic Energy Agency), as the high power e-beams can produce X-rays.

8.6.8 Most Important Parameters During e-Beam Crosslinking

The three most important parameters during the e-beam crosslinking process are:

- Voltage
- Amperage
- Radiation dose

8.6.8.1 Voltage

An e-beam is usually classified or identified by the maximum voltage the e-beam can achieve. A 3 MeV e-beam therefore will have the capability to deliver a voltage of maximum 3 million volt.

This does not necessarily mean that the e-beam is always running at that maximum voltage. The voltage of the beam of electrons, which is needed for crosslinking, is depending on the construction of the wire or cable.

The actual applied voltage of the e-beam is responsible for the final speed of the electrons, when exiting the scan horn at the titanium foil. If, for instance, a voltage of 1.0 million volt is applied, the electrons will have a speed which is $1/3^{rd}$, compared with the speed of electrons that will be accelerated with 3.0 million volt.

The penetration depth of a 1.0 MeV electron is lower than the penetration depth of a 3.0 MeV electron. In other words, the higher the wall thickness of the insulation or jacket, the higher the voltage that is needed to have the electrons penetrate through the complete wall and to create even crosslinking.

However, in practice it is not the actual wall thickness, which is of importance to set the voltage of the e-beam, but the so called "chordal length (CL)."

This is especially applicable for wires and cables, which can only pass under the e-beam a relatively low number of times and are irradiated from one direction only. One-sided irradiation systems use an UBHS, which will allow that the wires and cables will be irradiated on the bottom and top side, for instance, in a so-called "figure-8" set up of the UBHS, as shown in Figure 8.14. Should the wire or cable not twist during the process, the maximum penetration depth of the electrons need to be equal to the CL, as can be seen in Figure 8.21.

By using the ***Pythagoras Theorem***, we know that:

$$(R_1)^2 = (\mathrm{CL})^2 + (r_2)^2$$

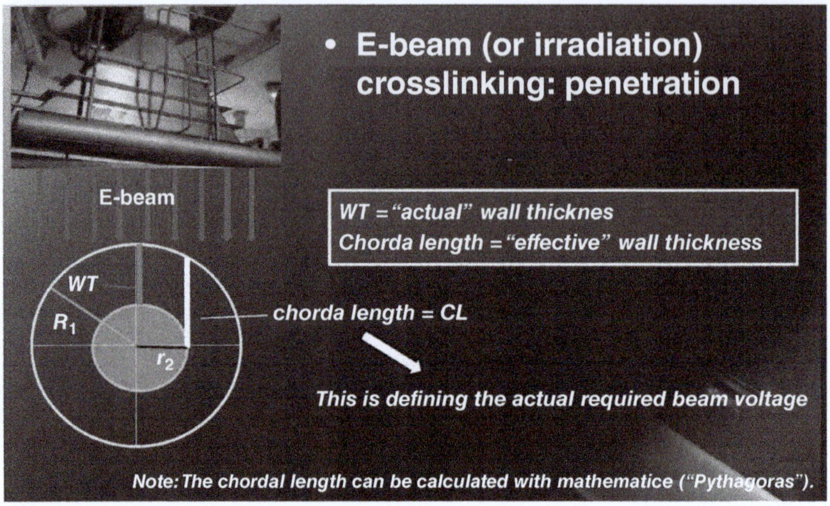

Figure 8.21 Chordal length is the chordal length. This is different from the actual wall thickness of an insulated wire or jacket cable.

where

R_1	$1/2 \times$ OD
OD	The outer diameter of the cable
WT	The wall thickness, to be crosslinked
r_2	Half the diameter of the core or $1/2 \times (\text{OD} - 2 \times \text{WT})$

Therefore:

$$\text{CL} = \sqrt{\{(R_1)^2 - (r_2)^2\}}$$

Not only is the required voltage depending on the wall thickness, but is also depending on the density of the material that shall be crosslinked. The higher the density, the more difficult for an electron to penetrate through the material.

A generally used "calculation" for the required voltage reads as follows:

$$E = \{197 \times d \times \sqrt{[\text{WT} \times (\text{OD} - \text{WT})]}\} + 325$$

where

E	The required voltage (in keV), which is required to crosslink a certain wall thickness
d	The density (in g/cm^3) of the material to be crosslinked
WT	The wall thickness (in mm) to be crosslinked
OD	The outer diameter (in mm) of the wire or cable

The following example shows us that totally different voltages are needed for different size cables, having the same insulation wall thickness.

Let's take:

- An insulated wire, having a wall thickness of 1.5 mm and an OD of 4 mm, and
- An insulated wire, having a wall thickness of 1.5 mm, but having an OD of 20 mm,

and that both use the same insulation compound, with a density of 1.25 g/cm^3.

(1) Calculation for the cable with an OD = 4.0 mm teaches us that we will need a voltage of $E = \{197 \times 1.25 \times \sqrt{}[1.5 \times (4 - 1.5)]\} + 325$ = around 800 keV (=0.8 MeV)
(2) Calculation for the cable with an OD = 20.0 mm teaches us that we will need a voltage of $E = \{197 \times 1.25 \times \sqrt{}[1.5 \times (20 - 1.5)]\} + 325$ = around 1600 keV (=1.6 MeV).

In general, the operating voltage range (*minimum to maximum voltage*) of an e-beam is part of the design of the unit. e-beam manufacturers provide a voltage adjustment device that permits customers to operate the e-beam at less than the maximum rated voltage.

This will make the e-beam more economic for both thin wall products and products, which use lower density compounds for insulation or jacketing applications. Important to know: when selecting an e-beam it is crucial to know which sizes of wires and/or cables need to be irradiated (Figure 8.22).

8.6.8.2 Amperage

The amperage, also known as e-beam current, is directly connected to the number of electrons in the beam of electrons. The higher the amount of electrons, which are able to hit the wire or cable, the sooner the total dose for crosslinking the wire or cable will be achieved.

The number of passes under the scan horn, the amperage of the e-beam, and the speed of the wires under the e-beam are all three related to the dose. Reducing the processing speed of the wire under the e-beam will increase the dose, as the time that the wire is hit by the e-beam is increasing. Reducing the number of passes will decrease the dose, as the total time that the wire is hit by the e-beam is reduced as well.

8.6.8.3 Radiation Dose

8.6.8.3.1 Total Dose The total absorption of a beam of electrons by a wire or cable is called the dose. This dose can be compared with the amount of raindrops hitting an umbrella during a rain shower (as shown in Figure 8.23).

In e-beam crosslinking the "*amount of energy*," created by the electrons, which is absorbed by the wire or cable is often expressed in *megarads* (Mrad), wherein 1 rad equals 100 ergs/g of absorbed energy. One megarad is also equal to 10 *kilograys* (=10 J/g), also known as kGy. The gray is the internationally used unit of dosimetry for e-beam.

The crosslinking of wire and cable requires a dose exposure of somewhere typically between 5 and 20 Mrad (*50–200 kGy*), whereas the curing of most coatings, inks, and adhesives requires a dose exposure in the range of 1–3 Mrad (*10–30 kGy*) to exhibit full cure.

Some useful dosimetry relationships are known as follows:

Dose = energy absorbed per unit mass
Rad = unit of dose equal to absorption of 100 ergs/g

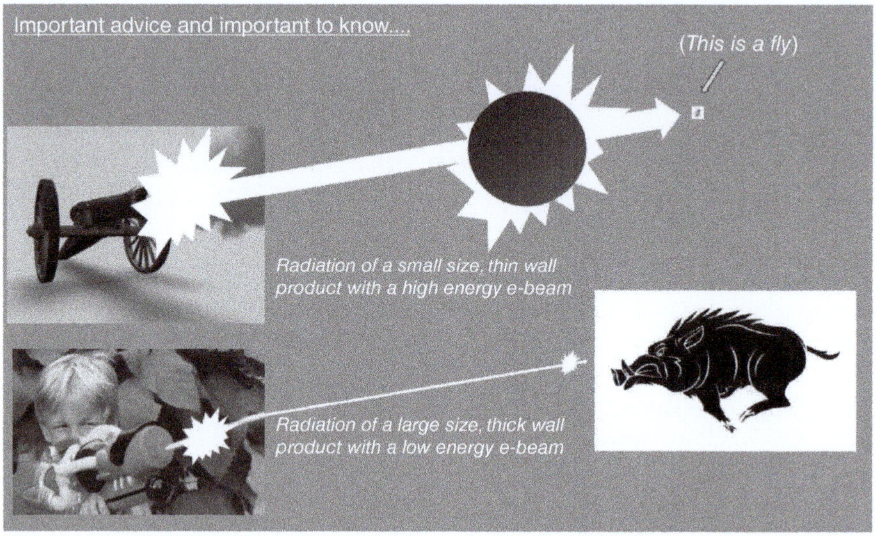

(a)

(b)

Figure 8.22 Selecting the right power of an e-beam is important for the range of products to be crosslinked. Source: (a) Fun Way Illustration/Shutterstock.com, (b) Pederk/E+/Getty images.

Megarad = Mrad = 1.000.000 rads = 10 kGy = 10 J/g = 2.39 calories/g = 4.3 BTU per pound = 10 W·s/g = 4.54 kW·s per pound (lbs)
Gray = unit of dose = absorption of 100 rads
Kilogray = kGy = 1.000 grays = 0.1 Mrad = 1 J/g

Each polymer or compound of a wire or cable will need a different dose to achieve a crosslink degree that will meet the requirements of the applicable cable specification.

8.6.8.3.2 Dose Distribution Probably the most "difficult part" of e-beam crosslinking is to understand the so called "dose distribution."

The reason for this is that the dose within the insulation or jacket is not perfectly uniform.

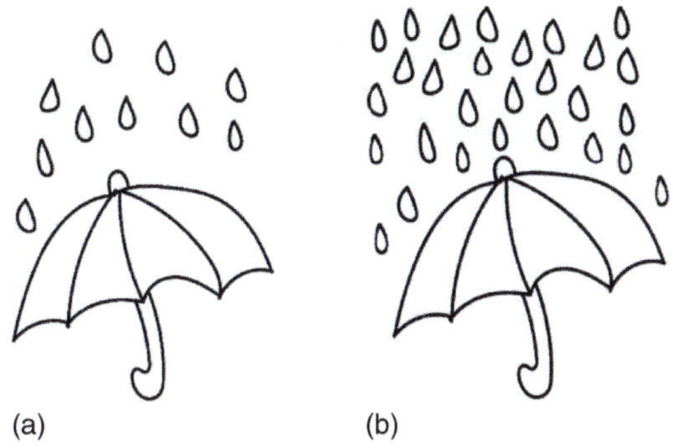

Figure 8.23 Umbrella receiving a low dose (a) and a high dose (b) during a rain shower.

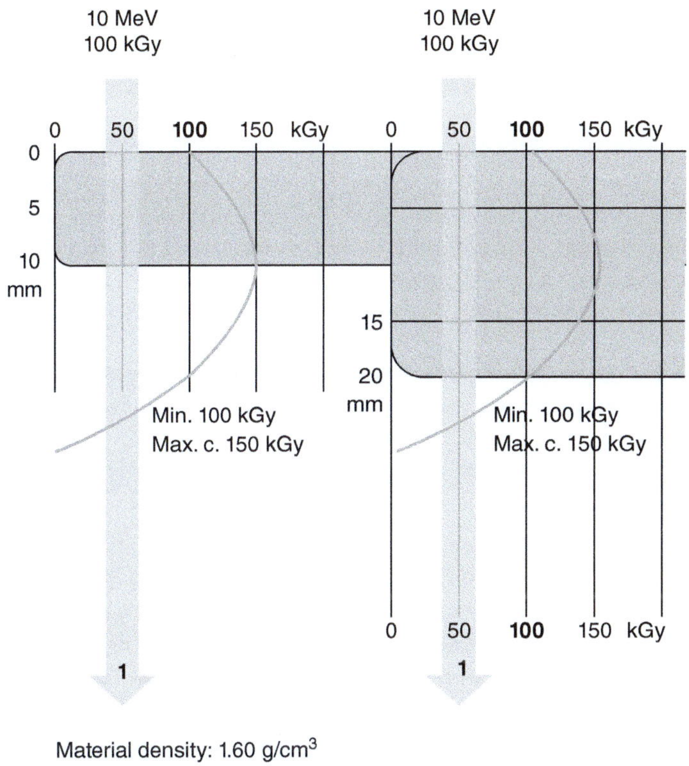

Figure 8.24 The dose distribution curve, resulting from irradiation of two samples (density = 1.60 g/cm^3) having different wall thicknesses, under a 10 MeV e-beam with 100 kGy.

Usually wires and cables are irradiated through *one-sided irradiation*, as the beam of electrons hit the product from one direction. The curve on the left side in Figure 8.24 shows that most of the energy (dose) is not deposited in the 10 mm thick sample. The grey colored area represents the sample. For this one-sided irradiation, a dose distribution curve as shown on the right side of Figure 8.24 is often used as an optimum dose distribution.

Table 8.1 Calculated ideal wall thickness at three different energies (voltages) of the e-beam, for a jacket with a density of 1.60 g/cm^3.

Voltage	1.0 MeV	2.0 MeV	3.0 MeV
D(opt) in mm	1.4	4.0	6.5
D(50e) in mm	1.9	4.8	7.7
D(50) in mm	1.7	4.3	6.9

In this curve we see that most of the energy (dose) is not deposited in the sample. For this sample 10 MeV is apparently too high.

The green curve on the right side shows the dose distribution curve, resulting from irradiation of a 20 mm thick sample under a 10 MeV e-beam with 100 kGy. In this curve we see that the exit dose is equal to the entrance dose (100 kGy). The 10 MeV energy fits to the thickness of the sample.

In general the dose initially tends to *increase* with depth in the material to about 150% of the value of the surface dose. This increase can be explained as follows:

As the free electrons travel through the irradiated material, these free electrons do interact with atoms, which can be found in the material and are known as "Coulomb force interactions." These interactions are the result of collisions with the orbital electrons or even the nuclei of atoms in the irradiated material.

(*Note*: Many studies have been made and are still being made, as e-beam technology is not only used for crosslinking, but also used in equipment for cancer therapy, where dose distribution in human tissue plays an important factor as well.)

After reaching its maximum the dose decreases to nearly zero at a depth where the electrons have lost all their kinetic energy.

There are a few interpretations or options of getting to the "optimum dose distribution" in a material:

1) **D(opt)** is the depth in the irradiated sample, where the exit dose is equal to the entrance (=*surface*) dose. This is shown in the right-side curve in Figure 8.24.
2) **D(50e)** is the depth in the irradiated sample, where the exit dose is equal to 50% of the entrance dose.
3) **D(50)** is the depth in the irradiated sample, where the exit dose is equal to 50% of the maximum dose.

Note: The speed of the electrons (in other words: the electron energy) should be sufficient to achieve one of the above optimum dose distributions.

For a halogen free sheathing compound, having a density of around 1.60 g/cm^3, the optimum wall thicknesses of the jacket, which is irradiated at 1.0, 2.0, or 3.0 MeV, is shown in Table 8.1.

The absorbed dose distribution depends on the thickness, density, and atomic composition of the insulation as well as the diameter and composition of the metallic conductor. Because of the cylindrical geometry and the shielding and backscattering effects of the conductor, accurate calculations of the dose distribution (*such as the Monte Carlo probability codes*) are not simple and not user-friendly.

However by measuring the

- Gel content, or the
- Hot-set elongation

on well selected samples, the quality of the crosslinking can be measured and interpreted in practice.

8.6.9 Capacity of an e-Beam

Productivity is an important issue in cable production and therefore it is important to know the potential output of an e-beam. In other words, how many kilograms can an e-beam crosslink per hour, day, or year.

8.6.9.1 Functional Capacity

The functional capacity of an e-beam is depending on three factors:

- The power of the e-beam (in kW)
- The efficiency of the e-beam, and
- The required dose needed to meet the specifications of the cable

It is clear that the higher the power of the e-beam, the more energy per kilograms can be transferred to material.

8.6.9.2 Efficiency

The efficiency of the e-beam depends on the beam efficiency factor, as not all produced electrons hit the cable.

This is due to several issues:

- The fact that a part of the electrons are "absorbed" by the tungsten foil at the exit of the scan horn.
- The gap between the layers of cable on the UBHS. The larger the gap, the more electrons will be "lost" for the crosslink process and thus reduce the efficiency of the e-beam.

We also have the productivity factor, which does influence the efficiency of the e-beam. It is clear that the more hours the e-beam is actually running, the higher the efficiency. Therefore larger cable drums, containing more meters on the drum, will have a positive effect on the efficiency, as drum changing may require to stop the e-beam and thus may reduce the efficiency.

8.6.9.3 Dose

Some polymers or compounds are less sensitive to e-beam irradiation than others. The higher the sensitivity, the less dose is needed to achieve sufficient crosslinking. The selection of polymers and materials has an important effect on the sensitivity, as described in Section 8.6.11 (Compound Design).

In practice the functional capacity of an e-beam can be calculated as defined in Figure 8.25.

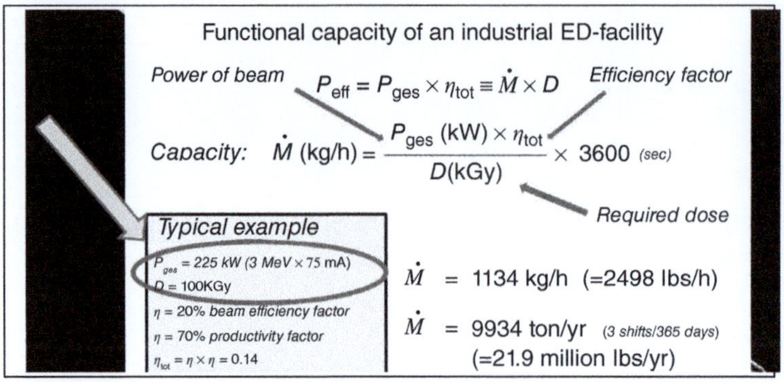

Figure 8.25 Example of a calculation of the capacity of an industrial e-beam.

8.6.10 Temperature Rise

The thermal capacity of polymeric insulating material is typically about half that of water, e.g. about 2 J/g/°C. With this value, a dose of 100 kGy absorbed quickly in a single pass of the wire through the e-beam would cause the temperature of the insulation to rise about 50 °C. However, with a multiple-pass method in a UBHS, which could have more than 100 passes, the dose per pass would be less than 1 kGy and the temperature rise per pass would be less than half a degree Celsius. In practice it appears that in such a multiple-pass system most of the heat is dissipated between the passes.

8.6.11 Compound Design

8.6.11.1 Polymers

Whereas the silane and peroxide crosslinkable systems have a limited range of polymers that can be selected, a large range of polymers and elastomers can be crosslinked by e-beam crosslinking.

Some of these polymers and elastomers are listed in Table 8.2. A visualization is shown in Figure 8.26.

Very soft rubbers, which tend to be sticky and may deform after winding on a drum in the extrusion process, may not be the best choice for silane- and e-beam

Table 8.2 List of some of the polymers and elastomers, which can be crosslinked by e-beam irradiation.

PVC	LLDPE	LDPE	MDPE	HDPE
POE	EVA	EEA	EMA	EVM
PVDF	FKM	CPE	Chloroprene	PBT
PA	SBR	TPE-U	TPE-O	TPE-V
TPE-E	TPE-U	CSPE	EPR	EPDM
ACM	IIR	(H)NBR	SiR	ETFE

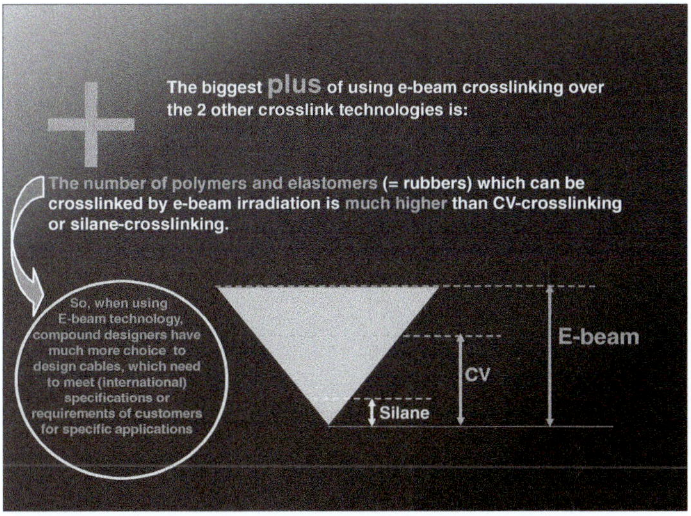

Figure 8.26 The large range of polymers and elastomers, available for e-beam crosslinking, give compound designers more choice to develop compounds for insulation or sheathing of wires and cables.

crosslinking, whereas these soft rubbers may still be an option for the CV-curing process.

(1) *Co-agents*: The addition of coagents such as, for instance, TAC, TAIC, or TMPTMA improves the crosslinking density in all cases significantly and improves all properties aligned with this. In some cases the coagents are essential to enable crosslinking. The coagents also reduce the radiation dose required for a certain crosslinking degree and by this save radiation time and cost.
(2) *Other raw materials*: The selection of other raw materials for e-beam crosslinkable compounds can be compared with the range, which is available for thermoplastic compound formulations.

8.6.12 Costs Indications of e-Beam Crosslinking

8.6.12.1 Annual Costs

The initial investment in an e-beam facility (*including e-beam, UBHSs, cable handling system [take-off and take-up], a concrete bunker [for e-beams of >1.5 MeV]*) may be considered as high, but more interesting to know is: What are the crosslinking costs per kilograms?

A well maintained e-beam will easily be productive for 30–40 years and more.

Based on an actual case, the following example will give an indication of the crosslinking costs.

Let's assume an investment of US$ 4.000.000 for a 4 MeV, 35 mA e-beam facility, including bunker. In that case it will come down to following:

1.	Depreciation per year:	US$ 200.000,– is based on 20 years depreciation
2.	Maintenance per year:	US$ 50.000,–
3.	Energy costs (*):	US$ 175.000,–
4.	Direct labor costs (**):	US$ 165.000,–
	Total annual costs:	US$ 590.000,–

Note: * A 4 MeV, 35 mA beam, having 140 kW will typically consume around 200 kW, as also auxiliary equipment "around" the e-beam will need power, such as the safety systems, winding and unwinding equipment, under-beam-handling-system, and control room. 200 kW/h comes down to 1.752.000 kW/yr. At US$ 0.10/kWh this gives a total energy costs of US$ 175.000,–.

** Three shifts. Labor costs per shift: 1 operator of US$ 40.000,– plus 25% of a production supervisor (US$ 60.000,– per year) is US$ 15.000,–. Total over three shifts: US$ 165.000,– per year.

8.6.12.2 Annual Throughput and Costs per Kilograms

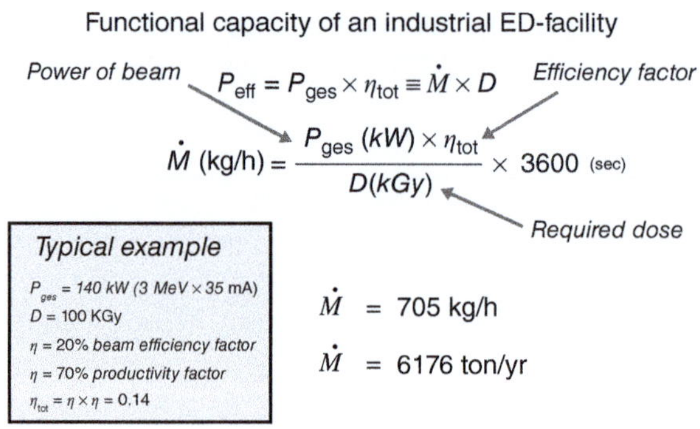

The 4 MeV, 35 mA e-beam with a total efficiency of 14% (*beam efficiency of 20% and a productivity in three shifts of 70%*) will be able to irradiate some 705 kg of polymeric material per hour at a dose of 100 kGy. This comes down to 6176 ton/yr. See the calculation in Figure 8.25.

Based on the annual costs of US$ 590.000, (*as per Section* 8.6.12.1) the average costs will therefore be **US$ 0.10/kg.**

And even when calculating with *10 years of depreciation* (*instead of 20 years*), increasing the energy costs to US$ 0.15/kWh (*instead of US$ 0.10/kWh)* and reducing the *productivity to 50%* (*instead of 70%*), the crosslinking costs for the e-beam process will be around: **US$ 0.20/kg.**

8.6.13 Advantages and Disadvantages of e-Beam Crosslinking

Advantages:

- No scorching or premature crosslinking during extrusion.
- Extruder can be stopped at all times for the change of colors or cable sizes, without crosslinking inside the extruder.

- Extrusion waste material can be re-used, as it is still thermoplastic.
- Reduced costs for waste disposal.
- Highest crosslinking speed compared with the peroxide and silane crosslinking process.
- Highest output in extrusion compared with the peroxide and silane crosslinking process.
- Cleanest and most environmentally friendly method of the three crosslink processes, as no chemicals are added to split the carbon–hydrogen bonds to facilitate crosslinking.
- Highly reliable and used in the cable industry for over 60 years.
- Crosslinking happens in split seconds.
- High throughputs possible, processing speeds of 1200–1500 m/min for small size wires possible.
- Dry process. No water or steam needed.
- Crosslinking takes place at room temperature, which contributes to a reduced "heat history."
- Easier compound design compared to the chemically reactive silane and peroxide crosslinking, so R&D teams will have more options to develop a suitable compound.
- Easy to optimize the crosslink degree of insulation or jacket.
- A wide range of polymers is available for this crosslink process.
- Nowadays wire processing parameter traceability systems can be integrated with data systems from other manufacturing processes in the cable factory.

Disadvantages:
- High initial investment (*however: see Section* 8.6.12).
- Some (*partly scientific*) knowledge will be needed for operating the e-beam and to understand the shielding requirements for X-ray protection.
- Once installed an e-beam cannot easily be moved (*except for self-shielded e-beams*).
- Very soft and therefore sticky rubbers may show deformation, prior to being crosslinked.

8.7 Conclusions

Each of the three crosslinking processes: silane crosslinking, peroxide crosslinking, and e-beam crosslinking has their own advantages and disadvantages.

However, an overall look learns us that e-beam crosslinking is the most versatile crosslinking system, as it is applicable to the widest range of polymers, it is the fastest process for crosslinking and if the capacity of the e-beam is well used the overall production costs per meter of cable will even be the lowest, when e-beam crosslinking is used in the manufacturing process of crosslinked cable.

Further Reading

Following articles, chapters, studies, patents, and books are considered to be very valuable to get a better understanding of the fundamentals of crosslinking, curing, and vulcanization.

Silane Crosslinking

Moshedian, J. and Hoseinpour, P.M. (2009). Polyethylene cross-linking by two step silane method. A review. *Iranian Polymer Journal* 18 (2): 103–128.

Jiao, C., Wang, Z., Gui, Z., and Hu, Y. (2005). Silane grafting and crosslinking of ethylene–octene copolymer. *European Polymer Journal* 41 (6): 1204–1211.

Shuhei Yasuda (2012). Silane crosslinked polyolefin insulated wire. Hitachi Cable, Ltd. Japanese Patent 129198, filed 20 March 2012. https://patents.justia.com/patent/20120312580.

Christele Kensicher (Lyons), Linda Boudiaf (Grigny) (2009). Method of cross-linking a filled polymer based on polyethylene. Patent Grant No. 7632574, filed 18 October 2006. https://patents.justia.com/patent/7632574.

Carlsson, R., Sultan, B.-A. and Jungkvist, J. (2004). Developments in reactor designed moisture curable polyethylenes. *AMI conference called 'Cables'*, Cologne (Germany)

Peroxide Crosslinking

Henning, S.K. and Costin, R. (2005). Fundamentals of curing elastomers with peroxides and coagents I: coagent structure–property relationships. *Presented at the Spring 167th Technical Meeting of the Rubber Division, American Chemical Society*, San Antonio, TX (16–18 May).

(2010). *Use of Polybutadiene Coagents in Peroxide Cured Elastomers for Wire and Cable Compounds*. Exton, PA, USA: Cray Valley USA LLC http://www.crayvalley.com/docs/technical-paper/polybutadiene-coagents-in-peroxide-cured-elastomers-for-w-c.pdf?sfvrsn=2.

Cheremisinoff, N.P. (ed.) (1989). Sulfur vulcanization. In: *Handbook of Polymer Science and Technology: Volume 2 – Performance Properties of Plastics and Elastomers*. New York, NY: Marcel Dekker Inc.

Cheremisinoff, N.P. (2001). *Condensed Encyclopedia of Polymer Engineering Terms*. Butterworth-Heinemann. ISBN: 978-0-08-050282-3 https://onlinelibrary.wiley.com/doi/abs/10.1002/pi.4980230319.

Mark, J.E., Erman, B., and Roland, C.M. (eds.) (2013). *The Science and Technology of Rubber*, 4e. Academic Press. ISBN: 978-0-12-3945846.

Kruzelak, J., Sykora, R., and Hudec, I. (2017). Vulcanization of rubber compounds with peroxide curing systems. *Rubber Chemistry and Technology* 90 (1): 60–88.

Electron-Beam Crosslinking

Wu, G., Zhai, M., and Wang, M. (eds.) (2018). *Radiation Technology for Advanced Materials: From Basic to Modern Applications*. Academic Press. ISBN: 978-12-814017-8.

Tikku, V. (2017). Cross-linking of wires and cables with electron beam (EB) – a new technology of future. *Wire and Cable India Journal* http://www.wirecable.in/2017/01/cross-linking-of-wires-cables-with-electron-beam-eb–-a-new-technology-of-future/.

(1993). Van de Graaff's Generator. In: *Electrical Engineering Handbook*. Boca Raton, Florida USA: CRC Press. ISBN: 0-8493-0185-8 https://en.wikipedia.org/wiki/Van_de_Graaff_generator.

Sanders, R. (2006). *Electron Beam Accelerators*. ESI, Printers Guide, Radtech (Association for UV & EB technology) https://www.ebeam.com/news-archive/archive/2017/electron-beam-accelerators.

Sabharwal, S.. Electron beam irradiation applications. *Proceedings of PAC2013*, Pasadena, CA, USA. https://accelconf.web.cern.ch/PAC2013/papers/weyb3.pdf.

Cleland, M.R. and Galloway, R.A. *Electron Beam Crosslinking of Wire and Cable Insulation*. White Paper and Technical Information Series TIS 01812. Edgewood, NY, USA: IBA Industrial, Inc.

Anon, V. (1982). *Ethylene-Propylene Elastomers, Radiation Cured Electrical Materials*. Product Report 82 ET-B 348, 16–19. Essochem Europe, Inc.

Zamore, A., Hackett, J., and Wald, S.E. (1993). *Successful Applications of Electron Beam Processing in Wire and Cable Manufacturing*. Technical Information Series TIS 01416. Edgewood, NY, USA: IBA Industrial, Inc.

Frounchi, M., Dadbin, S., and Panahinia, F. (2006). Comparison between electron-beam and chemical crosslinking of silicone rubber. *Nuclear Instruments and Methods in Physics Research Section B: Beam Interactions with Materials and Atoms* 243 (2): 354–358.

Cleland, M.R. and Galloway, R.A. (2009). *Accelerator Requirements for Electron Beam Crosslinking of Wire and Cable Insulation*. Technical Information Series TIS 01813. Edgewood, NY: IBA Industrial, Inc.

Drobny, J.G. (2003). *Radiation Technology for Polymers*. Boca Raton, FL, USA: CRC Press.

Lyons, B.J. (1995). Radiation crosslinking of fluoropolymers – a review. *Radiation Physics and Chemistry* 45 (2): 159–174.

Spenadel, L. (1985). Electron beam crosslinking of ethylene propylene electrical compounds. *Journal of Industrial Irradiation Technology* 3 (1): 7–39.

Cleland, M., Galloway, R., Genin, F., and Lindholm, M. (2002). The use of dose and charge distributions in electron beam processing. *Radiation Physics and Chemistry* 63 (3–6): 729–733.

Boye, W.M. (2008). Utilizing coagents in the electron beam cure of elastomers. 57th IWCS Conference: Proceedings of the International Wire & Cable Symposium (IWCS), Inc. (9–12 November 2008), Rhode Island Convention Center, Providence, Rhode Island, USA.

Radiation Safety of Gamma, Electron and X Ray Irradiation Facilities. SSG-8. Safety guide. IAEA, Vienna. https://www.iaea.org/publications/8401/radiation-safety-of-gamma-electron-and-x-ray-irradiation-facilities.

(2017). Radiation. Enhancing plastics properties by irradiation. Wiehl, Germany: BGS. https://en.bgs.eu/wp-content/uploads/2017/02/BGS_radiation_crosslinking_en-1.pdf.

Hot-Set-Elongation

Jacob Steendam (2009). Measuring the crosslinking level by means of hot-set-elongation (Hot creep). Wire & Cable Technology International: PTL. https://www.ptl-bv.com/hot-set-creep-ovens.

9
Nuclear Power Station Cables

Frank Liu

Jiangsu Shangshang Cable Group Co., Ltd., Shangshang Road, Liyang City, Jiangsu Province 213300, China

9.1 Development of Nuclear Power in the World

The development of nuclear power in the world originates from the development of nuclear fission energy technology. This technology was developed in the 1940s with the research mainly focused on the use of large amounts of energy released by atomic nuclear fission of specific nuclides to make weapons. By the 1950s, nuclear fission energy technology began to turn to peaceful use, i.e. for nuclear power generation. Nowadays, nuclear power plays a decisive role in the world's power energy. Of this, civil nuclear power has more than 16 500 reactor years of operation experience, and accounts for 11.5% of the world's electricity energy supply.

9.1.1 Development Stage of Nuclear Power in the World

When the United States first used nuclear energy to generate electricity in 1951, it had been nearly 70 years since human beings started using nuclear energy to generate electricity. The development of nuclear power has gone through four stages: verification and demonstration, high-speed development, stagnant development, and rational development.

Verification and demonstration stage: In 1942, the University of Chicago built the world's first nuclear reactor, which demonstrated the possibility of achieving controlled fission chain reaction. But it was during the Second World War that nuclear energy was mainly used for military purposes. Since the early 1950s, the United States, Soviet Union, United Kingdom, France, and other countries began to turn nuclear energy to civilian use, using existing nuclear technology to develop and build nuclear reactors for the purpose of generating electricity, thus entering the stage of verification and demonstration. In 1957, the United States built a pressurized water reactor (PWR) nuclear power plant, which opened the way for the development of light water reactor nuclear power. In 1956, United Kingdom built a graphite gas-cooled reactor nuclear power plant for both plutonium production and power generation. The Soviet Union

The Global Cable Industry: Materials, Markets, Products,
First Edition. Edited by Günter Beyer.

built a pressure tube graphite water cooled reactor nuclear power plant in 1954. Canada built a natural uranium heavy water reactor nuclear power plant in 1962. These nuclear power plants show relatively mature technology and low costs of power generation, which laid the foundation for the commercial promotion of nuclear power.

High-speed development stage: At the end of the 1960s, the developed countries entered the stage of high-speed economic development and electricity demand grew rapidly. For fear of scarcity of strategic resources such as oil, the United States, Soviet Union, United Kingdom, France, Germany, Japan, and other countries formulated long-term nuclear power development plans. Some developing countries also started to build nuclear power plants by purchasing complete sets of equipment. After the verification of the economy of the US light water reactor nuclear power, the first climax of nuclear power plant construction occurred. In 1967, the order of nuclear power plant reached 25.6 GW. Since 1969, the total installed capacity of nuclear power in the United States had exceeded that in United Kingdom, ranking first in the world. After the oil crisis in 1973, in order to get rid of the dependence on Middle East oil, the second climax of nuclear power plant construction came into being. Since then, by exporting light water reactor technology to other countries, the United States made light water reactor the mainstream type of nuclear power plant construction in the world. Under the situation of the great development of nuclear power, the United States, United Kingdom, France, Germany, and other countries also actively developed fast neutron breeder reactor and high temperature gas-cooled reactor and built numerous experimental reactors and prototype reactors.

Stagnant development stage: In 1979, the second oil crisis occurred in the world. Since then, economic development rate slowed down in various countries, and the growth rate of electricity demand had dropped dramatically with large-scale energy-saving measures and adjustment of industrial structure. Construction of many new nuclear power plants was suspended or delayed. Three Mile Island nuclear power plant accident occurred in 1979 in the United States and Chernobyl nuclear power plant accident occurred in 1986 in the Soviet Union. These incidents had a significant impact on the development of nuclear power in the world. The problem of public acceptance became the main obstacle to the development of nuclear power. Some countries temporarily suspended the development of nuclear power, such as Switzerland, Italy, and Austria. In order to ensure the safety of nuclear power, the United States took measures to improve safety after the Three Mile Island accident, making the construction period of nuclear power plants prolonged; investment increased, and the economic competitiveness of power generation of nuclear power plants decreased. The uncertainty of investment risk hindered the continued development of nuclear power especially.

Rational development stage: Since the end of 1980s, developed countries worked actively for the recovery of nuclear power construction, and started to formulate safer and more economical nuclear power design standards. The United States first formulated the Advanced Light Water Reactor Power Company utility requirement document (URD) and sorted out the safety approval procedures for nuclear power plants. The European utility requirement document (EUR) was also formulated in Western European countries, and similar documents were formulated in Japan and Korea too. According to URD requirements, nuclear power equipment suppliers carried out research and development on safer and more economical light water reactor type. General Electric Company in the United States, and Toshiba and Hitachi in Japan jointly developed the modified boiling water reactor of advanced boiling water reactor (ABWR); The US ABB-CE developed modified PWR system 80+; Westinghouse Company developed passive safety PWR of AP-600; Framatome in France and Siemens in Germany jointly developed modified European PWR of ethylene propylene rubber (EPR), etc. The development of nuclear power plants entered a rational development stage with emphasis on safety and environmental protection.

9.1.2 Current Status of Nuclear Power Development All Over the World

The world's first commercial nuclear power plant began in the 1950s. There are 445 commercial nuclear power reactors operating in 31 countries, with a total installed capacity of 387 GW and 64 reactors under construction. As the sustainable and reliable low-carbon energy source, nuclear power has provided more than 11% of the world's electricity. In addition, there are about 240 research reactors operating in 56 countries and 180 power reactors providing power for about 140 ships and submarines.

India: According to the national energy policy, India's nuclear power development goal is to achieve 14.5 GW installed by 2020, including light water reactors, heavy water reactors, and fast reactors. At present, in addition to 21 units already in operation, six units are under construction in India, including domestic and imported designs, as well as a prototype fast neutron breeder of 500 MWe (megawatts of electricity, unit of power for nuclear power), which will push India's "thorium project" to the second stage, paving the way for the ultimate utilization of India's abundant thorium resources as reactor fuel (Figure 9.1).

Russia: Russia plans to increase its installed capacity to 30.5 GW by 2020, mainly relying on its own advanced light water reactor. In addition, a large fast neutron breeder reactor has been started, and the main objective is to export. Construction of an initial floating nuclear power plant has begun and has been delivered in 2017. In addition, Russia is also very active in building and financing nuclear power plants in other countries (Figure 9.2).

Europe: Finland and France are expanding their 1650 MWe PWR technology EPR from AREVA, two of which have been built in Taishan, China. Besides, Germany and Belgium have politically decided to step out from the nuclear technology and to shut down stations in a few years. There are also several countries in Eastern Europe that have started or have clear plans to build new nuclear power units, mainly including Bulgaria, Czech, Hungary, Romania, Slovakia, Slovenia, and Turkey (Figure 9.3).

According to the annotation of the government's energy book in mid-2006, aged nuclear power units in the country are replaced by newly built nuclear

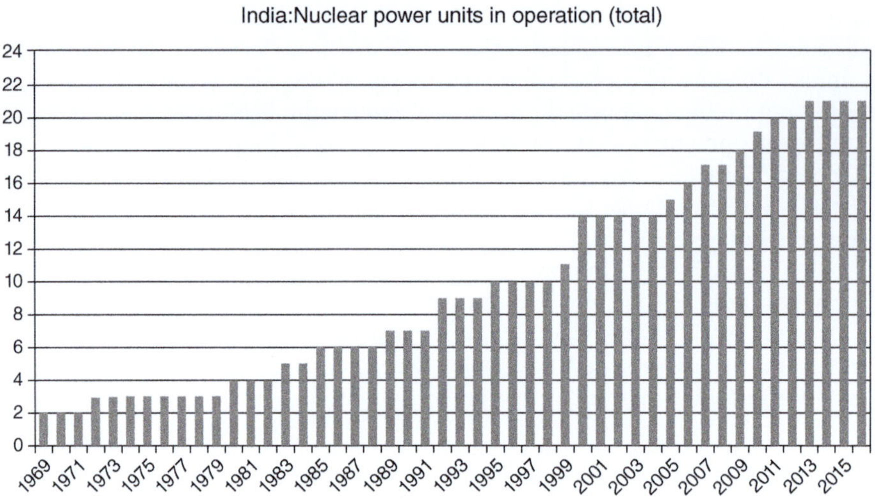

Figure 9.1 Nuclear power development in India. Source: Data from www.nuclear.net.sn.

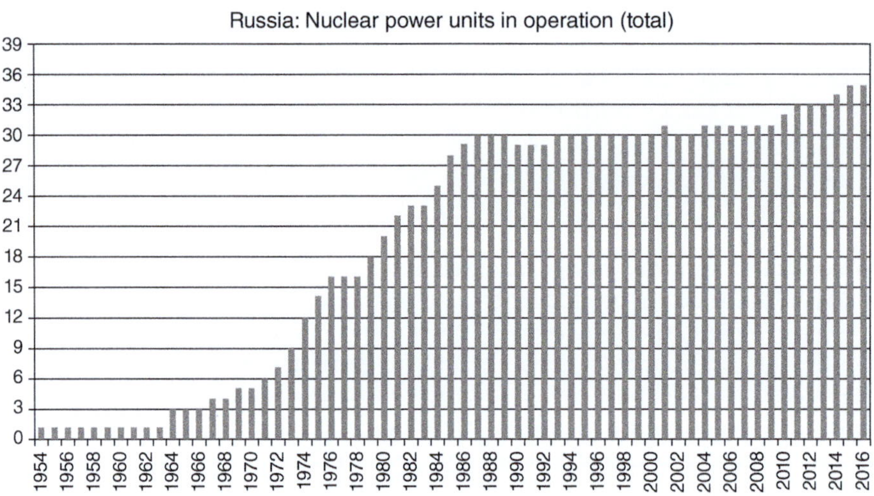

Figure 9.2 Nuclear power development in Russia. Source: Data from www.nuclear.net.sn.

power units. Four 1600 MWe French units are scheduled to operate in 2023. The development goal of Britain's nuclear power is that by 2030, new nuclear power units will be in operation at 16 GWe.

The United States: At present, four nuclear power units are under construction in the United States adopting AP1000, the advanced PWR technology of the United States. So far, an important reason for the interruption of new nuclear power projects in the United States is that the United States has made very successful progress in nuclear power operation and maintenance strategy. Over the past decade, innovation has increased the utilization of nuclear power

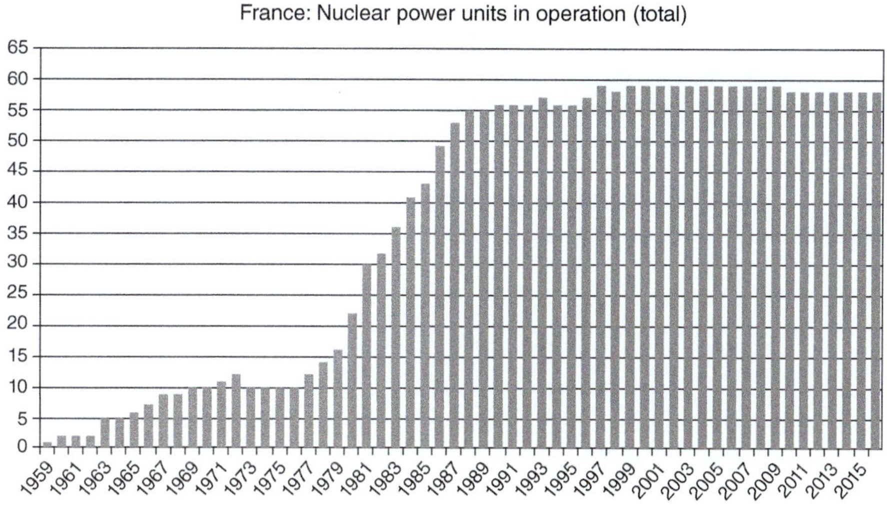

Figure 9.3 Nuclear power development in France. Source: Data from www.nuclear.net.sn.

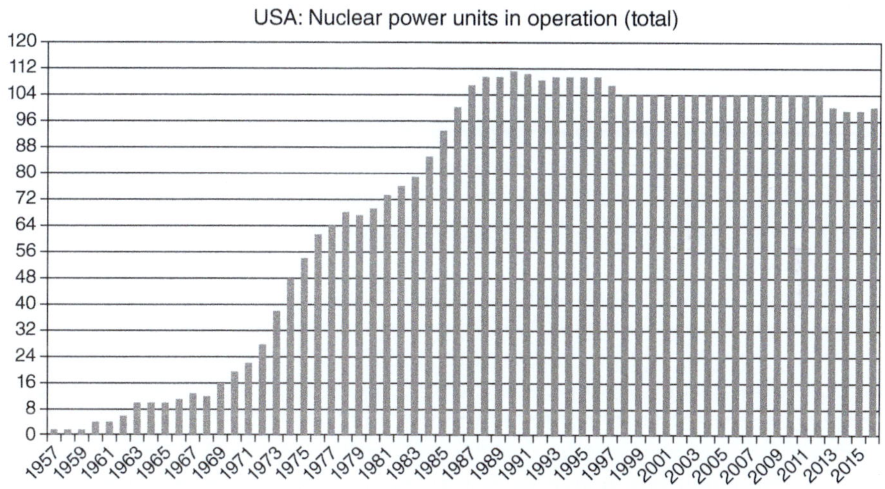

Figure 9.4 Nuclear power development in USA. Source: Data from www.nuclear.net.sn.

plants in the United States, resulting in a significant increase in power output, equivalent to 19 new 1000 MWe nuclear power units (Figure 9.4).

Africa: South Africa is the only nuclear power country in Africa. It operates two 930 MWe PWR units at the Koeberg nuclear power plant. South Africa indicated that it will further implement the nuclear power development plan (9600 MWe). Nigeria has sought support from the International Atomic Energy Agency (IAEA) for its plans to develop two 1000 MWe nuclear power plants.

Other emerging countries: The IAEA expects emerging countries such as Lithuania, the United Arab Emirates, Turkey, Belarus, Vietnam, Poland, and Bangladesh to develop nuclear power in the short term.

Development of nuclear power in China: China has defined for the first time the mid- and long-term target of installed capacity and annual power generation of nuclear power since 2007. So far, nuclear power plant has been constructed more than 10 years. In 2014, the work report of the two sessions of the Chinese government clearly put forward the idea of "starting a batch of nuclear power," which greatly promoted the construction of nuclear power shelved by the Fukushima nuclear power accident in Japan, and the related supporting industries also ushered in new opportunities for development.

Since 2015, China has ushered in a new round of nuclear power construction peak. With an average annual start-up rate of five to six nuclear power units, it is estimated that the annual investment in new nuclear power will be about 100 billion Yuan, of which nuclear island equipment accounts for about 20 billion Yuan and is expected to show an annual growth. By the end of December 2017, the number of nuclear power reactors in service in China had reached 38. In 2016, the power generation capacity reached 210 TWh, accounting for 8.45% of the total global nuclear power generation; in the past three years, China's nuclear power generation has maintained an average annual growth rate of more than 30%, which is much higher than the global growth level.

According to the current plan, all the units to be built in the future in China are the third generation of nuclear power plants. China's third-generation nuclear power technology has gone through the initial complete import, to the delivery of HPR1000, the first MW PWR nuclear power unit developed and manufactured independently, entering a critical stage of import substitution and not going abroad. Whether the key equipment and materials of nuclear reactor core can be localized or not has become the most important technical bottleneck.

The first CAP1400 demonstration unit in China will reach about 80%, and the localization rate of the first HPR1000 demonstration unit will not be less than 85%. In principle, equipment that can be localized in China's nuclear power construction needs to be purchased from domestic enterprises. At present, besides the reactor core, all the equipment and materials related to the newly built nuclear power plant have achieved import substitution, while only the related equipment and materials of the reactor core still rely on imports. The core is the most difficult part of nuclear reactor technology, and the control rod is the "brake" of nuclear reactions. Nuclear power control rods absorb thermal neutrons through silver, indium, and cadmium materials to adjust the reactor speed or to achieve emergency shutdown control. Control components are part of the consumables of nuclear reactors. In China, the replacement of control components for civil nuclear reactors takes 10 years as a cycle. At present, manufacturing nuclear reactor core components locally is the key to realizing independent construction of China's nuclear energy industry and not to go abroad. The core of a nuclear power reactor consists of fuel assembly, control rod assembly, neutron source assembly, combustible poison assembly, and blocking assembly. The fuel assembly and control assembly are the core

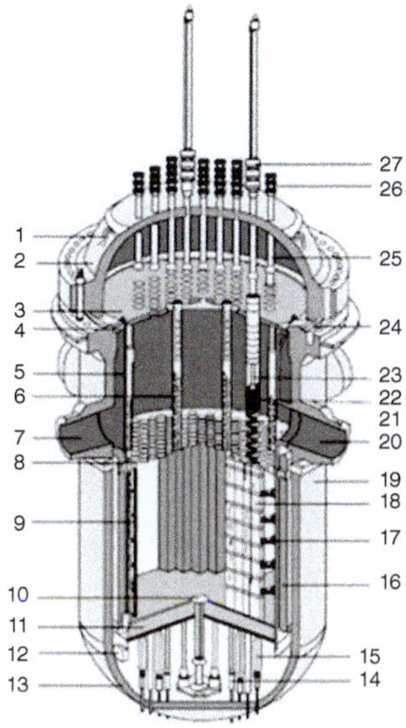

Note

1. Instrument lead tube
2. Core support column
3. Heat shield
4. Coaming
5. Fuel assembly
6. Reactor pressure vessel
7. Outlet sleeve
8. Control-rod bank (cluster)
9. Control rod guide thimble
10. Control rod drive assembly
11. Pressure spring
12. Insulation sleeve
13. Meter lead pipe inlet
14. Control rod driving mechanism
15. Lifting earring
16. Pressure vessel head
17. Guide tube support plate
18. Interior support ledge
19. Core support barrel
20. Upper support column
21. Inlet sleeve
22. Top core grid plate
23. Coaming
24. Access hole
25. Bottom core grid plate
26. Radial support pressuer
27. Vessel bottom head

Figure 9.5 Fuel system rods for nuclear power plants.

working components, and the other three components are related components. Fuel assemblies release thermal neutrons and heat through nuclear fission. The thermal neutrons keep the nuclear fission going and the heat drives generators to generate electricity. The control module absorbs thermal neutrons through silver, indium, and cadmium control rods, thus changing the number of neutrons in the core to achieve power regulation and emergency shutdown control of the nuclear reactor, to ensure nuclear safety (Figure 9.5).

At present, the fuel pellets for core fuel assemblies of CAP1400 and HPR1000 nuclear power plants in China have been locally manufactured by CNPC North Nuclear Fuel Element Co., Ltd. and CNPC Jianzhong Nuclear Fuel Components Co., Ltd., respectively. Zirconium alloy tubes for fuel cladding have also been manufactured locally by Western Energy Material Technologies Company and State Nuclear Baoti Zirconium Industry Company. Western materials are the main undertakers of the local manufacture of nuclear reactor core control components (including Ag–In–Cd control rods, neutron absorber materials, stainless steel, and Inconel 718 alloy materials for related components, etc.).

In addition to commercial nuclear power plants, there are 240 research reactors in operation and more under construction in 56 countries. These reactors are useful for many purposes, including research and manufacture of isotopes for medical and industrial purposes. Reactors for ship power are limited to the main Navy and have played an important role in powering submarines and large surface ships over the past 50 years. At present, at least 140 hulls, mostly submarines, are driven by about 180 nuclear reactors and have more than 13 000 reactor years of operational experience for submarine reactors.

9.1.3 Technical Characteristics and Differences of Nuclear Power Plants

9.1.3.1 Technical Characteristics of Various Reactors of Nuclear Power Plants

There are many kinds of nuclear power plants. The type of reactor is usually named by the moderator that reduces the speed of neutron movement in the reactor. Reactors using light water (H_2O) as moderator are called light water reactors; those using heavy water (D_2O, deuterium oxide) as moderator are called heavy water reactors; graphite reactors use graphite moderator. In addition, light water reactor can be divided into boiling water reactor and PWR according to the state of water in the reactor. The state of water in the boiling water reactor is boiling, while PWR keeps high temperature water liquid by pressurizing water. In addition to the above, there are also nuclear reactor called fast neutron breeder reactor that do not slow down neutrons, and are referred to as fast reactors. These different types of nuclear power plants not only have different neutron speeds, but also have different utilization efficiency and requirements for nuclear fuel.

Pressurized water reactor nuclear power plant: It is the nuclear power plant with PWR as its heat source. It consists mainly of nuclear and conventional islands. The four major components in the nuclear island of PWR nuclear power plant are steam generator, pressure regulator, main pump, and reactor core. Its working principle is as follows: the main pump feeds 120–160 atmospheric primary loop cooling water into the reactor core, takes the heat energy released from nuclear fuel out of the reactor core, and then enters the steam generator, transfers heat to the secondary loop water through the heat transfer tube, makes it boil, and generates steam; when the temperature of the primary loop cooling water drops, it enters the reactor core and completes the primary loop water cycle; the high-pressure steam generated in the second loop drives the steam turbine to generate electricity, and then enters the steam generator through the condenser and preheater to complete the water circulation in the second loop. PWR is easier to maintain because the primary loop and steam system are separate and the steam turbine is not polluted by radioactivity.

Boiling water reactor nuclear power plant: It is the nuclear power plant with boiling water reactor as heat source. Boiling water reactor is a power reactor that uses boiling light water as moderator and coolant and produces saturated steam directly in the reactor pressure vessel. Boiling water reactor and PWR belong to the category of light water reactors, which have the advantages of compact structure, safety and reliability, low construction cost, and strong load following ability. They all need to use low enriched uranium as fuel. The difference between boiling water reactor and PWR is that the cooling water is kept at a low pressure (about 70 atmospheric pressure), and the water passes through the reactor core into steam at about 285 °C, which is directly introduced into the steam turbine. Therefore, the boiling water reactor has only one loop, eliminating the steam generator, which is prone to leakage, so it seems simple. The boiling water reactor involves the direct entry of steam generated in the reactor into the steam turbine, and therefore the steam turbine is exposed to radioactive contamination. Therefore, the design and maintenance of boiling water reactor (BWR) are more troublesome than that of PWR.

Heavy water reactor nuclear power plant: It is the nuclear power plant with heavy water reactor as the heat source. Heavy water reactor is a reactor with heavy water as moderator. It can directly use natural uranium as nuclear fuel. It can be divided into pressure vessel type and pressure tube type. The coolant of heavy water reactor is unrestricted. Heavy water, light water, gas, or organic compound can be used to realize nonstop stacking of loading and unloading and continuous feeding, and the control rod to compensate fuel consumption can be omitted. Heavy water reactor is usually fueled by natural uranium, which is similar to PWR. However, because of the large grid pitch, the pressure vessel is much larger than that of PWR with the same power, so the maximum power of a single reactor can only be 300 000 kW. Heavy water reactor nuclear power plant is an early-developed nuclear power plant with various types, but only Canada reactor nuclear power plant developed in Canada has achieved industrial scale promotion.

Fast neutron breeder reactor: It is the nuclear power plant that converts heat released by fast neutron-induced chain fission reactions into electricity. In operation, a fast reactor not only consumes fissile materials but also produces new fissile materials. Moreover, it can produce more fissile materials than it consumes, realizing the proliferation of nuclear fissile materials. The commercial nuclear power plants in the world, such as PWR, boiling water reactor, heavy water reactor, and graphite gas-cooled reactor, are all non-breeder reactor types. They mainly use nuclear fission fuel. Even if the converted fissile materials, such as plutonium-239, are reused, the utilization rate of uranium resources is only 1–2%. However, in fast reactors, uranium-238 in principle can be converted into plutonium-239 and used. But considering all kinds of losses, the utilization rate of uranium resources can be increased to 60–70% by fast reactors. However, the development of fast reactor is still lagging behind, and Japan's Manjusri fast reactors as well as other fast reactors under development are not functioning properly.

9.1.3.2 Technical Development of Nuclear Power Plant

From the 1950s to the mid-1960s, the first-generation of nuclear power generation is the primary stage of nuclear fission power generation from military service to peaceful use, mainly developing early prototype nuclear power plants.

The second-generation of nuclear power generation is a period of great development for commercial nuclear power plants, from the mid-1960s to the late 1990s. The second-generation nuclear power plant includes PWR, boiling water reactor, heavy water reactor (Candu), gas-cooled reactor, and pressure tube graphite water cooled reactor nuclear power plants.

The design principle of the third-generation nuclear power unit is based on the accumulated technical reserve and operation experience of the second-generation nuclear power unit, aiming at its shortcomings, further adopting the new technology that has been developed and verified to be feasible, in order to significantly improve its safety and economy. At present, third-generation nuclear power is the mainstream nuclear power generation technology in the world.

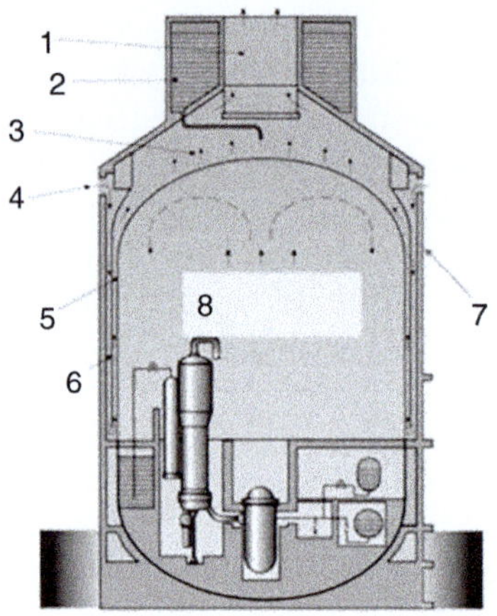

Note:
1. Natural convection air outlet
2. Passive containment cooling system water tank
3. Water film steamer
4. External cooling air inlet
5. Steel containment
6. Air baffle
7. Concreate shield building
8. Internal condensation and natural circulation

Figure 9.6 Containment for AP1000 nuclear power plant.

The fourth-generation nuclear power plant is different from the third-generation nuclear power plant in PWR technology. Its evaporator can reach 560 °C, and the power generation efficiency is greatly improved. More importantly, the high temperature gas-cooled reactor nuclear power plant has good inherent safety, which can ensure that core melting and radioactive release will not occur in any accident. The possibility of core melting due to residual heat of the reactor is excluded. There is no need to take out-of-plant emergency measures technically to achieve the nuclear safety goal of the fourth-generation nuclear power system.

9.1.3.3 Technical Characteristics of Typical Nuclear Power Plants in the World

At present, the world's main construction is the third-generation nuclear power plants with higher safety and performance with 60 years of service life. The technical characteristics of the typical third-generation nuclear power plants are described as follows (Figure 9.6).

Characteristics of AP1000 nuclear power plant: AP1000 is a PWR nuclear power technology; the nuclear fuel made of uranium undergoes fission in the "reactor" equipment to produce a large amount of heat energy, and then uses water under high pressure to bring out the heat energy; steam is generated in the steam generator, which drives the turbine with generators to rotate together, and electricity is constantly generated. AP1000 is characterized by concise design, easy operation, and full use of many "passive safety systems," such as gravity theory, natural circulation, and polymerization reaction, which are much simpler and more effective than the traditional PWR safety system. At the same time, AP1000 further improves the safety of nuclear power plants, and also significantly reduces the cost of nuclear power unit construction and long-term operation. In addition, AP1000 can use modular technology to carry out construction in parallel, greatly shortening the construction period of nuclear power units.

The characteristics of EPR nuclear power plant: EPR is a single reactor with four-loop units; its power reaches 1525 MWe, and its design life is 60 years. A double-layer containment design is adopted. The outer layer is made of reinforced concrete shell to resist external disasters, and the inner layer is made of pre-stressed concrete. Its main design features include the following: (i) Serious accident prevention – EPR considers serious accidents such as high-pressure melting reactor, hydrogen combustion and explosion, steam explosion, reactor core melt, and heat discharge in containment. (ii) Design of instrument and control system and main control room – EPR instrument and control system and main control room adopt mature design, fully absorb the experience feedback of digitized instrument and control system and man–machine interface of the power plant, and absorb the advantages of advanced technology and equipment. The instrument and control are arranged in four rows, which are located in different areas of the safety workshop to avoid common mode failure. The main control room is the same as the highly computerized control room of the N4 unit. It has a special man–machine interface for maintenance and diagnostics.

The characteristics of HPR1000 nuclear power plant: HPR1000 is characterized by "177 sets of fuel assembly cores," "multiple redundant safety systems," and "combined active and passive safety measures." It adopts the world's highest safety requirements and the latest technology standards to meet the safety requirements of IAEA and three generations of technical standards of the United States and Europe. HPR1000 is equipped with three cooling water tanks with a total water capacity of about 3000 tons in the passive containment heat export system, which serves as the final heat trap for heat release in the containment after serious accidents. HPR1000 also has 12 heat exchangers in the containment and passive safety injection water tank in the emergency core cooling system. When the pressure of the reactor coolant system drops below a certain value, the safe injection tank automatically injects boron water into the reactor coolant system to ensure the core cooling. HPR1000 has adopted effective passive safety measures on the basis of active safety to effectively respond to the passive safety system with loss of power source as a supplement to the project-proven, highly efficient, mature and reliable active safety system. It provides a variety of means to meet the safety requirements. It is one of the third-generation nuclear power plants with the highest acceptance in the current nuclear power market.

Through the passive design concept of these safety systems and the high-level heat exchanger water tank equipped, HPR1000 unit can meet the principle of non-intervention 72 hours after the accident. The passive safety system will be put into operation automatically when the design benchmark accident or the over-design benchmark accident or even serious accident occurs, preventing the reactor core from melting down, ensuring the integrity of the pressure vessel after core melting down, providing secondary side cooling of evaporator, ensuring that containment does not exceed temperature and pressure, and eliminating the risk of hydrogen detonation and explosion, and other safety functions.

The characteristics of CAP1400 nuclear power plant: CAP1400 nuclear power plant is based on the digestion, absorption, and comprehensive mastery of the third-generation of advanced AP1000 passive technology of nuclear power introduced by China, and the National Nuclear Power Technology Corporation has developed large-scale passive PWR nuclear power units with China's independent intellectual property rights and greater power through re-innovation. It is also the core content of China's major nuclear power projects, one of the "16 national science and technology projects." CAP1400 is the largest passive PWR nuclear power plant and meets the highest safety standards of commercial nuclear power plants in the world.

At present, units 1–2 of Shandong Rongcheng CAP1400 demonstration project are applying CAP1400 technology in China, with a capacity of 1.4 million kW and a design life of 60 years. The project is expected to be approved. The preliminary work of the two units is progressing smoothly.

CAP1400 adopts the combined design of passive reactor core cooling system and passive containment cooling system. Its safety is 2 orders of magnitude higher than that of the second-generation nuclear power. At the same time, CAP1400 adopts a simplified design, which significantly reduces the number of components compared with traditional PWR. In this way, the pressure of maintenance and inspection is reduced, and the probability of failure is greatly reduced.

For the steel containment, which is the most important part of reactor safety assurance, the size of the containment is enlarged in CAP1400 design, achieving larger free volume, optimal layout, and larger analysis margin of internal pressure of the containment. In the design of shielding workshop, a steel plate concrete structure is adopted, which has the ability to resist malicious impact of large commercial aircraft, and the space layout is optimized to improve the accessibility of personnel and maintainability of equipment.

In terms of economy, compared with the current second-generation nuclear power, the economy of CAP1400 is mainly reflected in the performance parameters and construction cost. CAP1400 is more economical and has a design life of 60 years, which is 20 years longer than that of the second-generation nuclear power plant.

Based on AP1000 modularization technology, CAP1400 optimizes the module design. Modular construction shortens the construction period: the construction period of unit 1 of the demonstration project is 56 months, and that of unit 2 is 50 months.

9.1.3.4 Characteristics of CPR1000 Nuclear Power Plant

CPR1000 is an improved MW PWR nuclear power technology scheme launched by China Guangzhou Nuclear Corporation. While introducing, digesting, and absorbing foreign advanced technology, it combines the gradual improvement and independent innovation over the past 20 years to form the "second-generation plus" MW PWR nuclear power technology.

CPR1000 technology originates from M310 reactor type, a megawatt reactor type imported from France. The prototype of French MW nuclear power technology is Westinghouse's standard 312 reactor, which has been built into a standardized CPY technology by improving batch production. In order to improve the export competitiveness of French nuclear power, France's Farmaton Company has formed the M310 reactor with better safety and economy on the basis of CPY. The nuclear power technology has the following characteristics:

(1) Reduce the burden of operators, reduce human error, and be conducive to handling multiple accidents and interface with the rules for handling serious accidents.
(2) Reduce the number of refueling overhauls, reduce overhaul costs, fuel cycle costs, production of radioactive waste, neutron flow of reactor pressure vessels, and irradiated dose of workers, and improve the utilization rate and annual power generation of power plants.
(3) Prevent or delay the penetration of pressure vessel, prevent the reaction between core melt and concrete, prevent the penetration of containment bottom plate, inhibit the production of hydrogen in containment, and improve the probability of containment integrity.
(4) Have a low leakage design, reduce neutron irradiation to pressure vessel, adopt integral forgings with strict control of Cu, P, S, and Ni content, prolong the service life of pressure vessel, and ultimately prolong the service life of nuclear power plant.
(5) Simplify the system (such as eliminating the main pipe anti-flick baffle and reducing the main pipe damper), save cost, improve the accessibility of maintenance and in-service inspection, and reduce the irradiation dose to staff.
(6) Display the progress and status of construction progress directly on the three-dimensional model, check the construction sequence and plan, show the difference of progress and plan, and provide support and service for the arrangement and optimization of construction plan.
(7) Check the three-dimensional layout of the system, check whether the interface is self-consistent, or check the three-dimensional space layout, set the best path, and shorten the overhaul period.

9.2 Development of Cables for Nuclear Power Plants

9.2.1 Development History of Cables for Nuclear Power Plants

In 1729, Grey discovered that electricity can be transmitted along metal wires. From then on, man had the concept of conductors. Later, Franklin of the United States invented the lightning rod to connect the earth with wires, and wires were put into practice for the first time. Because of the limitation of power sources, electricity and wires were not used effectively by human beings at that time. It was not until 1799 that the Italian Volt invented the battery to obtain a continuous current, and the source of electric energy was controlled, that the use of metal for conducting electricity was really discovered and utilized by humans. In the early nineteenth century, Oster of Denmark, Faraday of England, Ohm of Germany, and Henry of the United States discovered and founded many theories of modern electricity and electromagnetism, and the equipment for generating and utilizing electricity also came into being. In 1875, the Paris North Railway Station was built as the first power plant in the world. The nuclear power plant was built in the early 1950s. The world's first grid-connected nuclear power plant was built in 1954 in Obninsk, the former Soviet Union. From these early developments then nuclear power plants and cables for nuclear power plants were born.

During the construction of the first-generation nuclear power plant, cables did not have detailed classifications and special performance requirements except that the conductor specifications are roughly distinguished according to the actual current being carried. But at that time, the damage to the insulation and protective layer of cable caused by nuclear power environment had attracted the attention of designers.

The second-generation of nuclear power plants were built during the great development of commercial nuclear power plants, that is, from the 1960s to the late 1990s. At that time, cable extrusion materials such as cross-linked polyethylene (XLPE), EPR, polyvinyl chloride, low smoke halogen-free polyolefin, and so on emerged endlessly and were also continuously introduced to nuclear power plants as insulation and sheath materials for nuclear power cables. As countries rushed to build nuclear power plants, accidents such as the Three Mile Island nuclear accident and the Chernobyl nuclear leakage occurred one after another. The pace of building nuclear power plants was slowed down, and nuclear safety was formally put on the agenda. Nuclear safety-related equipment used in nuclear power plants must undergo rigorous accidental test appraisal before it can be used. Nuclear power cable as an important equipment running through the containment is also included in the appraisal. With the construction of the second-generation nuclear power plants and the second-generation improved nuclear power plants, the test standards for nuclear power cables have been constantly improved. The IEEE 383 [1] standard for qualifying class 1E electric cables, field splices and connectors for nuclear power generating stations, and NFM 64-001 Qualification Procedures for Electrical Equipment in Pressurized Water Reactor Containment under Accident Conditions and other qualification standards have been formed in this era. The concept of class 1E was put forward in the design and construction of the second-generation nuclear power plant. It refers to the safety level of the electrical systems and equipment of the reactor and nuclear power plant. These systems and equipment are necessary to complete the functions of emergency shutdown of the reactor, containment isolation, emergency cooling of the reactor core, waste heat discharge of the reactor and reactor plant, as well as to prevent a large amount of radioactive material from being discharged to the surrounding environment. The service life, radiation resistance, and nuclear accident resistance of class 1E cables are regulated by simulation tests in various standards. Only through these tests can they be called "class 1E cables for nuclear power plants" related to safety. They are also called "nuclear grade cables" or "nuclear safety grade cables" in the nuclear power industry, commonly known as "nuclear power cables." Thus, nuclear grade cables are separated from common cables used in nuclear power plants and have been studied as a new category of cables.

9.2.2 Classification of Cables for Nuclear Power Plants

9.2.2.1 Classification by Use Function

Cables in nuclear power plants are mainly used in nuclear reactor buildings, nuclear auxiliary buildings, turbine buildings, etc.; ducts, conduit, trays, or direct burial installation are generally used for cable laying, which requires cables of reliable service life, thermal stability, moisture resistance, chemical stability, and radiation resistance. In order to ensure the high reliability of the system design and avoid serious consequences caused by equipment damage, repetitive multiple independent line systems and devices are usually used. Usually, two sets of independent line systems are used for power cables and three sets of independent line systems are used for control cables. Similar to the application classification of standard cables, nuclear power cables can also be divided into power cables, control cables, instrument cables, measurement cables, communication cables, etc.

Medium voltage power cable: Mainly used for connection between medium voltage distributions panels in nuclear power plants, between medium voltage distribution panels and motors, and between medium voltage distribution panels and transformers.

Low voltage power cable: Mainly used for the connection of 0.6/1 kV rated power transmission and distribution system of nuclear power plants.

Control cable: Mainly used for control and signal connection in control system network of nuclear power plants.

Instrument cable: Mainly used to connect sensors with electrical cabinets, computers, and data processing systems in control system of nuclear power plants.

Signal cable: Mainly used to send or receive signal data for computer and data processing system in control system of nuclear power plants. An example is its use for connecting a thermocouple to a recorder.

Communication cable: Mainly used for transmission of calls, documents, data, and other electrical signals within nuclear power plants.

9.2.2.2 Classification by Safety Level

In nuclear power plants, there are special requirements for the cables connecting 1E equipment in the same functional cables, so the cables used in nuclear power plants can be divided into safety cables and non-safety cables according to the requirements of the safety level. Among them, the safety level cable (1E cable) includes K1, K2, K3 cables, and the non-safety level cable (non-1E cable) mainly includes non-critical (blow-out prevention) cables.

K1 cables: Cables installed in the containment of nuclear reactors and capable of performing their specified functions under normal environmental conditions and under the seismic load of safe shutdown as well as during or after any accident.

K2 cables: Cables installed in the containment of nuclear reactors and capable of performing their specified functions under normal environmental conditions and under the seismic load of safe shutdown.

K3 cables: Cables installed outside the containment of nuclear reactors and capable of performing their specified functions under normal environmental conditions and under seismic loads of safe shutdown.

The operating environment of the above three types of cables varies greatly, among which the K1 cables have the worst operating environment and the most stringent performance requirements. They can only be put into operation through simulated loss of coolant accident (LOCA) test. According to the actual operating environment of cables, when LOCA occurs in nuclear power plants, cables inside and outside containment will be severely tested. Therefore, some experts believe that for all the cables that are installed inside the nuclear reactor plant, LOCA test simulation shall have to be carried out. In fact, the design of cable structure and performance shall be based on the specific operating conditions of reactor buildings and auxiliary buildings.

9.2.2.3 Classification by Technical Types of Nuclear Power Plants

The technical types of nuclear power plants have different characteristics; especially the environment in the containment is quite different, and so the cables that meet the requirements of these nuclear power plants also have different technical characteristics. Usually it can be divided into "cable for EPR type nuclear power plant," "cable for AP1000 type nuclear power plant," etc.

9.2.2.4 Classification by Use Occasion

The energy of a nuclear power plant comes from atomic energy. The electrical equipment used in the acquisition process of nuclear fission or fusion reaction energy usually has higher requirements and the cables connecting the specific equipment also have various requirements, which necessitate the designer to carry out special design for specific environmental conditions. These cables, which are specially designed for special occasions within nuclear power plants, are usually named according to the environment in which they are used. For example, the super-soft flat cable for ring crane is called "ring suspension cable," and the flame-retardant DTS (data transmission system) communication cable for DTS system is called "DTS cable," etc.

9.2.3 Characteristics of Cables for Nuclear Power Plants

The most in-demand category of cables in nuclear power plants is class 1E K3 cables, which are mainly installed and laid outside the containment of nuclear power plants. Under normal conditions and seismic loads, they can meet the three basic safety functions of implementing or supporting reactivity control, waste heat removal, and radioactive material inclusion, and can prevent and mitigate accidents. They have low smoke, are halogen-free, and have low toxicity, corrosion resistance, and excellent electrical performance. Their service life is 40–60 years at an operation temperature of 90 °C (Figure 9.7 and Table 9.1).

Nuclear power plants have very stringent requirements for the operational environment, safety level, and working conditions of cables, so the performance requirements of cables are also high. Besides the conventional electrical and mechanical properties, there are also the following technical requirements.

(a)

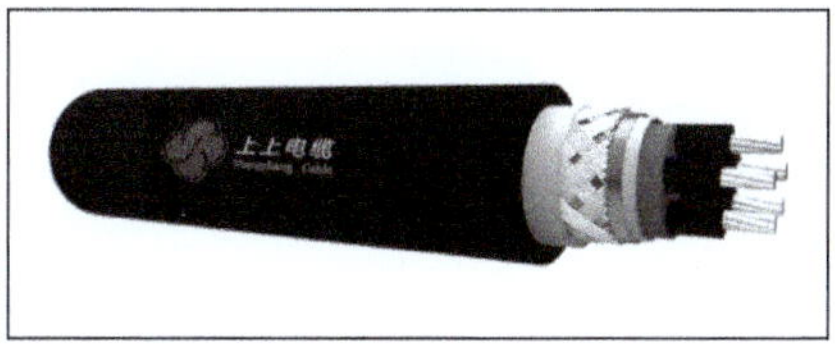

(b)

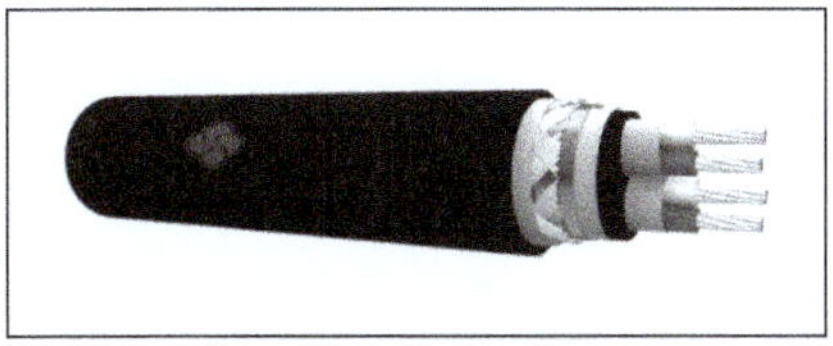

(c)

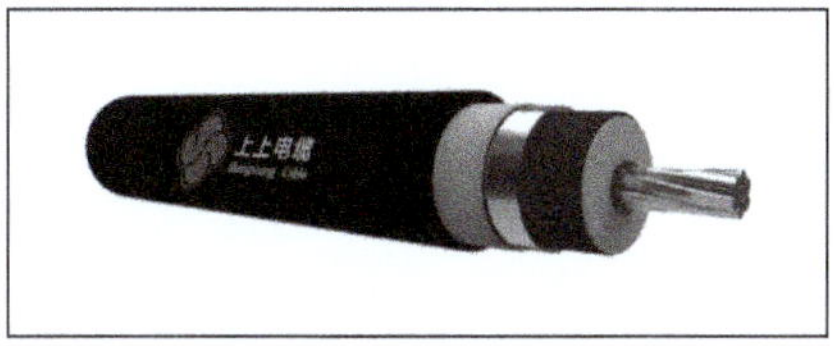

(d)

Figure 9.7 (a) Class 1E category K3 low voltage power cable, (b) Class 1E category K3 control cable, (c) Class 1E category K3 measurement power cable, (d) Class 1E category K3 medium voltage power cable. Source: © Jiangsu Shangshang Cable Group Co., Ltd.

(1) Combustion performance

According to the smoke density test method prescribed by IEC 61034, the light transmittance of the cable during combustion shall not be less than 60%. According to the test method for determining the pH value and conductivity of the aqueous solution prescribed by IEC 754-2, the pH value of the solution after combustion of the cable shall not be less than 4.3, and the conductivity shall not be less than 10 μS/mm. In addition, the cables shall pass the category B flame-retardant test specified in GB/T 18380.34 and the single vertical combustion test specified in GB/T 18380.12. Meanwhile, the insulation cores of some cables must pass the single vertical combustion test specified in GB/T 18380.12.

(2) Service life

The design life of nuclear power plants is generally 40 or 60 years, which requires that the service life of cables shall be at least 40 or 60 years. The thermal life evaluation tests specified in IEC 60216 and IEEE 383 standards are usually used for evaluating the service life of cables at different temperatures specific for various materials.

(3) Environment resistance

Heat resistance: Required to maintain normal operation at the highest ambient temperature and self-heating conditions. The normal working temperature of conductors required by cables in nuclear power plants is generally 90 °C (excluding cables inside AP1000 nuclear power plant containment). Therefore, cables need to have long-term heat-resistant performance.

Radiation resistance: During the operation of nuclear power plants, the cables will be exposed to sustained radiation dose. The cumulative dose of normal radiation is about 25 Mrad in 40 years. In the case of accident, the radiation dose can reach 60 Mrad (the radiation dose of third-generation nuclear power is higher in the case of accident and the radiation dose of

Table 9.1 Typical structure composition of cable for nuclear power plant.

Cable type	Power cable up to and including 6 kV	Power cable with voltage 1000 V	Control cable with voltage 1000 V	Measurement (instrument) cable	Communication cable
Conductor	Copper, tinned copper, or aluminum wire	Copper, tinned copper, or aluminum wire	Copper, tinned copper wire	Copper, tinned copper wire	Solid copper wire
Insulation	XLPE, EPR	Cross-linked polyolefin (XLPO), EPR	XLPO, EPR	XLPO, EPR	PE
Filler	Low smoke halogen-free flame-retardant material or glass fiber	Low smoke halogen-free flame-retardant material or glass fiber	Low smoke halogen-free flame-retardant material or glass fiber	Low smoke halogen-free flame-retardant material or glass fiber	—
Inner covering	Low smoke halogen-free flame-retardant material	Low smoke halogen-free flame-retardant material	Low smoke halogen-free flame-retardant material	Low smoke halogen-free flame-retardant material	Low smoke halogen-free flame-retardant material
Metallic screen	Copper tape wrapping	—	Copper, tinned copper wire braiding, or copper tape, copper–plastic tape wrapping	Copper, tinned copper wire braiding, or copper tape, copper–plastic tape wrapping	Copper, tinned copper wire braiding, or copper tape, copper–plastic tape wrapping
Armor	Copper tape (if any)	Steel tape (if any)	Steel tape (if any)	Steel tape (if any)	Steel tape (if any)
Outer sheath	Thermosetting or thermoplastic low smoke halogen-free flame-retardant material	Thermosetting or thermoplastic low smoke halogen-free flame-retardant material	Thermosetting or thermoplastic low smoke halogen-free flame-retardant material	Thermosetting or thermoplastic low smoke halogen-free flame-retardant material	Thermosetting or thermoplastic low smoke halogen-free flame-retardant material

HPR1000 in the case of accident will reach 145 Mrad). K1 cables shall also pass the LOCA test stipulated in IEEE 383 and other relevant national and international standards.

Humidity resistance: In the case of chemical spraying under accident, and in the case of loss of coolant, the cable will work under high temperature and high pressure, which requires the cable to work normally at 100% relative humidity and steam.

Chemical stability: Cables can withstand chemical environments such as resistance to oil, acid, alkali, and ozone.

(4) Low toxicity

Toxicity index is the summation of the toxicity factor (the concentration of gas produced under the conditions of the test, scaled up for 100 g of material, and recalculated as though the combustion products were diffused into a volume of 1 m^3/the concentration of a gas considered fatal to humans for a 30 minute exposure) of selected gases produced by the combustion of a material under the conditions specified in the test.

According to the toxicity index measurement method of the British military standard NES 713, it is required that the measured value shall be not more than 5.

9.2.4 Standards for Nuclear Cables

Owing to the different technical types of nuclear power plants, the manufacturing standards of nuclear cables in different countries have also formed different standards systems. Some laws and regulations, qualification procedures, and product test standards that can be used as reference standards for nuclear cable manufacturing are listed in Table 9.2.

9.2.5 Qualification and Test of Nuclear Cables

9.2.5.1 Significance of Class 1E Equipment Qualification in Nuclear Power Plants

The purpose of equipment qualification is to verify that it can perform its expected safety function under normal and accident loading. To maintain the safety of nuclear power plant and make its system and equipment have the necessary reliability, redundancy and diversity are adopted in the design of nuclear safety system. By providing a multichannel (redundant) design, random failures in one channel of the system will not prevent the realization of its security functions, because other independent channels can perform the same functions. This design method of nuclear power plants requires the assurance that redundant systems will not be unavailable at the same time by common cause failures. The so-called common cause failure refers to the equipment or system failure due to the same cause, which may arise from the design, manufacture, operation, environment, or human factors on the equipment.

An important way to prevent common cause faults induced by environmental factors (including normal operating conditions, LOCA operating conditions,

Table 9.2 Standards of nuclear cables (examples).

No.	Standard no.	Name of standard
Nuclear safety laws and regulations		
1.	HAF 003	Quality assurance safety regulations for nuclear power plants
2.	HAF 601	Regulations on design, manufacture, installation and NDI supervision and management of civil nuclear safety equipment
3.	RCC-E	Design and construction rules for electrical components of nuclear islands
Qualification procedure of nuclear (class 1E) cables		
4.	NFM 64-001	Procedure for qualification of electric equipment installed in containments for pressurized water reactors subject to accident conditions
5.	IEEE 383	Standard for qualifying class 1E electric cables and field splices for nuclear power generating stations
6.	IEC 60780	Nuclear power plants – electrical equipment of the safety system – qualification
7.	RG1.131	Qualification tests of electric cables, field splices, and connections for light water cooled nuclear power plants
8.	GB/T 22577	Class 1E cables for nuclear power generating stations – general requirements
9.	GB/T 12727	Nuclear power plants – electrical equipment of the safety system – qualification
10.	EJ/T 1197	Test method and environment condition for qualification of electrical equipment of the safety system in nuclear power plants
11.	EJ/T705	Type test of class 1E electric cables and field splices for nuclear power plants
Basis of cable product and testing standard		
12.	CST 74C 06800	Technical specification of electric cables for nuclear power plant
13.	GB/T 3956	Conductors of insulated cables
14.	GB/T 2951	Common test methods for insulating and sheathing materials of electric and optical cables
15.	GB/T 12706	Power cables with extruded insulation and their accessories for rated voltages from 1 kV ($U_m = 1.2$ kV) up to 35 kV ($U_m = 40.5$ kV)
16.	GB/T 3048	Test methods for electrical properties of electric cables and wires
17.	GB/T 18380.12	Test on electric and optical fiber cables under fire conditions – part 12: test for vertical flame propagation for a single insulated wire or cable – procedure for 1 kW pre-mixed flame

Table 9.2 (Continued)

18.	GB/T 18380.34	Test on electric and optical fiber cables under fire conditions – part 34: test for vertical flame spread of vertically mounted bunched wires or cables – category B
19.	GB/T 19216	Tests for electric cables under fire conditions – circuit integrity
20.	GB/T 17650	Test on gases evolved during combustion of materials from cables
21.	GB/T 17651	Measurement of smoke density of cables burning under defined conditions
22.	ASTM D2863	Standard Test Method for Measuring the Minimum Oxygen Concentration to Support Combustion of Plastics
23.	IEC 60502	Power cables with extruded insulation and their accessories for rated voltages from 1 kV(U_m = 1.2 kV) up to 30 kV (U_m = 40.5 kV)
24.	IEC 60332	Test for vertical flame spread of vertically mounted bunched wires or cables
25.	DIN VDE 0266	Power cables with improved characteristics in the case of fire – nominal voltages 0.6/1 kV
26.	DIN VDE 0472-814	Test of cables, wires, and flexible cords, continuance of isolation effect under fire conditions
27.	DIN VDE 0250-812	Cables, wires, and flexible cords for power installation Rubber insulated flexible cable

and earthquake operating conditions) is to qualify equipment related to nuclear safety.

9.2.5.2 Method of Class 1E Equipment Qualification in Nuclear Power Plants [2]

At present, the commonly used qualification methods of nuclear-grade equipment include testing, calculation, analysis, and a combination of all methods.

(1) *Qualification by testing*: The test can check the various performance of the equipment and verify the operation performance of nuclear grade equipment. The testing method is to carry out a series of tests on one or a class of typical equipment installed in the field. The parameters of these tests can simulate the environmental parameters of operating conditions, and can cause the same failure as those easily occurring in operation.
(2) *Qualification by calculation*: Calculation is an important way to obtain equipment parameters. In the process of qualifying specific equipment, a series of data calculations are is needed. It is acceptable to use this method to verify the impact of load on specific equipment. Generally, this method is mainly used for the qualification of mechanical equipment.
(3) *Qualification by analysis*: Qualitative or quantitative reasoning method is used to prove the ability of equipment to perform its safety function. This method is often used to qualify equipment that cannot be tested due to size. For example, the evaluation of the anti-seismic function of vertical pumps can accurately feedback the safety margin of pumps under earthquake

conditions by investigating the operability, integrity, and stability of pumps based on quantitative reasoning method.

(4) *Qualification by a combined method*: For some complicated nuclear grade equipment, if a single qualification method is used, it will not be able to complete all the equipment qualification, so it is necessary to integrate various methods for qualification. For example, a combination of test method and analysis method can be used for this kind of equipment. Some functions or components need the test method, while others need the analytical method.

(5) *Qualification by analogical method*: Logical reasoning is used to prove that the equipment to be qualified is similar to the equipment that has been previously qualified.

(6) *Qualification by experience*: This method involves inferring the ability of similar items used in nuclear power plants to perform their safety functions based on the performance records of a certain item in conventional industry. Usually, this method can be used for seismic qualification of equipment, because there are a large number of seismic records. Experience method is not blind subjective judgment; it must have sufficient data records for reference; commonly used data are production records, seismic records, fault records, etc.

9.2.5.3 Content and Difference of Class 1E Cable Qualification Test for Nuclear Power Plants

Class 1E cables used in nuclear power plants are usually qualified by test method and analogical method. The specific qualification test items and methods for class 1E cables generally used in nuclear power plants are listed in Table 9.3.

Qualification test items of class 1E cable for mainstream nuclear power plants are listed in Table 9.4.

9.2.5.4 Key Points for Qualification of Class 1E Nuclear Power Cables

The function of quality qualification is to ensure that class 1E cables can work normally under specified conditions, and there is no mechanical defect that can cause general faults. In China, class 1E nuclear power cables must be qualified by testing in accordance with the relevant procedures, and has promulgated the "National Nuclear Safety Administration of civil nuclear safety equipment design and manufacture license" according to the relevant laws and regulations of the state and the nuclear safety regulatory requirements. The quality qualification test of 1E cables used in nuclear power plants involves the key points of selecting representative test simulators, evaluating cable life aging performance, and qualification test sequence. The following is a brief description of each test point.

9.2.5.4.1 Representativeness of Cable Samples A cable or a group of cables is used during qualification testing to represent a number of cable styles. The representative cables shall contain the following characteristics of the cable styles: (i) manufactured by a specific manufacturer using the same processes and controls; (ii) contains the same materials, including insulation, jacket, filler, binder tape, shields, and factory splices (if appropriate); (iii) the same or higher service rating; (iv) the same or higher service rating, (v) the same or higher volts per mil

Table 9.3 Qualification test items for class 1E cables.

No.	Test item	Test type	Test method (standards for reference)
Part 1: Characteristics and performance test of cable			
1.	Cable construction and dimension	R,T	GB/T2951
2.	Conductor resistance	R,T	GB/T3048.4
3.	Working capacitance for instrumentation cable	R,T	GB 5441.2
4.	Unbalance capacitance for instrumentation cable	R,T	GB 5441.3
5.	Partial discharge test for medium voltage power cable	R,T	GB/T3048.12
6.	Voltage test	R,T	GB/T3048.8
7.	Bending test, followed by a partial discharge test for medium voltage power cable	T	GB/T 12706.2
8.	Tanδ measurement for medium voltage power cable	T	GB/T 12706.2
9.	Heating cycle test, followed by a partial discharge test for medium voltage power cable	T	GB/T 12706.2
10.	Impulse test, followed by a power frequency voltage test for medium voltage power cable	T	GB/T 3048.8
11.	Voltage test for 4 h	T	GB/T 12706.2
12.	Resistivity of semi-conducting screen for medium voltage power cable	T	GB/T 12706.2
13.	Strippability of insulation screen for medium voltage power cable	T	GB/T 12706.2
14.	Tests for determining the mechanical properties of insulation before and after aging	S,T	GB/T 2951.11
15.	Hot set test of insulation	S,T	GB/T 2951.21
16.	Aging test of insulation in air bomb for rubber material	T	GB/T 2951.12
17.	Ozone resistance test of insulation for rubber material	T	GB/T 2951.21
18.	Water absorption test on insulation material	T	GB/T 2951.13
19.	Tests for determining the mechanical properties of sheath before and after aging	S,T	GB/T 2951.11
20.	Oil immersion test of sheath	T	GB/T 2951.21
21.	Water absorption test of sheath	T	GB/T 2951.13
22.	Hot set test of sheath	S,T	GB/T 2951.21
23.	Aging test of sheath in air bomb for rubber material cable	T	GB/T 2951.12
24.	Pressure test at high temperature on sheath	T	GB/T 2951.31

(*continued*)

Table 9.3 (Continued)

No.	Test item	Test type	Test method (standards for reference)
25.	Tests on sheath at low temperatures	T	GB/T 2951.14
26.	Crack resistance test of sheath	T	GB/T 2951.31
27.	Toxicity index of insulation and sheath	T	Def. stan. 02-713
28.	Differences between oxygen index	T	ASTM D2863
29.	Smoke release test	T	GB/T 17651
30.	Fluorine content test	T	IEC 60684-2
31.	Test on gases evolved during combustion	T	GB/T 17650
32.	Vertical flame test of insulated cores	S,T	GB/T 18380.12
33.	Bunched flame test	S,T	GB/T 18380.34
34.	Durability of printed marking	T	GB/T 6995.1
35.	Cable additional aging test (compatibility test)	T	GB/T 2951.12
Part 2: Class 1E functional verification test of cables			
36.	Accelerated aging test for equivalent life operation for K1, K2, K3 cables	T	IEEE 383 GB/T 22577 GB/T 18380.34
37.	Conventional radiation aging test for K1 and K2 cables	T	IEEE 383 GB/T 22577
38.	Accident radiation aging test for K1 cables	T	IEEE 383 GB/T 22577
39.	Accident simulation test for K1 cables	T	IEEE 383 GB/T 22577
40.	Immersion test (if any) for K1 cables	T	IEEE 383 GB/T 22577 Detailed technical specification

operating level; and (vi) construction or configuration features (e.g. number or type of conductors).

In addition, the selection of cable samples should also refer to the supply scope, nonmetallic content, combustible content, and other parameters. It should cover all the supply scope and include the principles of maximum core number, minimum core number, maximum cross-section area, minimum cross-section area, and maximum combustible content.

9.2.5.4.2 Cable Lifetime Aging Performance Evaluation

Basic Principles of Accelerated Aging of Materials Since the Fukushima accident, the standard of safe operation of nuclear power plants has become higher. Therefore, the National Nuclear Safety Administration (NNSA) has strict requirements on the design and manufacture of various nuclear power equipment and materials. Among them, the conductor, insulation, and sheath of cables used in nuclear power plants are required to be proved to maintain their mechanical and electrical properties just after they have been put into operation and after

Table 9.4 Qualification test item of class 1E cable for mainstream nuclear power plant.

Qualification test item	EPR	AP1000	M310	Water–water energetic reactor (VVER)
Cable characteristics	(1) Cable major requirements are based on CST 74 C06800 (2) Design life 60 yr (3) Low smoke halogen free (4) Flame retardant (bunched category B) (5) Classified to K1, K2, K3	(1) Cable major requirements are based on IEC 0502, combined with the requirement in IEEE 383 (2) Design life 60 yr (3) Low smoke halogen free (4) Flame retardant (bunched category B) (5) Divided into outside containment and inside containment	(1) Cable major requirements are based on GB/T 12706, combined with the requirement in RCC-E (2) Design life 40 yr (3) Low smoke halogen free (4) Flame retardant (bunched category B) (5) Classified to K1, K2, K3	(1) Cable major requirements are based on IEC 0502, combined with the requirement in IEEE 383 (2) Design life 40 yr (3) Low smoke halogen free (4) Flame retardant (bunched category B) (5) Classified to K0, K1, K2, K3
Design specification and standard	(1) CST 74 C06800 (2) RCC-E-2005 (3) Technical specification	(1) IEC 60502-2009 (2) IEEE 383-1974 (3) Technical specification	(1) GB/T 12706-2008 (2) GB/T 18380-2008 (3) RCCE-1993 (4) Technical specification	(1) IEC 60502 (2) IEC 60332 (3) IEC 60331 (4) IEEE 383 (5) Technical specification
Standard for conventional test item	IEC 60502	IEC 60502	GB/T 12706	IEC 60502
Standard for material life evaluation	IEC 60216	IEC 60216	IEC 60216	IEC 60216
Standard for thermal aging	RCC-E fixed condition	IEEE 383 need to deduce by oneself	RCC-E fixed condition	IEEE 383 need to deduce by oneself
Normal radiation aging	RCCE	IEEE 383	RCCE	IEEE 383
Accident radiation aging	RCCE	IEEE 383	RCCE	IEEE 383
Design basis accident (DBA)	RCCE	Technical specification	RCCE	Technical specification
Immersion	No	Yes	No	No

40 years (60 years) of use. If it is not possible to verify the mechanical and electrical properties of the materials after 40 or 60 years of cable construction, the NNSA will not agree to the construction or operation of nuclear power plants.

Since it is impossible to wait for cables to be inspected after decades of operation, it is necessary to simulate the aging condition of cables in a specific way, and then carry out the above tests on the aged cables to prove whether the cables can meet the requirements. The performance guarantee of nuclear grade cables mainly depends on the insulation material. Therefore, the accelerated aging of insulation is simulated by Arrhenius equation, and then tested. From the elongation at break the extrapolated lifetime of cable insulation can be calculated. If the cable can meet the requirements after accelerated aging, the durability of the cable can be proved.

For cable insulation material, its life mainly consists of three basic factors: (i) Polymer structure: cable insulation material is a mixture of organic polymers, including the most basic polymers, copolymers, and additives, which complicate the material characteristics. (ii) Environmental conditions for storage and operation of polymers: the activation energy, thermal aging rate, and radiation tolerance dose of mixtures prepared with various components are different, and the environmental conditions for operation will be different with different laying positions and modes. (iii) Working hours: long-term operating conditions of cables, environmental stress pulse intensity, and frequency under accident conditions, etc.

The main manifestations of material aging are (i) decrease of elongation at break and tensile strength, (ii) increase of hardness or modulus of elasticity, (iii) increase of density, and (iv) increase of dielectric loss. In most cases, the change in electrical performance is very small, and the change in mechanical performance results in failure of the cable, and insulation cracking occurs prior to the electrical failure.

Elongation at break (EAB) is used to measure the flexibility of materials and is defined as the percentage of elongation growth of materials from tension to fracture. Many years of experience at home and abroad have shown that EAB of cable insulation material is the most reliable characteristic parameter reflecting the life state of a cable. Generally, when the elongation at break of the material after aging is 50% of the absolute elongation before thermal exposure, the material reaches the end of life, and the lifetime of the material is estimated by the least square method. IEC 60216.2 specifies the thermal aging of polymer materials, and its theoretical basis is generally based on Arrhenius equation, which is used to carry out aging tests at aging temperature. Four temperature aging points are selected according to Arrhenius equation to carry out experiments, and then the Arrhenius diagram of 1E class cable is obtained by connecting the points in the experiment (as follows). According to this principle, the complete life of cable can be calculated and the residual life can also be calculated (Figure 9.8) [3].

The unit of transverse axis is the aging time, and the unit of longitudinal axis is the breaking elongation of cable. The aging failure can be judged by 50%.

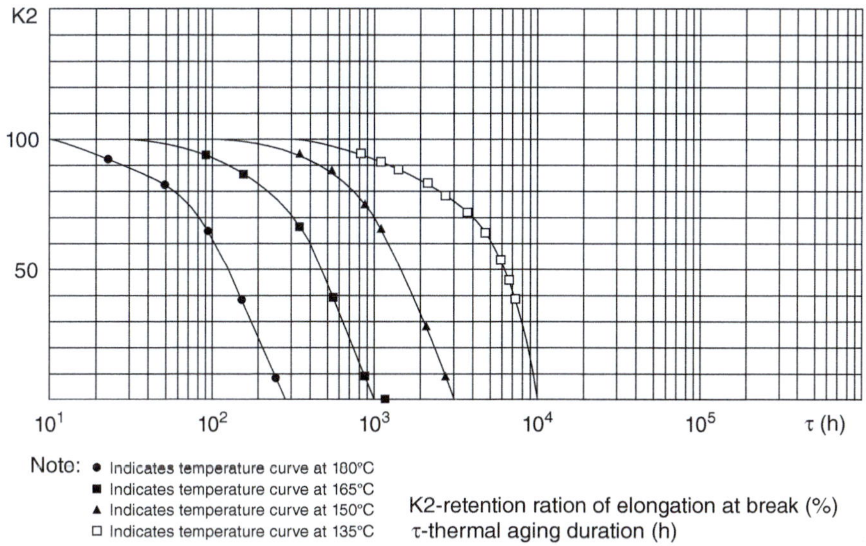

Figure 9.8 Aging life evaluation cable materials.

The aging temperature selection principle:

(1) When calculating the thermal life, the lowest aging temperature should make the average failure time not less than 5000 hours, and the extrapolation of the heat-resistant wire should not exceed 25 K.
(2) The maximum aging temperature should be the temperature at which the average failure time measured is not less than 100 hours.
(3) If the aging mechanism is expected to be the same in the whole aging range, the temperature difference between adjacent aging temperatures should be equivalent, usually 15 K.

Cable Thermal Aging Evaluation Method Select the material's accelerated aging temperature, and then calculate the accelerated aging duration under the determined accelerated aging temperature. Generally, cable insulation or sheath material of smaller activation energy is selected for cable thermal aging evaluation.

(1) Arrhenius equation

$$\tau = A \times \exp\left(\frac{E_a}{RH}\right) \tag{9.1}$$

where

τ	Duration for material change
H	$\theta + \mathrm{H}_0$ insulation temperature
θ	Celsius temperature (°C)
H_0	273.15 K
A	Constant related to material and diagnosis test
E_a	Material activation energy (eV)
R	Boltzmann constant, 0.000 086 17

Equation (9.1) is changed to

$$\ln \tau = \ln A + \frac{E_a}{RH} \tag{9.2}$$

Substitute the qualification lifetime τ_2 and working temperature H_2 in 4.2:

$$\ln \tau_2 = \ln A + \frac{E_a}{RH_2} \tag{9.3}$$

Substitute the accelerated aging time τ_1 and working temperature H_1 in 4.2:

$$\ln \tau_1 = \ln A + \frac{E_a}{RH_1} \tag{9.4}$$

Subtracting Eq. (9.3) from Eq. (9.4),

$$\ln \tau_1 - \ln \tau_2 = \frac{E_a}{RH_1} - \frac{E_a}{RH_2} \tag{9.5}$$

Equation (9.5) can be changed to

$$\tau_1 = \tau_2 \times \exp\left(\frac{E_a}{RH_1} - \frac{E_a}{RH_2}\right) \tag{9.6}$$

(2) Material activation energy data

$$E_a = M\ (\text{kJ/mol}) = M/96.485\ (\text{eV})$$

$$1\ \text{eV} = 96.485\ (\text{kJ/mol});\ 1\ \text{eV} = 1.602 \times 10^{-22}\ (\text{kJ}) \times 6.023 \times 10^{23}$$

$$6.023 \times 10^{23}\ \text{per mol}$$

Activation energy data of insulating materials is provided by the material suppliers:

$$E_a = \text{M}\ (\text{kJ/mol}) = M/96.485\ (\text{eV})$$

$$1\ \text{eV} = 96.485\ (\text{kJ/mol});\ 1\ \text{eV} = 1.602 \times 10^{-22}\ (\text{kJ}) \times 6.023 \times 10^{23}$$

6.023×10^{23} is the Avogadro constant, which refer to the

quantity of atoms per mol.

(3) Other parameters of cable aging

Material working temperature $T_2, H_2 = 273.15 + T_2$

Accelerated aging temperature $T_1, H_1 = 273.15 + T_1$

(Accelerated aging temperature shall be reasonably selected according to heat resistance and test time of materials.)

Material life to be qualified is 60 years, $\tau_2 = 60 \times 365 \times 24 = 525\,600$ hours

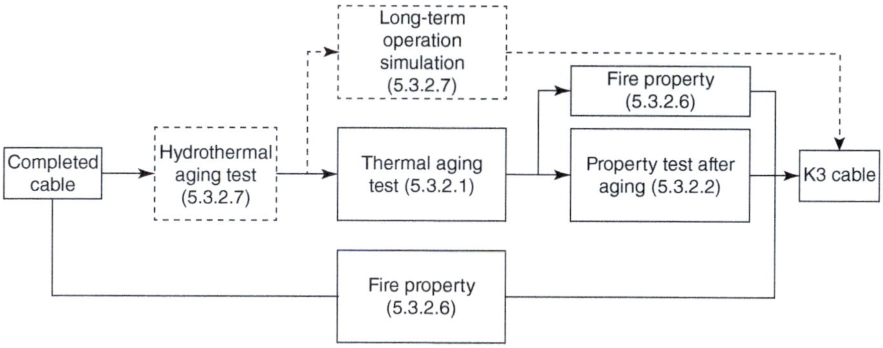

Figure 9.9 Quality qualification test process for category K3 cable.

(4) Calculation of accelerated aging time
Substitute material activation energy data and other parameters of cable aging in Eq. (9.6)

$$\tau_1 = 525\,600 \times \exp(E_a \cdot - 4.855)$$

60-year life : 155 °C accelerated aging duration h,

$$\tau_1 = 525\,600 \times \exp(E_a \cdot - 4.855)$$

$$\tau_1 = 525\,600 \times \exp(E_a \cdot - 4.535)$$

60-year life : 150 °C accelerated aging duration h,

$$\tau_1 = 525\,600 \times \exp(E_a \cdot - 4.535)$$

9.2.5.4.3 *Qualification Tests Sequence* Qualification tests for class 1E cable sample are divided into cable characteristics, performance verification tests, and function verification tests. Characteristics and performance verification tests mainly include routine performance tests such as routine mechanical and physical properties, electrical properties, and low smoke halogen-free flame-retardant properties. Function verification tests mainly include thermal aging, irradiation aging, and Data Base environment (DBE; Rated Continuous Current during the Most Severe DBE Environmental Conditions For rating of conductors, refer to NFPA-70 [4] and IEEE 317 [5]) (Figures 9.9–9.11).

9.2.6 Research on Nuclear Cable Compounds

9.2.6.1 Application of Available Compounds

The performance of nuclear cable compounds requires low smoke, halogen-free, and flame retardant characteristics, and thus polyvinyl chloride (PVC) materials often used in civil cables cannot be used in nuclear cables. It is because chloride reacts with hydrogen in the air to form gases such as hydrogen chloride (HCl) when PVC cables are on fire. When people inhale these gases, they will

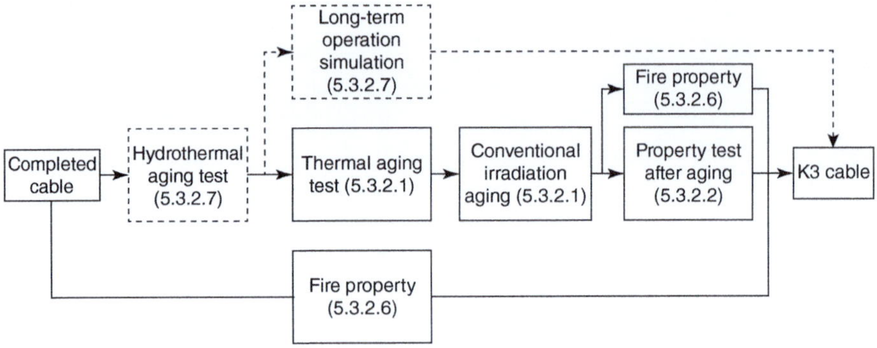

Figure 9.10 Quality qualification test process for category K2 cable.

suffocate to death. In addition, HCl is an acid substance that will cause losses to electrical equipment and steel structures of buildings by causing corrosion, and then by failure further increase the hazards of fire. Particularly, HCl will impact communication systems and control systems, causing greater losses. The safety protection systems such as reactor protection system and distributed control system (DCS) control system play an important role in nuclear power plants. Once the safety protection systems lose their protection functions, the safety of nuclear power plants will not be guaranteed.

According to the above introduction, the following characteristics for the compounds used in nuclear cables must be tested: smoke density, toxicity, corrosiveness, flame retardancy, etc., with nuclear cables being designed per those properties.

(1) *Smoke density*: As the name implies, it is a quantitative measured value of smoke generated during combustion. The value can be obtained by measuring the smoke generated during cable combustion by using specific light and the light transmission reduction after 1 m of cable being burnt in a specific space, such as a room of 100 m^2. Generally, the total accumulation of smoke – D_m is used to express this value. If D_m exceeds a certain range, it shows that the performance of nuclear cable cannot meet the requirements. If the cable is manufactured with this raw material, it will cause a lot of smoke in the case of fire, which will cause loss of life through smoke inhalation and equipment contamination.
According to the above analysis, the main reason for smoke density is HCl. Therefore, in order to ensure smoke density of nuclear cables to meet the design requirements, halogen-free compounds should be selected as insulation materials.
(2) Toxicity and corrosiveness: According to the above analysis, toxicity and corrosiveness are also caused by HCl, so HCl shall be eliminated in the latest design of nuclear cables.
(3) *Flame retardant*: The guarantee of flame-retardant performance is to ensure that nuclear cables do not support further growth of fire in the event of fire. It should be noted here that flame retardancy does not represent fire resistance and nuclear cables may not be required as fire survival cables; nuclear cables

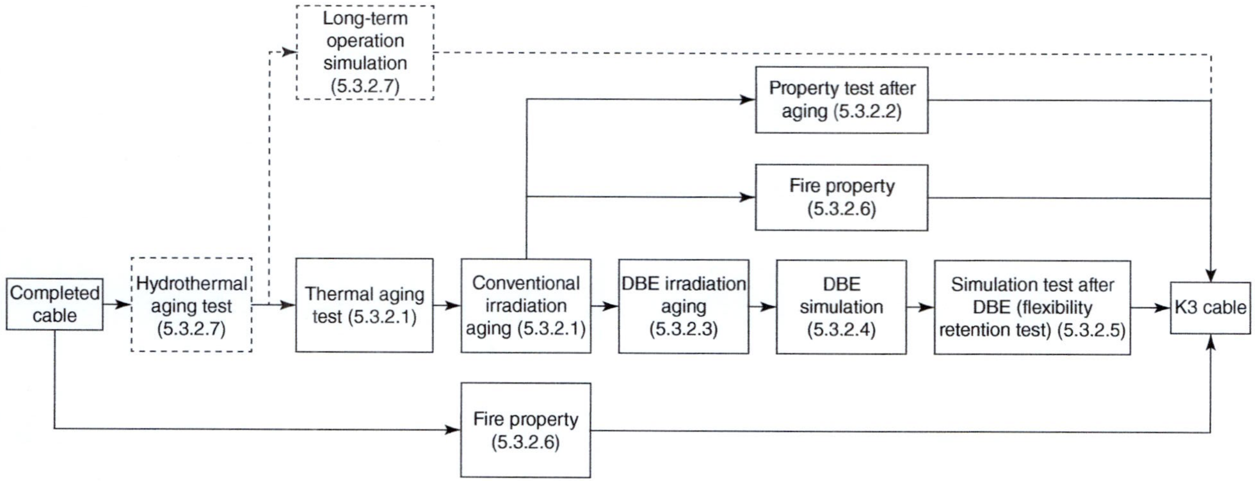

Figure 9.11 Quality qualification test process for category K1 cable.

are permitted to burn but the fire must not spread by flaming droplets from the cable. Therefore, the flame retardancy of nuclear cables is enabled by the addition of flame retardants.

(4) *Thermal resistance*: Because of the design of nuclear power plant, the temperature of the primary loop of nuclear power is above 200 °C. At such high temperatures, the nuclear cable must keep normal operation, and meanwhile, it must have a service life of 40 or 60 years. This characteristic must be considered in the design stage.
(5) *Irradiation resistance*: Once an accident occurs in a nuclear power plant, the greatest threat is that it will produce a large number of nuclear radiations, which will have an inestimable impact on people and the environment. The nuclear cables laid inside the nuclear power plant are subjected to a large amount of radiation at all times, which will make the materials brittle, and mechanical properties such as elongation at break and tensile strength will decline. Therefore, the materials used for nuclear cables must be able to work under long-term nuclear radiation, that is to say, anti-irradiation additives (anti-rads) should be added to cable compounds.
(6) *Loss of cooling accident LOCA resistance*: Generally, the most important thing in nuclear power plants is the primary loop. If there is no accident in the loop, the safety of nuclear power will not be a major problem, and the most common accident in the loop is the LOCA. When LOCA occurs, the original high temperature and high pressure water vapor in the primary loop will leak out. In this water vapor, there are also various chemicals with strong reactive elements and high corrosivity. At the same time, the nuclear fuel in the nuclear power plant may leak out, which will cause extremely high radiation dose, far beyond that in normal operating conditions. At this time, the nuclear cables must be capable of maintaining the performance of operation to start all safety protection measures. This LOCA survival requirement must be taken as a key target of nuclear cables.

On the basis of the operation of various nuclear power plants at home and abroad, various cable research institutes in China have carried out an analysis and research on the compounds that can be used to manufacture nuclear cables. After several rounds of screening and comparison, rubber or plastic compounds are usually used as cable materials in nuclear power projects that are running and under construction. So far, EPR and XLPE are used in nuclear cables in China.

Ethylene propylene rubber is a synthetic rubber that has high chemical stability, and can be used as a mature industrial product after various tests. It has good performance of resistance to thermal aging, radiation, and high-pressure vapor under LOCA accident. Meanwhile, EPR is suitable for insulation materials for its strong ability to withstand voltage with regard to the electrical performance of insulation. However, the rubber EPR itself is not flame retardant, so according to the above analysis, it is necessary to add flame-retardant agents to rubber and make it into flame-retardant products to meet the above performance requirements.

Cross-linked polyethylene is a polymer in which flame-retardant and anti-irradiation agents are added, and then processed by a technology to make the polyethylene cross-linked, so as to achieve the performance and standards of nuclear cables. With the development of China in the past 20 years, the manufacturing process of XLPE is the most advanced in the world. The nuclear cables produced abroad cannot satisfy the single vertical combustion test. At present, only those from China can satisfy this requirement.

9.2.6.2 Discussion on Future Materials

(1) Polyetheretherketone
Poly-ether-ether-ketone (PEEK) is one of the most important commercial poly-aromatic ether resins. Because of the rigid phenyl ring in its molecular structure, PEEK has excellent high temperature properties, mechanical properties, electrical insulation properties, and radiation and chemicals resistant properties. The ether bonds in the molecular structure make it flexible and easy to shape and process. Increasing the ratio of keto group to ether group in molecular structure can significantly improve the heat resistance grade of PEEK. A key point for PEEK is that it has excellent radiation resistant property, which can withstand a high radiation dose and its products can maintain excellent performance after 1100 Mrad irradiation, which is far superior to other general radiation resistant materials. Verified by the National Cable Research Institutes, the main companies that can manufacture PEEK that can meet the future development needs are Victrex, Solvey, and Evonik (Table 9.5).

(2) Polyamide
Polyimide (PI) is an aromatic heterocyclic polymer with repetitive imide groups in the main chain of molecules. PI has excellent thermal stability, excellent insulation performance, good mechanical properties, good environmental stability, and radiation resistance, and is one of the ideal insulation materials. The main varieties of PI suitable for melt extrusion are thermoplastic polyimide (TPI), polyetherimide (PEI), polyamide imide (PAI), etc.

(a) Thermoplastic polyimide
Because of its linear molecular structure, TPI can be melted and processed, which enlarges the application scope of PI. At present, the commercialized TPI is mainly composed of Mitsui's Auram series resins and SABIC's Extem series resins. Auram is a crystalline PI with high glass transition temperature (T_g 250 °C) and high melting point (T_m 388 °C). Extem series resins produced by SABIC are amorphous polymers. The highest T_g grade is Extem XH series for which T_g is 267 °C, the softening temperature is 260 °C, and the T_m is between 380 and 410 °C (Table 9.6).

(b) Polyetherimide
Polyetherimide has the conventional properties of PI resin, because of the introduction of flexible ether bond groups in the main chain, which weakens the intermolecular force and reduces the steric hindrance, providing it with more fluidity and processability than ordinary PI. At the same time, the flexibility of PEI is improved by introducing block copolymerization

Table 9.5 Basic performance table of commercial PEEK.

No.	Items	Testing method	Units	VICTREX® 450G	Solvey KetaSpire® KT-851	Evonik Vestakeep® 1000G
1	Proportion	ASTM D792	—	1.30	1.30	1.30
2	Tensile strength	ASTM D638	MPa	100	96	100
	Tensile modulus	ASTM D638	MPa	4100	3600	3700
3	Elongation at break	ASTM D638	%	45	30	30
4	Coefficient of thermal expansion	ASTM D696	1/°C	4.5e−5	4.3e−5	6e−5
5	dielectric constant, 1kHz	ASTM D150	—	3.2	—	2.9
6	Dielectric loss angle, 1 kHz	ASTM D150	—	0.003	—	0.003
7	Volume resistivity	ASTM D257	Ω cm	e16	e17	e15
	Dielectric strength, 2 mm	ASTM D149	kV/mm	23	—	16
8	Limit oxygen index	ASTM D2863	%	35	36.8	38
9	Flame retardancy	UL-94	—	V-0	V-0	V-0

of silicon and oxygen bonds into the main chain of PEI, and the application of PEI in cable is expanded. PEI is the most widely used PI plastics in the cable field. At present, the commercial PEI resin products are mainly ULTEM series resin of SABIC Company (Table 9.7).

(c) Polyamide imide

The molecular structure of PAI has both heat-resistant aromatic heteroamine group and flexible amide group. The Tg of PAI ranges from 278 °C to 285 °C. Compared with most engineering plastics, PAI has higher strength and is suitable for temperature change from low temperature to high temperature of about 232 °C. PAI has excellent properties of heat resistance and flame resistance. It also has high thermal deformation temperature, reaching 274 °C, and low linear expansion coefficient. At present, the commercialized PAI resin products are mainly Torlon® series resin of Solvey Company (Table 9.8).

(3) Other materials

In addition to special polymer materials such as polyaryl ether ketone and PI, polyphenylsulfone (PPSU) in polysulfone materials has good mechanical properties, high temperature resistance, steam resistance, radiation resistance, and flame-retardant properties due to its molecular structure of phenylene, sulfonyl, and ether groups.

Table 9.6 Basic performance tables of two commercialized TPI.

No.	Items	Testing method	Units	AURUM® PL450C	SABIC EXTEM™ XH1015
1	Proportion	ASTM D792	—	1.33	1.31
2	Tensile strength	ASTM D638	MPa	92	96
3	Elongation at break	ASTM D638	%	90	10
4	Coefficient of thermal expansion	ASTM D696	1 per °C	5.5e−5	5e−5
5	Dielectric constant, 1 kHz	ASTM D150	—	3.2	3.1
6	Dielectric loss angle, 1 kHz	ASTM D150	—	0.0009	0.02
7	Volume resistivity	ASTM D257	Ω cm	e17–e18	e16
8	Limit oxygen index	ASTM D2863	%	47	47
9	Flame retardancy	UL-94	—	V-0	V-0

Table 9.7 Basic performance table of PEI resin for cable.

No.	Items	Testing method	Units	SABIC ULTEM™ STM1500
1	Proportion	ASTM D792	—	1.18
2	Tensile strength	ASTM D638	MPa	28
3	Elongation at break	ASTM D638	%	100
4	Coefficient of thermal expansion	ASTM D696	1 per °C	5e−5
5	Melt index (295 °C/6.6 kg)	ASTM D1238	g/10 min	12
6	Dielectric loss angle, 100 kHz	ASTM D150	—	0.0056
7	Volume resistivity	ASTM D257	Ω cm	4.1e16
8	Dielectric strength, 3.2 mm	ASTM D149	kV/mm	16.3
9	Limit oxygen index	ASTM D2863	%	46

Table 9.8 Basic performance table of PAI for cable.

No.	Items	Testing method	Units	Solvey Torlon® 4203L
1	Proportion	ASTM D792	—	1.42
2	Tensile strength	ASTM D1708	MPa	192
3	Tensile modulus	ASTM D1708	MPa	4900
4	Elongation at break	ASTM D1708	%	15
5	Coefficient of thermal expansion	ASTM E831	1 per °C	3.1e−5
6	Melt index (295 °C/6.6 kg)	ASTM D1238	g/10 min	12
7	Dielectric constant, 1 MHz	ASTM D150	—	3.90
8	Dielectric loss angle, 1 MHz	ASTM D150	—	0.031
9	Volume resistivity	ASTM D257	Ω cm	2.0e17
10	Dielectric strength, 1 mm	ASTM D149	kV/mm	23.6
11	Limit oxygen index	ASTM D2863	%	45
12	Flame retardancy	UL-94	—	V-0

9.3 Future Developments of Nuclear Cables

9.3.1 Development Trend of Global Nuclear Power

In 2014, the top 10 countries of nuclear power consumption in the world were the United States, France, Russia, South Korea, China, Canada, Germany, Ukraine, Sweden, and the United Kingdom, among which the United States ranked first with 189.8 million tons of oil equivalent consumption in the world, while China ranked fifth with 28.6 million tons of oil equivalent consumption. After the accident at Fukushima Daiichi Nuclear Power Plant in Japan, the pace of nuclear power development has been slowed down in all countries in the world and some countries have even announced plans to abandon nuclear power. However, the United States has not changed its established nuclear power development plan due to the Fukushima nuclear power plant accident and Russia has also announced plans to build 38 more nuclear power units between 2020 and 2030, in addition to guaranteeing the operation of 33 units in 10 nuclear power plants. There are different attitudes toward nuclear power development in Europe. A few countries announced that they would abandon nuclear power. Germany and Switzerland decided to abandon nuclear power completely before 2022 and 2034 respectively. France, Britain, Finland, and Czech Republic indicated that they would expand the installed capacity of nuclear power. After the Fukushima nuclear power plant accident in Japan, China also suspended approval of new nuclear power projects and conducted comprehensive security inspection of nuclear power plants throughout the country. China's large-scale nuclear power development strategy was firmly shelved. However, the development of nuclear power is of great significance to meet the growing energy demand of China's economic and social development, to ensure energy security supply, to protect the environment, to optimize the structure of the power industry, to achieve sustainable development, and to enhance China's comprehensive economic strength, industrial technology, and international status. China will strategically promote the development of nuclear power on the premise of enhancing the safety of nuclear energy.

Generally speaking, the idea of large-scale nuclear abandonment is not realistic. Even Japan cannot abandon nuclear power indefinitely. It is an unrealistic and short-sighted behavior to abandon nuclear power because it will increase the per unit electricity cost and lead to the decline of industrial competitiveness. Even if Japan rapidly increases the domestic power generation capacity of natural gas thermal power plants and solar, tidal, and wind energy, it will be difficult to solve the current situation of domestic power shortage. Therefore, the global nuclear power industry will continue to move slowly forward in the future.

9.3.2 Development Trend of Nuclear Power in China

In China, it is the general trend to develop nuclear power, a clean energy source with low carbon and environmental protection. In the process of power generation, nuclear power does not produce air pollutants such as sulfur dioxide, nitrogen oxides, and soot, and the emission of carbon dioxide is much lower than that of thermal power. It can replace the role of thermal power plant. In the power system, the electric load below the lowest charge in a given time is called the basic load. Constrained by natural conditions, power generation of hydropower, wind power, and photovoltaics has obvious fluctuation and intermittence, which makes it difficult to play the role of base-load power plant. As a kind of high-density energy, nuclear power has a large capacity and can effectively guarantee power quality. In France, the proportion of installed capacity of nuclear power is as high as 72.28% of the total power supply. Nuclear power has replaced the status of thermal power plant in majority in France, which cannot be achieved by other clean energy sources (Figure 9.12).

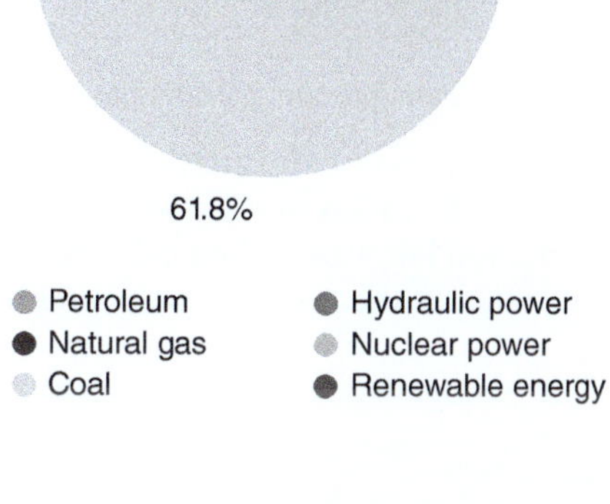

Figure 9.12 Structure of the primary energy consumption in China.

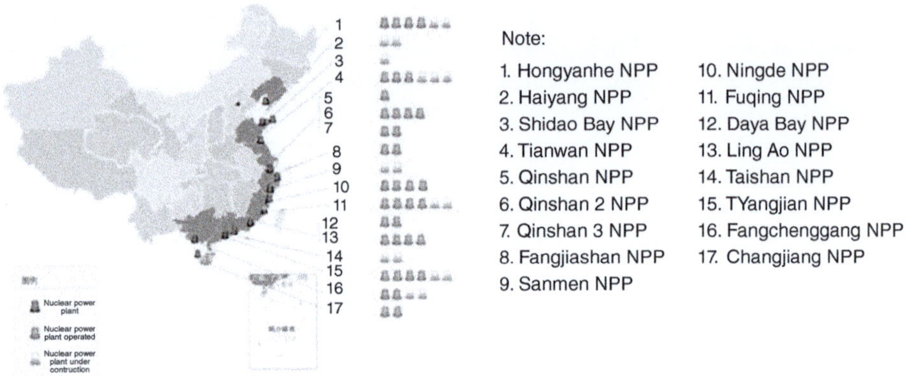

Figure 9.13 Distribution of nuclear power plant operated and under construction in China.

The coastal market totals nearly 500 billion RMB in China. By the end of 2017, 37 nuclear power units have been put into operation with an installed capacity of 35.81 million kW and 19 units under construction with a scale of 22.004 million kW. According to the 13th Five-Year Plan, China will achieve the target of 58 million kW in operation and 30 million kW in construction by 2020, but there is still a shortage of 30.18 million kW. With an average cost of 16 000 RMB/kW for third generation of units, the total coastal nuclear power market is estimated to be nearly 500 billion RMB. In order to achieve the target, the compound growth rate of nuclear power installed in 2018–2020 should reach 17.43%. In the long run, China has great potential for nuclear power development. According to the statistics of the World Nuclear Association, by the end of 2017, the total amount of China's preparatory and reserve projects was about 164 million kW (Figure 9.13).

Inland nuclear power is worth looking forward to, with an exploitable capacity of about 62 million kW. There is little difference in technical requirements between inland nuclear power plants and coastal nuclear power plants. At present, all nuclear power plants in operation and under construction in China are located in coastal areas, while the nuclear power plants of several major nuclear power countries in the world are mainly located inland. There are more than 440 nuclear power plants worldwide, of which more than 50% are located in inland areas. China has completed the preliminary feasibility study and examined 31 inland reserve sites. Conservatively, assuming an average of two units per site with an installed capacity of 1 million kW, China's inland nuclear power can develop about 62 million kW (Figure 9.14).

9.3.3 Development Trend of Nuclear Cables

Nuclear power plants have been rapidly developed and a total of 2700–2900 km of nuclear cables are needed for a nuclear power station unit with a capacity of a million kilowatts. The nuclear cables needed for nuclear units that will be constructed in the next few years and the nuclear cables for special projects will be

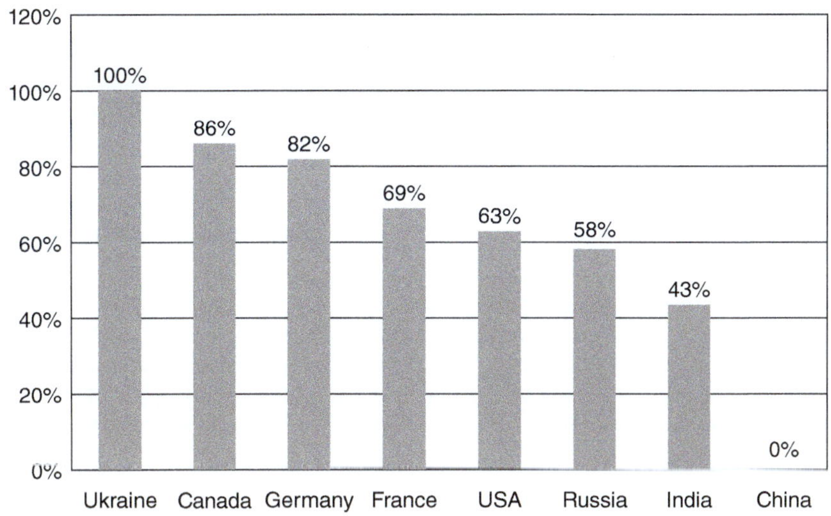

Figure 9.14 The proportion of inland nuclear power units in major nuclear power countries.

about 25 000–27 000 km/yr, which will bring huge business opportunities for the cable industry. The development of cables for nuclear power plants is showing trends of safety and environmental protection.

(1) Safety
In order to ensure the high reliability of nuclear power plant design, it is not only required to ensure the implementation of the three basic safety functions during normal operation and start up and shutdown of nuclear power plant, but also to ensure the implementation of the three basic safety functions under various extreme events.
The extreme events considered in the safety design of nuclear power plants are mainly caused by equipment failure, such as pipeline rupture, pump shaft jamming, valve malfunction, and flooding, injection, humidity, pressure, radiation, and fire caused by electrical failure. These failures can be considered as low as 10^{-6} per reactor per year. In order to avoid these faults effectively and ensure the safe and reliable shutdown of power plants after such accidents, the design process requires that the cables used in nuclear power plants (especially class 1E cables) shall have better fire resistance, higher radiation resistance, longer corrosion resistance in acid and alkali solution, and high temperature and high pressure impact resistance. It not only puts forward higher requirements for cable construction and technology, but also puts forward nearly strict requirements for the use of cable materials.
(2) Environmental protection
Green environmental protection products are the development trend of all industries, and environmental protection cable is the goal of wire and cable industry. Environment-friendly nuclear cables have also become the

direction of cable developers for nuclear power plants. In the standard GB/T 22577 "General Requirements for Class 1E Cables for Nuclear Power Plants," it has been clearly stipulated that class 1E cables for nuclear power plants shall be halogen free, low smoke, low toxicity, and flame retardant. Other countries have similar requirements in class 1E qualification standards, but most of the cables used in nuclear power plants are low smoke halogen-free and flame-retardant cables. The concept is confined to "smoke" and "halogen" taking PVC cables as reference, while "halogen" is confined to HCl (hydrogen chloride) gases, and does not refer to HF (hydrogen fluoride) gases, and does not involve heavy metal content control, nor does it concern the recyclability of materials. So, what kind of cable is an environment-friendly nuclear cable? Typical environment-friendly nuclear cables shall have the following characteristics besides good electrical performance, good mechanical performance, and properties of long-term radiation resistance, long-term environmental aging resistance, and LOCA resistance, which are unique to nuclear power plants.

1. *Halogen free*: Apply green environmental protection insulation, sheath, and special oxygen isolating layer materials not only to ensure cables have good electrical performance, and physical and mechanical properties, but also to ensure that the products do not contain halogen, solving the "secondary pollution" formed during combustion, and avoiding the carcinogenic "dioxin" substances produced during the combustion of traditional PVC wires.
2. *Excellent flame retardancy*: The environmental protection cable fully guarantees its performance for buildings with high requirements for fire protection. It is not easy to burn in fire, and can prevent the spread of flame and the expansion of disaster after combustion.
3. *Low toxicity*: There are no heavy metals harmful to people such as lead and cadmium in the insulation and sheath, so it will not pollute soil and water sources when the cable is used and discarded. In addition, mice were safe under the prescribed experimental conditions after rigorous toxicity tests.
4. *No corrosive gases*: The new type of special coating material without pollution to the environment will not produce toxic and corrosive gases such as HCl during production, use, and combustion, and will emit very little acid gas. It will do little harm to personnel, equipment, and instruments, and show more environmental protection features.
5. Waterproof and ultraviolet proof: Green environmental protection material with special molecular structure ensures ultralow water absorption. Special ultraviolet absorbance makes the product have good anti-ultraviolet function. It guarantees the safety of this kind of product and prolongs its service lifetime.
6. *High transmittance*: When the cable burns, the smoke generation is low, which is conducive to evacuation and fire-fighting. The light transmittance of the product is more than 60%, which is far higher than the standard of less than 20% for traditional flame-retardant cable.

References

1 IEEE 383. (2003). Type test of class 1E cables, field joints and connectors for nuclear power stations.
2 Kong, L. (2012). Classification and Qualification of Safety Level of Nuclear Equipment. *Technological Development of Enterprise* 31 (22).
3 Wang S (2016). The study on the Nuclear Class Cable Material for the Safe Operation of Nuclear Power, Master degree thesis in Zhejiang University of Technology.
4 NFPA 70 (2011). National Electrical Code.
5 IEEE 317 (2013). Standard for Electric Penetration Assemblies in Containment Structures for Nuclear Power Generating Stations.

10

Submarine Cables

George Georgallis

Hellenic Cables S.A., 33 Amarousiou-Halandriou Street, Maroussi, Athens 15125, Greece

10.1 Introduction

One of the earliest applications of submarine cables was for the transmission of power using stronger electrical currents with the intent of detonating powder mines across the river Neva during the Napoleonic wars in 1812.

Since then submarine power cables have gradually increased their power transmission capability underwater over long lengths and for many applications.

However, the initial commercial interest in submarine cables, which were first installed in the mid-nineteenth century, was not for power transmission but for telegraph cables used for communication purposes.

A hundred years later, as communication technology further developed, telephone submarine cables were first installed and by the early 1980s submarine fiber-optic (FO) cables were used, enabling data transmission.

10.2 Submarine Power Cable Applications

Over the years, submarine power cables have been used to transmit power across rivers and lakes, and between destinations separated by water such as the following:

- Oil and gas platform connections to the mainland
- Interconnection between offshore wind generator platforms and power export from wind farm substations to shore
- Connection of islands to main grid or interconnection between islands
- Interconnection between power grids of neighboring countries
- Power transmission from tidal and wave energy generation installations

Installation depths may vary from a few meters to over 1500 m and cable route lengths from a few hundred meters to 400 km or more.

Cable rated voltages may vary from 6 kV up to 500 kV alternating current (AC) or up to 640 kV direct current (DC) with the maximum transmitted power capability in the range of Giga Watts.

The Global Cable Industry: Materials, Markets, Products,
First Edition. Edited by Günter Beyer.

10.3 Submarine Power Cable Design Overview

Modern Submarine Power cables are characterized by their robust mechanical design, normally equipped with a metallic wire armor layer for mechanical protection and as a strain-bearing member, which accommodates a large part of the installation tension.

Submarine power cable designs include single-core or three-core cables for AC systems or single-core, two-core, or concentric cables for DC systems.

The three-core design for AC cables provides a less costly solution overall than the manufacture, installation, and protection of three separate single-core cables. This requires however larger and more complex manufacturing facilities and larger cable laying vessels, in particular for long high voltage (HV) or extra high voltage cables.

High voltage AC and DC cables usually have an extruded lead sheath over an extruded or lapped insulation core as a radial moisture barrier.

Many three-core AC submarine cables incorporate fiber-optic (FO) elements in their construction, usually between the outer interstices of the laid up cores for data or communications transmission or for thermal or strain sensing and monitoring purposes (Figure 10.1).

Submarine power cables manufacturing plants require access to docking facilities that are capable of loading the cable on transportation or cable laying vessels.

When designing a submarine cable for a power transmission system, the following factors need to be considered:

- Electric power system characteristics
- Transmitted power requirements
- Route lengths
- Installation depth
- Installation conditions
- Static or dynamic operation

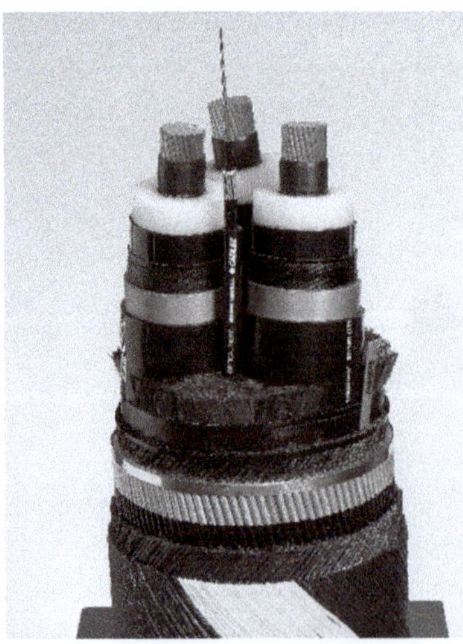

Figure 10.1 Three-core HVAC XLPE insulated submarine cable. Source: Courtesy of Hellenic Cables SA.

- Project-specific cable design, manufacturing, and testing considerations
- Submarine cable accessories
- Means of installation, protection, and maintenance

The above are briefly discussed in the text that follows.

10.4 Power System Considerations

The power system employed could be a three-phase AC or a DC system.

Alternating current submarine cable systems ranging from medium to high or extra high voltage are used for a wide range of transmitted power values and varying route lengths and in general do not have the large capital expenditure costs or large footprints associated with high voltage direct current (HVDC) converter stations.

High voltage direct current transmission compared to AC is suitable for higher transmitted power over longer distances and exhibits reduced power losses and hence lower operational costs over the system service life.

The power transmission requirements for a given project for which the electric power system is specified will consequently determine the sizing of the cable conductor. Other factors with an impact on the cable size are the installation conditions along the route on the choice of insulation system and conductor material.

Submarine power cables operating in an AC system act as capacitors, which for longer lengths and higher voltages significantly limit the transmitted active power capability. The use of shunt reactors at one or both route ends compensates for the cable capacitive power impact and can improve the active power transmission capability of the system.

10.5 Submarine Power Cable Elements

10.5.1 Cable Conductor

The cable conductor material, just as in land cables, can be either copper or aluminum.

Copper is a material that has a higher electrical conductivity, mechanical strength, and corrosion resistance than aluminum and has been for many years the preferred choice as the conductor material for submarine cables. However, in recent years aluminum, although with only 60% of the electrical conductivity of copper, is chosen for many medium and high voltage submarine cable applications. The reason is that it leads to more cost-effective and lightweight cable designs for the same power transmission than copper, although with the drawback that the cables have a larger cross section and volume, which may be a storage limiting factor in some instances.

The conductor designs can be of the stranded wire type or, in particular for aluminum, of the solid type, which may be stiffer but is naturally water blocked and provides a slightly more compact and less costly cable.

Stranded conductors for cables with extruded insulation – lapped insulation cables have conductors that are water blocked by the impregnating fluid – require sealing against the longitudinal propagation of water at any installation depth, the most demanding situation being at the deepest point of the route.

Water sealing is achieved in most applications by applying a suitable compound over each conductor wire layer during the manufacturing process.

10.5.2 Cable Insulation

The cable insulation can be extruded, usually cross-linked polyethylene (XLPE), or in some cases ethylene propylene rubber (EPR) or lapped paper or paper laminate combined with polypropylene that is oil/fluid filled (self-contained oil filled [SCOF]) or mass impregnated non-draining (MIND).

An indication of the ratio of the various insulation systems in service for submarine cables is provided in the statistics from CIGRE TB 379 [1] for HV (60–500 kV) submarine systems in operation at the end of 2005.

The quantity of cables in service for AC systems is 3697 circuit km in total, of which 67% are oil filled and 33% XLPE insulated, and for DC systems the values are 3366 circuit km in total, of which 13% is oil filled, 80% MIND, and 7% XLPE.

10.5.2.1 Lapped Insulation

Lapped insulation for submarine cables is currently used mostly in the production of HVDC cables and to a lesser degree for HV/extra high voltage (EHV) AC cables.

Self-contained fluid filled (SCFF) or SCOF cable designs have been used earlier for high voltage alternating current (HVAC) or HVDC submarine cable systems.

Lapped insulation for such cables consists of multiple layers of cellulose paper tapes impregnated with an appropriate oil or fluid with a low viscosity permitting the flow of fluid along the whole cable length. The fluid is maintained at a positive pressure by using fluid reservoirs at the cable ends to accommodate the fluid volume change caused by the ambient or cable operating temperature changes. These cables have maximum length limitations around 40–50 km, imposed by the need to achieve sufficient fluid flow to the whole cable length.

Examples of such systems are the Vancouver island to Canadian Mainland through Texada island 525 kV interconnection in 1984 [2] or, for HVDC submarine interconnections, the Kii Channel Honshu–Shikoku island ±500 kV cable in 2000 [3]. In the latter case, the insulation for the cable was polypropylene paper laminate (PPL) permitting a higher continuous operating temperature.

The insulation of MIND cables also consists of multiple layers of cellulose paper tapes but in this case they are impregnated with an oil or fluid that exhibits high viscosity at the operating temperature, which does not permit drainage.

Many high power HVDC submarine cables above 320 kV and up to 500 kV have MIND insulations that do not have the cable length limitations of the SCFF types and do not have the environmental risk of oil or fluid leakage in case the cable integrity is compromised for some reason.

Mass impregnated non-draining insulation has been used for medium voltage (MV) submarine cable applications in earlier years.

Certain types of MIND HVDC cables have lapped insulation consisting of PPL tapes restricting drainage and thus permitting higher maximum continuous operating temperatures of 85 °C than the simple MI cables of 50–55 °C.

Examples of 500 kV MI HVDC submarine cable installations are the Italy Sardinia interconnector (SAPEI – 2011) and the Sweden–Finland interconnection (Fennoscan 2 – 2011) [4].

10.5.2.2 Extruded Insulation

10.5.2.2.1 AC Cables – Medium and High/Extra Voltage Currently, XLPE is the most widely used insulation material for AC medium and high voltage submarine cable applications up to 400 kV and voltage source converter (VSC) HVDC cables up to 320 kV.

Ethylene propylene rubber is used in medium voltage applications requiring higher cable flexibility and mechanical resilience. Although EPR submarine cables up to 150 kV AC have been produced, XLPE is usually preferred for extruded HVAC insulation since it has lower dielectric losses permitting higher current ratings with lower risk of insulation thermal runaway.

Cross-linked polyethylene for insulating medium and high voltage submarine cable cores is applied by extrusion and the cross-linking is done by using peroxides as agents to achieve the cross-linking/vulcanizing of low density polyethylene molecule chains in inert nitrogen atmosphere pressurized tubes with the application of heat to form three-dimensional molecular structures.

Peroxide is usually incorporated in the insulation compound prepared in pellet form by the XLPE compound provider along with antioxidants to prevent premature cross-linking or pre-curing of the insulation material during the extrusion process and to protect the material during its service life from thermo-oxidative aging.

During the cross-linking process volatile by-products are generated such as water vapor, methane, acetophenone, and cumylalcohol, which can cause void formation in the still non-solidified insulation around the conductor if the pressure in the tube is not appropriately controlled.

The insulated cores are degassed in suitable heated chambers permitting diffusion of the by-products through insulation. The degassing temperature usually does not exceed 70 °C to permit timely diffusion without risking insulation deformation of the cores.

Degassing is required not only for XLPE insulated high voltage AC and DC cable cores but also for medium voltage AC cable cores for submarine cable applications in order to avoid gaseous by-products diffusing through the insulation during operation.

Gas in the insulation voids can mask the partial discharge (PD) measurements, thus posing a risk of not detecting accurately potentially damaging insulation defects during routine partial discharge tests of smaller lengths or on samples from larger ones.

10.5.2.3 HVDC Cables

For HVDC XLPE insulated cables the electric field distribution within the insulation is controlled by the electrical resistance of the insulation, which in turn is highly dependent on the temperature and on the electric stress.

Use of XLPE as HVDC cable insulation material requires the development of compounds with a much higher electrical resistance compared to the classic XLPE for HVAC insulation.

These materials can be unfilled with a high level of physical and chemical cleanliness and with a lower peroxide content than standard XLPE or of a nanocomposite filled type.

The degassing requirements for HVDC XLPE insulation are even stricter since many of the cross-linking by-products are polar, thus contributing to space charge increase in the insulation [5–8].

10.5.3 Metallic Sheaths

HV/EHV AC and HVDC submarine cables of almost all insulation types have a metallic barrier protecting the insulation against radial ingress of water, which eventually would cause damage to the insulation and considerably shorten the cable service life.

The practically universal metallic barrier for these cables consists of a seamless lead alloy sheath applied by extrusion over the insulated core, also acting as a metallic radial shield. The lead sheaths of submarine cables have polymeric oversheaths, usually PE applied in tandem just after the lead extrusion, which provide mechanical protection to the underlying lead during the subsequent handling and manufacturing processes until the cable is completed.

European Standard EN 50307 "Lead and Lead Alloys – Lead and Lead Alloy Sheaths and Sleeves of Electric Cables" [9] specifies the designation and requirements, and in addition provides information on the use of lead alloys as sheath material for submarine and land cables.

Dynamic HV submarine cables cannot have a lead alloy sheath for moisture protection because although modern lead alloys exhibit adequate fatigue performance for most land applications, submarine dynamic cable cores are subjected to much higher strains – bending, tension, compression, and torsion – in their service life. For such applications the preferred material is copper, copper alloy, or stainless steel, usually longitudinally welded – smooth or corrugated – exhibiting a much higher fatigue resistance at the cost of lower corrosion resistance compared to lead, again with the protective PE jacket.

Medium voltage cables with extruded XLPE insulation up to and including 66 kV may have a lighter and less costly metallic sheath such as an aluminum or copper laminate foil of 0.1–0.2 mm thickness longitudinally applied over the insulated core and bonded to an extruded medium or high density polyethylene jacket.

In some cases just the PE jacket is applied without the metal barrier, a "semi-wet" design, in which case the permeation of water molecules through the jacket, reaching and eventually saturating the insulation, is delayed but not avoided.

Certain cable designs usually with an EPR insulation employ a fully "wet" design with no metallic or PE radial moisture barrier at all.

With the exception of the fully wet designs for EPR insulations, all HV and MV cables with extruded insulation have semi-conductive water blocking tapes applied over the insulated cores, which restricts the longitudinal water propagation in case the sheath integrity is compromised. For MV cables with aluminum or copper laminate foils or even semi-wet designs with just extruded sheaths the water blocking tapes also serve as moisture absorbers for any moisture diffusing through the jackets, delaying the saturation of the insulation with water.

10.5.4 Armor Layers

For submarine cables the armor normally consists of one or two layers of metallic wires helically applied over the laid up cores for three-core AC cables or over the sheathed core for single-core cables, in all cases with a suitable bedding layer underneath such as bitumen soaked polypropylene yarns. For three-core AC cables or for single-core HVDC cables, zinc-coated round steel wires are commonly used.

European Standard EN 10257-2 "Zinc or Zinc Alloy Coated Non-alloy Steel Wire for Armoring Either Power Cables or Telecommunication Cables – Part 2: Submarine Cables" specifies galvanized steel wires for submarine cables [10].

The zinc coating assisted by the bitumen soaking of the polypropylene bedding and serving layers provide corrosion protection to the steel wires limited by factors such as the installation environment, water salinity, burial of cable, possible erosion, or electrochemical effects by induced or stray currents.

For single-core AC cable designs where the use of a magnetic material would cause high power losses in the armor and consequently have a negative impact on the cable current rating, copper, aluminum, or stainless steel wires are used.

Single- or three-core submarine cables with a single armor layer are in general "coilable," meaning that they can be coiled in non-rotating turntables or static tanks provided the direction of coiling is such that the coiled cable armor opens during coiling.

For each coil turn, the cable will twist around its axis by 360° and the coil should have adequate cable length to be able to accommodate the torsion evenly without cable damage.

Since there is a higher risk of damage to the thicker insulations of high voltage cables, most coiled cable applications refer to medium voltage cables.

The coilability of a cable is tested with a coiling test as part of the mechanical type testing and should include flexible joints of the power cores if applicable.

10.5.5 Other Cable Elements

A three-core submarine power cable will have material filling up the space between the outer interstices of the laid up cores, providing a round shape, which in turn permits an even application of the armoring wires, smoother handling of the cable, and more uniform distribution of mechanical stresses from handing and installation.

The filler material could be polypropylene yarns of different diameters to achieve optimum filling of the interstices or extruded polyolefin or polyvinyl chloride (PVC) based profiles permitting the incorporation of any integrated fiber-optic units.

Fiber-optic units, armored or unarmored, are frequently integrated in three-core submarine power cables for the purpose of transmitting data or for measuring the temperature distribution in the cable during operation or the strain distribution during installation.

10.6 Submarine Cable Manufacturing

Submarine cable manufacturing plants are generally capable of producing, testing, and storing the complete cable lengths required for a particular project.

Submarine cable semi-finished components and final cables are usually produced in long lengths and are transported to intermediate or storage turntables respectively by means of roller systems whereas for land cables the norm is that intermediate production components and final lengths are wound on drums.

Submarine cable manufacturing plants require access to facilities capable of directly loading the vessels that will either install the cables or transport them elsewhere to eventually be loaded on a cable laying vessel (CLV) to be installed. In many cases, the limiting factor for the maximum finished cable length produced is imposed by the maximum weight or volume loading capacity of the cable laying vessel (Figure 10.2).

Conductors are manufactured in lengths corresponding to the core production lengths and are coiled in turntables before proceeding to the insulation

Figure 10.2 Submarine cable manufacturing plant with docking facility. Source: Hellenic Cables.

process. Stranded conductors for cables with extruded insulation of dry or semi-dry application are water-blocked using suitable compounds that limit the longitudinal ingress of water even at the deepest installation depths.

The insulation application process involves the production of long insulated core lengths limited by the manufacturing process.

For extruded XLPE insulated cables, the limiting factor for the production lengths would be the insulation line throughput limits for producing high quality insulation and for lapped paper insulated MIND cables the limit would be the capacity of the insulation impregnating vessel.

Triple extruded XLPE insulated cores are degassed in degassing turntables before being further processed.

Other elements such as semiconductive tape layers, metallic screen in the form of wires or tapes, or lead sheaths in tandem with a PE jacket – semiconductive for long cable lengths – are applied on the manufactured lengths in the manufacturing processes that follow.

The application equipment is similar to those of the land cables, the difference being usually in the long lengths and the intermediate turntables in place of intermediate drums.

For long lengths of completed cables, it may be necessary to connect with factory joints production lengths of insulated and sheathed cores at the manufacturing plant. These are normally of the flexible type, which permit the same handling of the cable with joints as the rest of the cable during manufacture and installation.

It is desirable to produce long insulated core lengths not only for cost and manufacturing capacity improvement reasons but also to improve the reliability of the cable system by minimizing the number of factory joints.

The jointing is done during the laying up process for three-core cables and in any case before the armoring process for all cable types by connecting manufactured insulated cores of the same or different cross sections or conductor materials.

The conductors are connected normally by welding and the insulation and semiconductive layers are restored usually by taping and vulcanizing, followed by the metallic sheath and PE jacket restoration. At the same time, the polypropylene yarn or extruded fillers are inserted along with any FO units that are integrated in the outer interstices (Figures 10.3 and 10.4).

The number of factory joints depends on the final cable length to be delivered.

The armor layer or layers, along with the bitumen soaked polypropylene yarn layers under and over the armor wires, are applied during the armoring process on the complete final cable length.

10.7 Submarine Power Cable Accessories

The cable may have one or more sea or field joints, usually of the rigid type.

These are scheduled when smaller cable lengths need to be connected in the field as is the case when the overall route lengths exceed the cable laying vessel

Figure 10.3 Laying up of submarine cable. Source: Hellenic Cables.

Figure 10.4 Tensile bending type test of submarine cable. Source: Hellenic Cables.

loading capabilities. They may be unscheduled if a damaged cable section needs to be replaced during the initial laying operations or within the scope of replacing a section of cable that has been damaged in operation.

For cost and risk optimization reasons it is preferable to install complete cable lengths, minimizing the number of field joints or completely avoiding them.

An interconnecting or export submarine cable may connect with a land cable via a transition joint at the landing area or can appropriately connect to a substation or an overhead line if the distance from the landing area is very short.

Cables for medium or high voltage offshore applications can connect to turbine or transformer platforms using joints, gas insulated switchgear (GIS) type terminations, or employing connectors, which in many cases have a quick connect–disconnect capability to facilitate maintenance or emergency situations.

10.8 Submarine Power Cable Testing

Submarine cables compared to land cables may have

- long delivery lengths and respectively higher weights;
- a number of factory joints, generally fully flexible;
- field joints, rigid or flexible, connecting armored cables (during installation or following repair);
- armor layers.

These cables may be subjected to high water pressures when installed and should be able to perform as required.

For instance, if they are of the wet type design, the insulation needs to exhibit resistance to water treeing phenomena in the presence of moisture.

If they are of the dry type, the insulation needs to be protected from the radial ingress of water and if damaged their design needs to ensure limited longitudinal propagation of water in the conductor and under the metal sheath.

Submarine cables are tested as specified in the appropriate technical documentation, relevant standards where available and applicable, and technical brochures and recommendations.

Submarine cables compared to land cables are subjected to much more severe mechanical stresses and strains during manufacturing, handling, installation, and repair operations. Since repair or replacement operations are costly and require longer implementation times for submarine cables, proactively verifying that the cable system is fit for the purpose and thus reducing design risks has its merits.

Therefore, type tests for submarine cables of all voltage levels are system tests including accessories. These tests include realistic mechanical tests and water penetration tests, before the specified electric tests, as appropriate for the system, such as coiling tests for coilable cables, tensile bending tests including cable and – if foreseen for a project – factory joints, and tensile tests for rigid field joints. Project-specific tests may be carried out such as crushing or long-term pressure tests or bend restrictor tests for field joint casings.

Submarine power cable system qualification testing procedures and requirements reflect the necessity to ensure that the cable systems are fit for the purpose considering the above but also any specific project conditions.

In this sense, to relate to the actual handling and installation sequences encountered by a submarine cable, the appropriate mechanical and pressure or water penetration type tests are carried out before the electrical tests, as described in Cigre TB 490 [11] and Cigre TB 623 [12].

Tests can be categorized as follows:

- *Routine*: Tests on all manufactured components, such as cable lengths or accessories, to ensure that the components meet the specified requirements.
- *Sample*: Tests on samples of accessories or from complete cables at a specified sampling frequency to verify that the components meet the specified requirements.

Figure 10.5 Voltage test of submarine cable. Source: Hellenic Cables.

- *Type*: Tests made before supplying on a general commercial basis a type of submarine cable system in order to demonstrate that the performance characteristics meet the intended application. Submarine cables of all voltage levels are usually type tested as cable systems – cable and accessories (Figures 10.4 and 10.5).
- *Project-specific and development or special*: Tests that may be performed, if design values or if conditions (such as handling, storage, installation, or operation) are not covered by the type tests or if, for instance, information regarding materials or cable characterization properties for engineering purposes is required.
- *Pre-qualification*: Test applicable to HV/EHV submarine cables with extruded insulation, as described in Cigre TB 490, to validate the long-term performance of the complete cable system including the cable and accessories, the essential differentiating factor for submarine systems as opposed to land cable systems being the factory joint.
- *Accelerated wet aging*: Tests for AC cables with extruded insulation of wet or semi-wet design without a metal barrier over insulated core. These and other tests related to assess insulation performance of submarine cables (6–72.5 kV) for the offshore wind farm industry are introduced in Cigre TB 722 [13].
- *Sea trials*: A sea trial test is costly and relates to the interaction during installation and retrieval between the submarine cable including factory and field joints if applicable and the installation experience and equipment. Usually a cable supplier/installer and the customer judge if a sea trial test is necessary for a specific project.
- Factory acceptance (FAT): Tests carried out on the finished cable at the manufacturing facility before loading out on the cable transport or laying vessel to check that each length meets the specified requirements.
- *After installation or commissioning*: Tests made to demonstrate the integrity of the cable system as installed and usually involve high voltage testing as described in Cigre TB 490. A time domain reflectometry (TDR) measurement is recommended to provide a "fingerprint" of the wave propagation characteristics of the cable, facilitating fault location if required.

The first international submarine cable standard to be issued is IEC 63026 [14], expected in early 2020, with a scope applicable to cables with extruded insulation up to 72.5 kV.

With the above exception and in the absence of relevant international submarine cable standards, Cigre TB 623 "Recommendations for mechanical testing of submarine cables," which covers mechanical tests on AC/DC submarine cables of almost all types and voltage levels, and Cigre TB 490 "Recommendations for testing of long AC submarine cables with extruded insulation for system voltage above 30 (36) to 500 (550) kV" provide the foundation for testing submarine cable systems. These brochures refer accordingly to the relevant land cable standards, where applicable, such as IEC 60840 [15], IEC 62067 [16], or IEC 62895 [17].

Cigre TB 623 provides a simple analytical calculation method, adapted from Electra 171 [18], to determine the installation maximum tensile force exerted on the cable, for carrying out the mechanical tensile and tensile bending type tests. This tensile force depends on the maximum sea depth and includes the static and dynamic drag and inertial components. TB 623 recommends that a detailed dynamic analysis for deeper seas and for more demanding conditions is performed. This analysis should take into account the cable physical properties along with the cable laying vessel dynamic response characteristics and the available sea state information for a given project.

10.8.1 Dynamic Submarine Cable Testing

Submarine power cables may experience cyclic or varying loads and strain for shorter or longer periods of time or even during the whole of their service life, potentially leading to accumulation of fatigue damage of specific cable elements.

The lead sheath, in a cable of such design, although proven to be robust regarding thermal fatigue over many years of service, is usually the element most susceptible to fatigue damage from mechanical cyclic or varying loading.

When a submarine power cable is installed, the cable is continuously in linear movement as it is fed out from a static storage tank if coilable or from a turntable and therefore is not subjected to fatigue.

If a sea joint is required, either scheduled or not as in the case of a cable repair, the cable jointing operation requires the feeding to be halted for a few days and the cable hanging from the vessel may be subjected to repeated wave-induced ship movements, and therefore, fatigue could be an issue.

A suspended span of a cable installed on the sea bottom may be subjected to fatigue damaging vortex-induced vibrations (VIVs) for long time lengths, depending on the local sea currents and cable physical properties.

Submarine power cables may hang freely from a fixed or a floating wind or oil and gas platform; other applications for wave or tidal power generation may involve continuous flexing of the cables connected to the power generators and be the cause for fatigue damage.

A detailed fatigue analysis in situations similar to those above can provide information to assist in implementing mitigating measures such as avoiding extreme weather conditions or releasing cable length from time to time in field jointing operations or protecting cable spans when VIV is encountered or anticipated, without changing the cable design.

Dynamic submarine cables on the other hand are specifically designed to endure continuously varying mechanical stresses and strains throughout their service life.

Dynamic cables without a metallic radial water barrier as is the case for some inter-array applications should have an insulation system with water tree resisting performance. For high and extra high voltage cables, other metallic barriers with a greater fatigue endurance can be used such as copper, copper alloys, or stainless steel.

Dynamic submarine power cables should have their fatigue performance investigated and verified for the intended application taking into account the project-specific environmental conditions, vessel/platform characteristics and motion, and the cable configuration.

This is done initially through a global analysis to determine extreme loads, strains, and curvatures including the effect of related accessories/equipment such as bend stiffeners, hang-offs, or buoyancy modules followed by local analysis on the cable components to assess the imposed stress/strain impact.

Further analysis employing historical environmental data and expected cable service life can estimate the accumulated fatigue damage to the various cable components assuming linear cumulative damage (Palmgren–Miner rule). Then an assessment can be done if the cable design is fit for its intended application.

The component stress–strain curves can relate the anticipated number of cycles to failure and therefore service life with a suitable safety factor to the applied stress or strain of the cable components.

Component fatigue testing can provide the stress–strain curves of the individual components.

Global, local, and fatigue analyses, taking into account frictional forces between the cable layers and elements, can be implemented using commercially available finite element based software; however, analytical simplified calculations, if available, can provide quick initial estimates.

Finally, a full-scale fatigue test of a cable sample with the relevant end fittings and accessories attached, such as bend stiffener or bellmouth, provides the verification that the dynamic cable is considered as fit for purpose.

The cable system is subjected to at least 1.5 million bending cycles, as described in Cigre TB 623, or to an appropriate number as to represent the cable service life.

The cyclic bending of the full-scale fatigue test is typically divided into a number of 5–7 blocks with different bend radii and number of cycles, chosen to realistically represent the accumulated fatigue damage of the project service life. During the bending cycles, a representative constant tensile force is applied.

Apart from the visual inspection after the fatigue test, the cable is subjected to an electrical routine test before and after the test to verify the cable integrity.

If the cable contains optical fibers, their continuity is verified before and after the test.

Since no safety factor is mentioned for the full-scale fatigue test in Cigre TB 623, this is something that can be agreed to between the manufacturer and the customer for a specific project.

10.8.2 Submarine Power Cable Sizing and Other Design Considerations

The analytical steady state and transient cable rating methods described in IEC 60287 [19] and IEC 60853 [20] also apply to submarine cables that are solidly bonded at both ends but with a few considerations discussed below.

Although the armor losses in three-core armored AC submarine cables are generally considered to be overrated and many studies have been published on the issue [21], the analytical IEC 60287 methodology is useful for providing quickly and on the safe side initial current rating results.

However, steady state and transient rating calculations, which are based on finite element modeling simulations, using measured armor loss factors, are currently considered by many system designers, in particular for renewable offshore applications, as being able to provide realistic cable thermal performance providing a greater confidence in finding the optimum cable sizes and cost-effective solutions. In many cases, these cable systems need to be cost-effective, combining both capital expenditure and operating costs, and in addition may be required to operate for longer periods of time near their nominal power transmission capacity.

Appreciable and realistic capital expenditure savings can be achieved for export cable and inter-array cable sizing purposes and therefore cost optimization by considering historical load data and using, as previously discussed, advanced calculation tools for estimating cable transient rather than steady state thermal rating.

Another issue to consider with AC cables of long lengths is the capacitive charging current that contributes to the thermal rating of a power cable, thus reducing its active power transmission capability. However, the charging current can be compensated by using reactive coils from one or both ends of the solid bonded submarine cable system.

Projects with compensated HVAC cables of very long lengths exceeding 100 km route lengths have been implemented such as the 2 × 200 MVA·150 kV Lavrion–Syros interconnection with a route of 108 km and the 2 × 118 MVA·132 kV Ibiza–Mallorca project with a 118 km submarine cable length.

For very long submarine cable lengths, further optimization of the cable size can be achieved by calculating the thermal rating of the cable along thermally unfavorable parts of its route and if feasible by adjusting accordingly the conductor size or material or even the armor design:

- Sea bottom soil properties or specified burial depths may change for buried sections of a cable, thus enabling sizing variations.
- Shunt compensation of AC cables results in variation of the current affecting thermally the cable along parts of its route, consequently offering an opportunity of size adjustment.

The installation depth will have an impact on the design of the power cable mechanical load bearing elements, namely the conductor and the armor, since these elements essentially bear the cables' own static weight and dynamic tensile force during installation.

Since the cable conductor cross section is determined by the power rating – if the system voltage and conductor material have been chosen – the armor cross section may need to vary in order to accommodate the tensile forces. If the armor cross section needs to be adjusted for this purpose, this in turn may also affect the total cable weight.

For higher installation depths, torsionally neutral cables with two or more armor wire layers applied helically in alternating directions may be necessary.

If the submarine cables are subjected to long-term dynamic operation, such as cables connected to floating structures, they should be designed to withstand the associated cumulative fatigue damage during their service life.

10.8.3 Submarine Power Cable Installation, Protection, and Commissioning

Submarine cables have been and are installed with suitably adapted barges or supply vessels equipped with turntables or for coilable cables with static tanks. Smaller sized cables of shorter lengths can be installed from drums on ships. Installation of the larger and longer cables is normally implemented by dedicated CLVs with dynamic positioning (DP) capabilities and equipped with one or more turntables capable of storing in total up to 10 000 tons or more of cable.

An appropriate desktop study and a subsequent detailed marine survey of the scheduled cable route preceding the submarine cable installation can provide information permitting the safe installation and appropriate protection of the cable. Factors to consider are the selection of appropriate installation windows, selection of appropriate protective measures for the installed cable, consideration of other cable crossings, identification of unexploded ordnance (UXO) and suitability of equipment for trenching or jetting, avoidance of obstacles or free-hanging cable lengths, investigation of potential bottom stability, and VIV issues.

The successful outcome of a submarine cable turnkey installation project requires detailed preparation, which starts with a desktop study serving as a roadmap for the more detailed, costly, and resource-consuming survey related work to follow.

The aim of this study is to determine the optimum installation solution from a cost and technical point of view by identifying the appropriate route, ensuring safe installation and effective protection of the cable while enabling the cable system to operate as specified.

The desktop study will normally rely on any existing published survey data, available environmental information, marine activity, and possible restrictions of right of way on the sea bottom or at the landing area. It will investigate information related to existing submarine cables or lines that may need to be crossed and will try to determine the technical feasibility of the installation and most importantly to identify potential risks to the cable.

A desktop study may need to identify alternative cable installation areas/corridors and to recommend possible cable protection or burial approaches.

Finally, a desktop study should be able to produce a documented recommendation or guide for the marine survey to follow.

A marine survey of the proposed cable route(s) is carried out using fully equipped specialized survey vessels and provides details concerning depth and seabed profiles including slopes, ridges, and large obstacles using sonar-based equipment.

Sub-bottom profiling can provide information on the thickness and characteristics of the layers below the sea bottom, which are needed if the cable is to be buried for protection along the route or part of the route. Soil sampling in the form of cores can provide additional information concerning the properties of the layers under the sea bottom. Of particular interest are soil thermal resistivity values that are used for cable rating calculations and hardness and type of soil, which are useful in the selection of the appropriate cable burial methods.

Modern cable laying vessels are equipped with dynamic positioning (DP) systems permitting the vessels to maintain their positions precisely under various weather conditions and have large turntables capable of accommodating long lengths of cable up to 9–10.000 tons, thus minimizing or removing completely the need for scheduled field joints even for long cable routes.

Global and local cable analysis software can be used to proactively investigate various installation scenarios and prepare for or avoid situations that may pose increased risk during installation.

During installation, the cables are fed out in a controlled manner using frictional forces to counter the cable tension. Friction is applied either by applying pressure – "squeezing" the cable using linear machines equipped with wheels or treads in two or more rows of an appropriate length to avoid crushing the cable – or by using capstans without squeezing the cable.

In both methods the cable is subjected to combined mechanical stresses from tension and bending when leaving the sheave or capstan.

This can be mitigated to a degree by using vertical laying equipment with linear machines placed after the bending element.

Vessels with the capability of installing the cable from the middle of the vessel rather than the stern can achieve lower inertial tensional forces on the cable.

Coilable cables can be transported and installed by vessels equipped with static tanks.

Small cables with short lengths may use drums for transportation and laying purposes.

After the final cable survey has been performed, a risk assessment study can be performed along the route for the cable to be installed and suitable protection measures may be proposed and cost estimated.

Protection measures could consist of cable burial along selected parts of the route or use of externally applied protection shells, mattresses, or rock dumping over the cable or the use of directional drilling at landing areas both to avoid difficult or restricted areas and to provide cable protection.

During cable installation, remotely controlled vehicles (ROVs) can be employed to monitor by video the cable at touchdown. The monitored laying angle and cable tension can be compared during laying with the theoretical values.

Figure 10.6 Site acceptance testing of HVAC submarine cable. Source: Hellenic Cables.

The position of the cable after installation can be checked by video or sonar or by divers at smaller depths.

A high voltage routine test as specified in the relevant land cable standard, or if this is not feasible a lower voltage for a longer duration, is recommended after a new installation (Cigre TB 490) (Figure 10.6). As a minimum requirement for AC cables a "soaking" test of the system is acceptable with application of the system voltage at no load for 24 hours.

A TDR test performed on the installed cable is recommended in order to "fingerprint" the cable wave propagation characteristics.

Periodical maintenance surveys of the whole or part of the cable route may provide information on the cable condition during its service life.

10.9 Submarine Power Cables – What Next?

We can expect the development of dynamic HV/EHV AC and HVDC cables for deep-sea floating wind generation applications as the offshore wind generator power increases and the introduction of new insulating high-performance polypropylene thermoplastic elastomer (HPTE) materials for both HVAC and HVDC submarine cables.

Lighter cables for deep waters with total or partial replacement of heavy steel armor with lighter high strength synthetic armor wires will permit installation of three-core HV/EHV power cables in depths exceeding 1000 m.

References

1 CIGRE TB 397 (2009). *Update of Service Experience of HV Underground and Submarine Cable Systems*. CIGRE.

2 Foxall, R.G., Bjorlow-Larsen, K., and Bazzi, G., et al. (1984). Design, manufacture and installation of a 525 kV alternating current submarine cable link from mainland Canada to Vancouver island. 21-04, JICABLE 84 (5–10 March 1984), Versailles, France.

3 Inque, T., Mitani, M., Shigetoshi, I., et al. (1999). Installation of 500 kV De PPLP-insulated oil-filled submarine cable. B7.1, JICABLE 99 (20–24 June 1999), Versailles, France.

4 Europacable (2011). *An Introduction to High Voltage Direct Current (HVDC) Underground Cables*. Europacable.

5 Mazzanti, G. and Marzinotto, M. (2013). *Extruded Cables For High-Voltage Direct-Current Transmission*. Wiley.

6 Billore, J., Auge, J.L., Allais A., et al. (2015). Study of the thermo-oxidative ageing of crosslinked polyethylene for HVDC applications. B6.1, JICABLE 15 (21–25 June 2015), Versailles, France.

7 Eriksson, V., Andersson, J., Englund, V., et al. 2015. Long term performance of XLPE insulation materials for HVDC cables. B6.2, JICABLE 15 (21–25 June 2015), Versailles, France.

8 Hoang, A.T., Wang, L., Serdyuk, Y.V., et al. 2015. DC electrical conductivity of LDPE-based nanocomposites. B6.5, JICABLE 15 (21–25 June 2015), Versailles, France.

9 EN 50307. (2002). *Lead and lead alloys – lead and lead alloy sheaths and sleeves of electric cables*. CENELEC.

10 EN 10257-2. (2011). *Zinc or zinc alloy coated non-alloy steel wire for armouring either power cables or telecommunication cables – Part 2: Submarine cable*. CENELEC.

11 CIGRE TB 490. (2012). *Recommendations for testing of long AC submarine cables with extruded insulation for system voltage above 30 (36) to 500 (550) kV*. CIGRE.

12 CIGRE TB 623. (2015). *Recommendations for mechanical testing of submarine cables*. CIGRE.

13 CIGRE TB 722. (2018). *Recommendations for additional testing for submarine cables from 6 kV (U_m = 7.2 kV) up to 60 kV (U_m = 72.5 kV)*. CIGRE.

14 IEC 63026 ED1. (2011). *Submarine power cables with extruded insulation and their accessories for rated voltages from 6 kV (U_m = 7.2 kV) up to 60 kV (U_m = 72.5 kV) – test methods and requirements*. IEC.

15 IEC 60840. (2011). *Power cables with extruded insulation and their accessories for rated voltages above 30 kV (U_m = 36 kV) up to 150 kV (U_m = 170 kV) – test methods and requirements*. IEC.

16 IEC 62067. (2011). *Power cables with extruded insulation and their accessories for rated voltages above 150 kV (U_m = 170 kV) up to 500 kV (U_m = 550 kV) – test methods and requirements*. IEC.

17 IEC 62895. (2017). *High voltage direct current (HVDC) power transmission – cables with extruded insulation and their accessories for rated voltages up to 320 kV for land applications – test methods and requirements*. IEC.

18 ELECTRA 171. (1997). *Recommendations for mechanical tests on submarine cables*. CENELEC.

19 IEC 60287. (2014). *Electric cables – calculation of the current rating – Part 1-1: Current rating equations (100% load factor) and calculation of losses – general (Note: IEC 60287 consists of a number of parts)*. IEC.

20 IEC 60853. (1985). *Calculation of the cyclic and emergency current rating of cables. Part 1: Cyclic rating factor for cables up to and including 18/30(36) kV (Note: IEC 6083 consists of a number of parts)*. IEC.

21 CIGRE TB 640. (2015). *A guide for current calculations of insulated cables*. CIGRE.

11

MV and HV Cables

Detlef Wald

Management, Eifelkabel Detlef Wald, Rebenhügelweg 8, Villmergen 5612, Switzerland

11.1 History of Cables

The first recorded cable was reported to have been used in Russia by a mining engineer in 1812. After this there were the famous rigid "Street Pipes" invented by Thomas Edison in 1880 and operated at 110 V, with direct current (DC). This was followed by Ferranti who invented paper cables in 1891. The cable in Pearl Street Substation, USA, has been in operation for 90 years at the time of writing. Most of the paper cables have an operation time of 65–75 years with their life being limited by the lead sheath or the conduction being too small where more power needs to be transmitted [1].

11.2 Today's Cables

Today, cables are mainly produced by extrusion, meaning using a solid dielectric. There is however still a significant amount of the so-called paper cables produced and installed around the world, even if the length is declining.

11.2.1 Paper Cables

There are different kinds of paper cables, with certain types still being produced in some parts of the world.

For medium voltage there are two different types of the so-called PILC (paper insulated lead cables). In North America these cables are mainly three-core oil filled cables covered with a lead shield. In Europe and Africa it is a mass impregnated paper that is used. However, in all areas the production facilities are closing down due to lower demand. There is still a big amount of mass impregnated non-draining (MIND) cables installed especially for medium voltage. Also, the highest voltage cable installed in 2018 was a polypropylene-paper laminated (PPL) cable with 600 kV DC.

The Global Cable Industry: Materials, Markets, Products,
First Edition. Edited by Günter Beyer.

11.2.2 Extruded Cables

Today, primarily solid extruded cables are used for new medium voltage cables. The insulation compound is polyethylene in its cross-linked form. Polyethylene was firstly used in thermoplastic form but quickly evolved into cross-linked form. In North America also HMWPE (high molecular weight polyethylene) was used. At the beginning the cross-linking was done in a CCV line using steam to heat up the polymer. After the failure of the first cables due to water trees, steam was replaced by nitrogen. This also gave the advantage of getting higher temperature and thus higher line speed.

11.3 Conductor Design

The conductor materials used are aluminum and copper; they can be either stranded or solid type. The latter is very seldom used as solid conductor. For extra high voltage cables the wires in the conductor can also use enamelled wire or oxidised ones in order to reduce the skin effect for bigger conductors.

The conductor is mostly first covered with semiconductive tapes, which can have different functions, such as water blocking and couching. Also water blocking yarns or powders are used for strand blocking. These tapes can contain a water-absorbing polymer, the so-called super absorbent polymers (SAPs), carbon black, and fibers.

11.4 Conductor Screen

The next layer is an extruded semiconductive polymeric cross-linked shield. Certain utilities also accept a non-cross-linked conductor shield. However, this will reduce the impulse strength and the resistance against the growth of water trees (Figure 11.1).

The conductor shields contain normally a furnace carbon black, peroxide, antioxidant, and an ethylene copolymer. Mostly the ethylene copolymer is either an ethylene-acrylate or an ethylene-vinyl-acetate. The latter gives better adhesion to aluminum. The first gives better heat stability as it is shown as heat loss in Figure 11.2.

In North America certain utilities require acetylene black for the conductor screen for medium voltage, whereas normally these compounds are used for extra high voltage or DC application (Figure 11.3).

In Figure 11.4 the impulse test results are compared on a so-called mini cable. The conductor size was 25 mm^2, stranded aluminum. Two different insulation and three different semiconductive conductor screens were used in this test. The so-called Supersmooth® one does not show better results than a good conventional one.

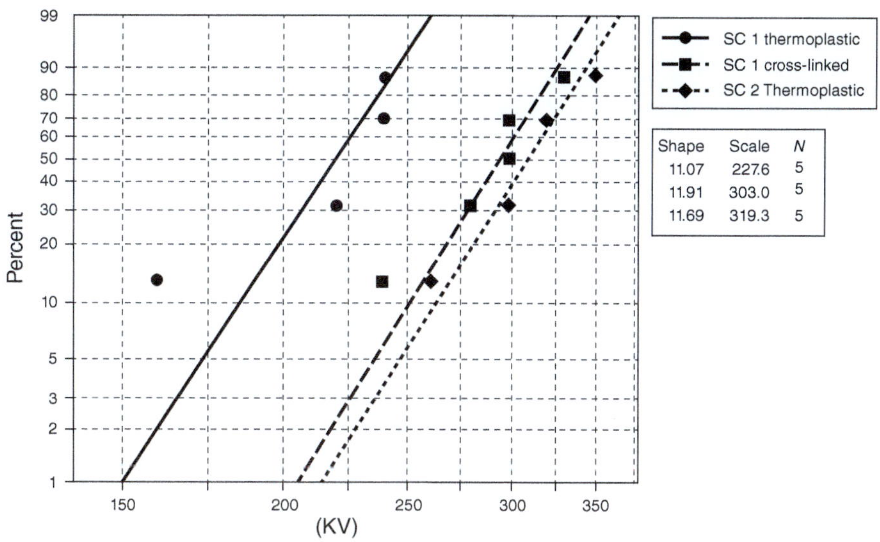

Figure 11.1 Impulse strength depending on cross-linking of the conductor screen.

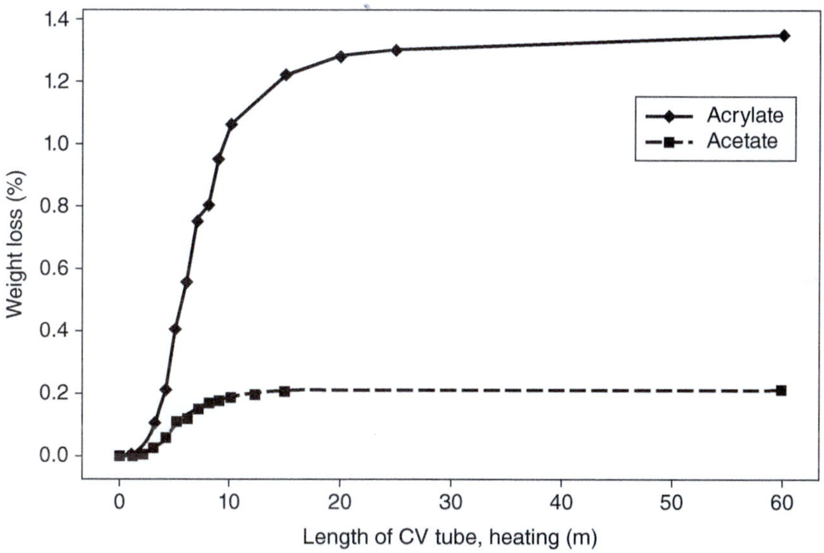

Figure 11.2 Loss of weight during the cross-linking process in a CV tube.

11.5 Insulation

Either cross-linked polyethylene (XLPE) or ethylene propylene rubber (EPR) is used as insulation for medium voltage cables. After the discovery of water trees in the 1970s the requirements for insulation have been increased. The cleanliness of the insulation has to be at a certain level in order to avoid the starting points of water trees.

(a) (b)

Figure 11.3 Surface finish of conventional (a) and Supersmooth® (b) screens.

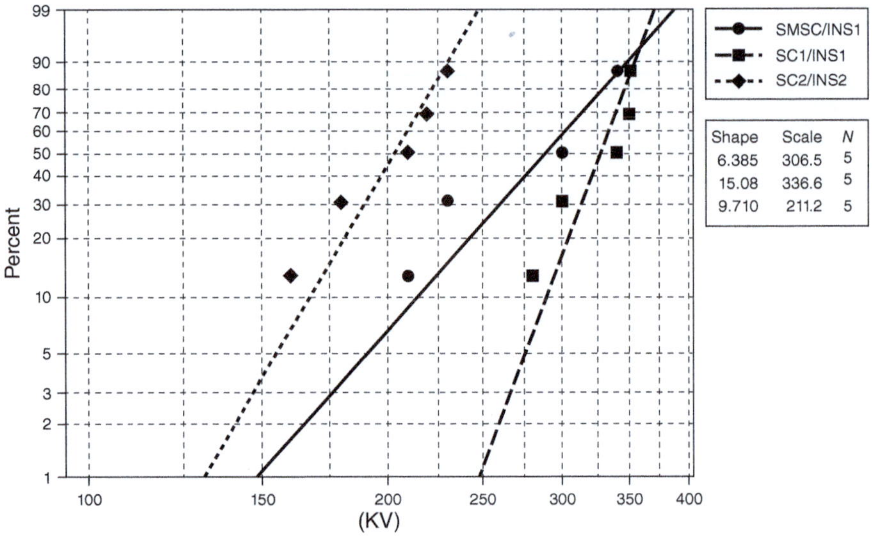

Figure 11.4 Impulse performance depending on the screen with medium voltage cables.

It should be highlighted that some of the properties of polyethylene vary in commercially available material. Included will be some reflections on the inherent variability of a material that is produced in production campaigns of many thousands of tonnes. We will also go into more detail about commonly used XLPE, which is the main material used for power cable applications [2].

Here we would like to describe the differences that exist in commercially produced polyethylene, specifically high pressure, low-density polyethylene (LDPE), since these are most commonly used in power cable applications as base resin. The basic "surprise" for people might be that LDPE, even if it is called hydrocarbon, contains not just these two basic atoms, hydrogen and carbon. We need to understand the production process in order to know that oxygen might be included into the chain. Oxygen might have a significant influence on the electrical properties – and in this case we are not speaking of "oxidized particles" that are generated later during the production process. This oxygen might come from the chain transfer agent (CTA) or the initiators that are used when producing LDPE in the reactor.

We need to further understand the difference between the two production methods that are used to produce LDPE: the autoclave and the tubular reactor, since this also might have an influence on the performance of the cables and its properties, especially cleanliness.

Insulations for cables, unlike almost all other polymer components, are required to have a very long operational life (>50 years) and are installed in complex locations that make their replacement problematic. Thus, as a consequence of the need for longevity the different antioxidants used will also influence the overall properties of the cable insulation. They will additionally influence the decomposition of the peroxide that is used for cross-linking of the polyethylene. The chemistry of the peroxide needs to be examined further. Generally, the compounder uses dicumyl-peroxide, which decomposes into several by-products (Table 11.1). The influence of these by-products themselves is evaluated under several conditions. The ratio of these by-products during the cable production nevertheless depends on the chemical composition of the XLPE itself, the production parameters, and the degassing process.

11.5.1 Production of Polyethylene

The first production of polyethylene (LDPE) by Pechmann only worked because the autoclave was leaking. Oxygen was unknowingly introduced into the process. We can see that during the production of LDPE we need oxygen in addition to hydrocarbon. Today, oxygen is introduced either directly or via peroxides. The type of peroxide might vary from producer to producer and with the product. This oxygen is incorporated in low amount into the polyethylene chain. The amount of oxygen in the chain has an influence on the electrical properties such as loss factor and resistance. Simultaneously, a CTA is added into the reactor. This agent regulates the chain length of the polyethylene and is added into the chain of LDPE [3].

11.5.1.1 Influence of the Chain Transfer Agent

During the polymerization of ethylene different types of CTAs such as alkanes, α-olefinens, or ketones are used. Figures 11.5 and 11.6 show the influence of commercially used CTAs on the loss factor of XLPE with a common antioxidant (AO1) for this process. The loss factor has been measured at low stress level with a Shering bridge using a low electrical field and a pressed plaque as test specimen. On a tested cable the loss factor can be significantly higher depending on an increased electrical field. The influence of antioxidants on the electrical properties will be explained.

Table 11.1 List of peroxide by-products.

Chemical name	Percentage (typical), not all substances identified
Ethylbenzene	0.09
Styrene	0.14
Cumene	2.58
Phenol	0.13
α-Methylstyrene	23.93
Methylacetophenones	0.52
C12 hydrocarbon	0.67
O-containing aromatic	0.04
C2-styrene	0.18
Acetophenone	45.3
Cumyl alcohol	10.6
2-Methoxy-2-phenlypropane	0.58
C5 benzyl derivative	0.18
C3 styrene	0.15
C6 benzyl derivative	0.38
Di-isopropenyl acetophenones	1.4
Diacetylbenzene	1.11

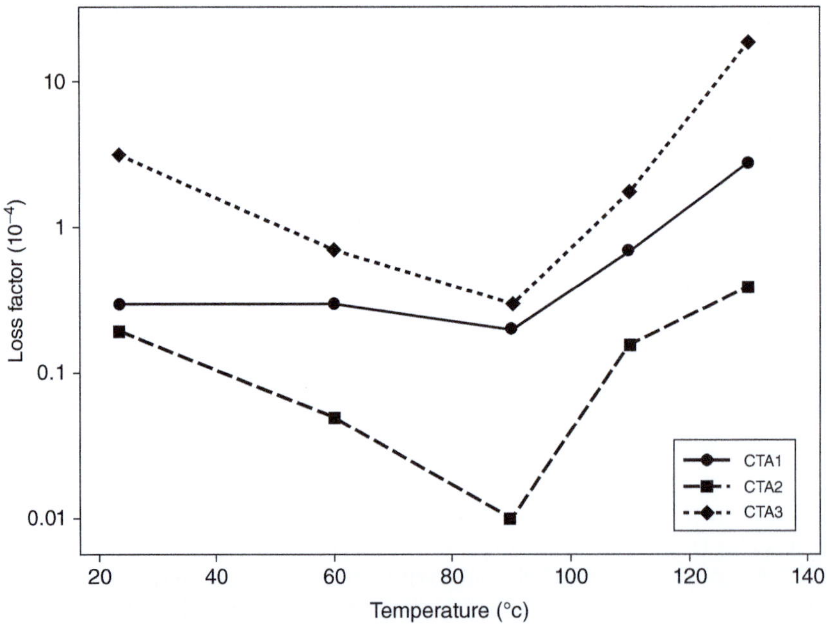

Figure 11.5 Loss factor depending on the type of CTA used [4].

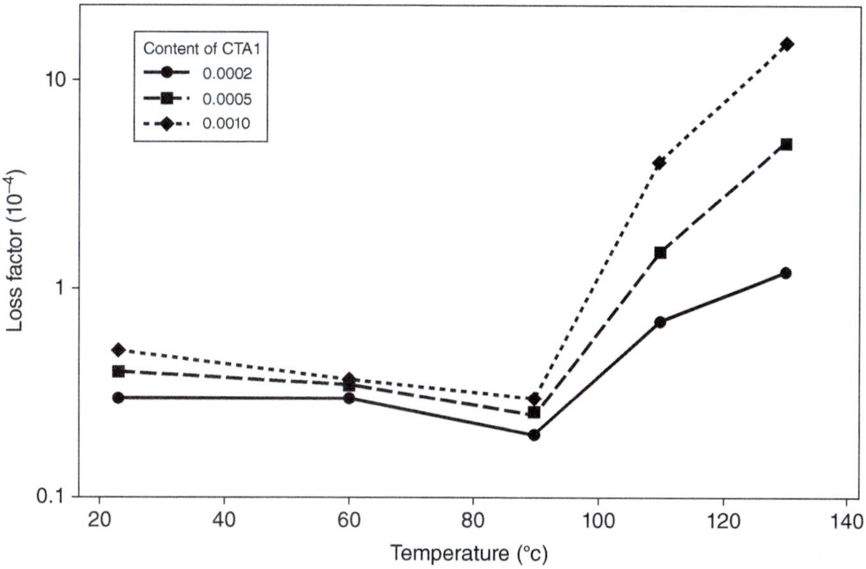

Figure 11.6 Loss factor depending on the temperature and amount of chain transfer agent CTA2 [5].

Depending on the process and the producer the amount of CTA could vary. Figure 11.6 shows the influence of the amount of CTA on the loss factor. As mentioned earlier this influence could increase with a high electrical field. Certain types of CTA will create a polyethylene that is not suitable for the cable application.

In a recent study by J.C. Fothergill et al., the influence of the CTA on the electrical resistivity was demonstrated [6]. It was mentioned in this study that decomposition products of peroxide will increase the differences between these two polymers. Additionally, if you further take the use of different types of peroxides into the equation, you can only estimate the variations in result in this test.

During the last few years, dienes have also been used as CTA [7, 8] to increase the reactivity of the polyethylene. The use of these dienes has much increased the reactivity of the polyethylene with peroxide. The total influence of the CTA on the properties of XLPE has not completely been studied and care should be taken if we transfer older knowledge to this cross-linking process since it is chemically not a first order reaction any more (Figure 11.7).

11.5.1.2 Cross-linked Polyethylene

11.5.1.2.1 Cross-linking Process After this short introduction of polyethylene we should now look more into XLPE, which is commonly used as insulation material in the cable industry. We will not discuss here the different processes of cross-linking [10, 11] since this would go too far. We will instead focus on the peroxide cross-linking process since this is the dominating process in the industry. The picture of XLPE is normally given very simplistically in the literature. However, recent studies by Smedberg et al. have shown that this

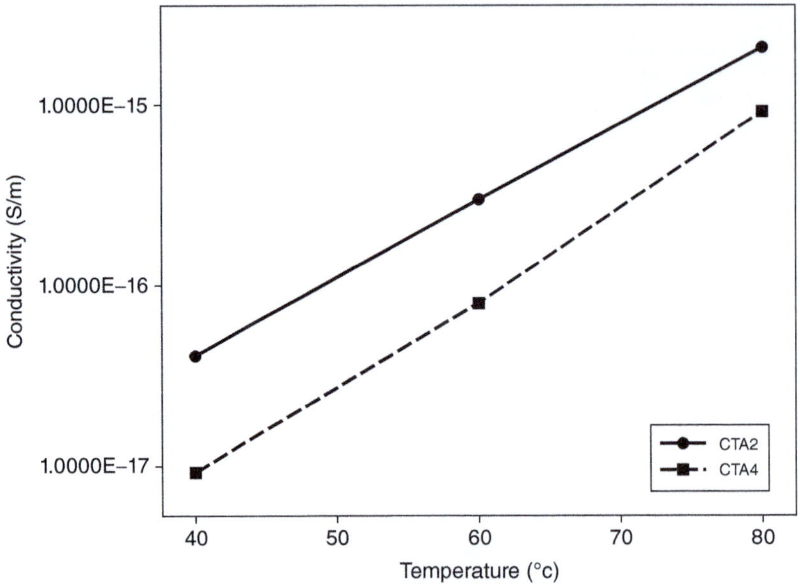

Figure 11.7 Resistivity depending on the type of CTA [9].

picture is wrong and XLPE looks more like a spider web with different types of cross-linking points [12]. In this research it is demonstrated that besides chemical cross-linking there are also physical cross-links and both are nearly equally strong and influence the structure of the XLPE. If we consider our previous knowledge of amorphous and crystalline structures, we need to understand what is happening with these structures during the cross-linking process and processing of the cable further on.

11.5.1.2.2 Influence of Peroxide and Antioxidant Polyethylene and especially XLPE are never used without peroxide and antioxidants. The choice of these additives will influence the morphology of the insulation. There are not too many studies evaluating the influence of different peroxides. This may be due to the fact that dicumyl-peroxide is normally used as a cross-linking agent. However, several studies have evaluated the morphology using different antioxidants.

The crystallization temperature (T_{cr}) of a pure LDPE was found to be 91.9 °C. This temperature can be influenced by additives such as antioxidants. Several combinations of antioxidants were mixed with polyethylene:

- 1,1′-Thiobis-(2-methyl-4-hydroxy-5-*tert*-butyl-methylphenol) (AO1)
- Pentaerythiritol-tetrakis-(3-3,5-di-*tert*-butyl-4-hydroxyphenyl)-propionate (AO2)
- Octadecyl-3-(3,5-di-*tert*-butyl-4-hydroxyphenyl)-propionate (AO3)
- 4,4-Thio-bis(2-*tert*-butyl-5-methyl-phenol) (AO4)
- Dioctadecyl 3,3′-thiopropionate (AO5)

The highest increase of Tcr was noted for the samples mixed with AO5 of 97 °C – a noteworthy increase in crystallization [13].

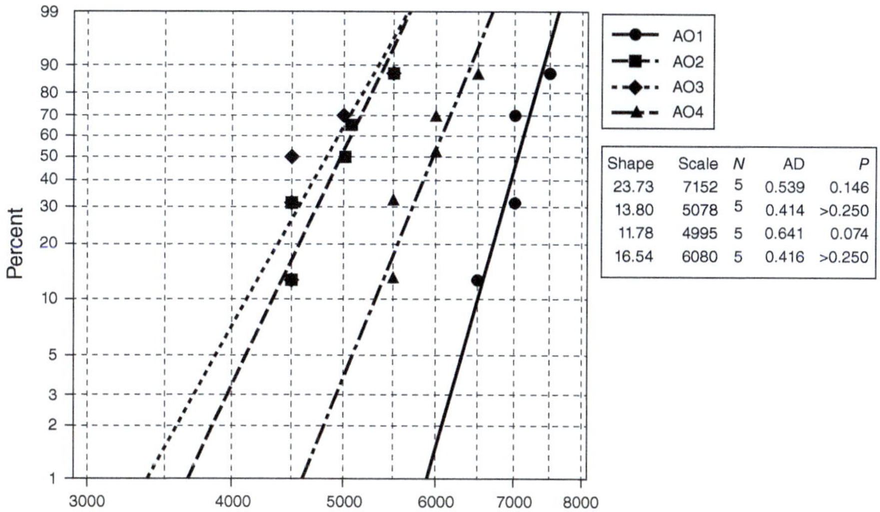

Figure 11.8 Tree inception voltage depending on antioxidant type.

11.5.1.2.3 The Influence of Antioxidants Several different kinds of mixtures of antioxidants are used in commercial XLPE formulation. The antioxidants are chosen both for their long-term protection (since cables are designed for a life time of 40 years) and for the interference or, better said, lack of interference in the cross-linking process. Several antioxidants are also used in combination. The influence of these antioxidants on the properties of XLPE has been evaluated in many studies and here we will just highlight some of them. A. Campus et al. demonstrate that an LDPE with methyl-ethyl-ketone (MEK) or butanon-1 as CTA has large fully rounded spherulites, which are present only in LDPE base resin without additives. Antioxidants such as AO1 will influence these structures [14]. The advantage of this antioxidant is that it is grafted during the cross-linking process at the polyethylene backbone and therefore cannot be removed by solvents or other physical processes [15]. In a study by K. Fukunaga [16] the influence of different antioxidants on the crystallinity and conductivity of polyethylene was evaluated. The polyethylene used is produced on a tube reactor with propane as CTA. This used AO5 together with a phenolic antioxidant AO2 and showed that the conductivity was dependent on the concentration and type of antioxidants (Figure 11.8).

A. Campus et al. confirmed [17] that the cross-linked segments of XLPE are situated in the amorphous phase and therefore do not modify the crystalline structure on the level of the elementary cells. They also confirmed that with AO4, the crystallinity decreases from 48% to 45% for this particular LDPE [18]. The electrical properties such as tree inception voltage were studied by J. Kjellqvist et al. [19]. They used as a base polymer a polyethylene that had been produced on a tubular reactor. More details about the polymer were not revealed in this study. However, several different antioxidants and combinations were mixed and their influence on the tree inception voltage compared to a virgin polymer. It is worthwhile to mention that this, like earlier, expressed only one of the parameters for how to choose the additives.

11.5.1.2.4 The Influence of Peroxide and Cross-linking Parameters The decomposition products of peroxide and the way they are created are described in several publications, e.g. [20, 21], and are mainly acetophenone (AP), α-methyl-styrene (MS), cumyl alcohol (CA), methane, and water. Methane should be removed out of the cable for safety reasons. However, also the other cross-linking by-products will influence the electrical properties under AC and DC conditions. The creation of acetophenone and cumyl alcohol is a competing reaction and, depending on the cross-linking temperature, one or the other is preferred. Cumyl alcohol is reacting to α-methyl-styerene and water [22]. In this study the amount of chemically created water was measured depending on the curing time and temperature. It clearly demonstrates that today's focus on speed might not be optimal for a good cable, but unfortunately economics force the cable maker in this direction. This study was done on a polymer that was produced on a tabular reactor with CTA1 and AO1. The amount of water created has been calculated by the amount of α-methyl-styrene, since there is also each time a small amount of "other" water inside the polymer and this gives the smallest error in the calculation.

It is known that acetophenone increases the AC tree inception voltage, decreases the insulation resistance, and increases the space charge after a DC voltage application. In addition, it has recently been pointed out that the dielectric loss factor of PE considerably increases at high temperature and high electrical stress region because of conductivity enhancement, in spite of PE being a nonpolar polymer. This study was done on a polyethylene with a melt index of 3, produced on a tubular reactor. The authors soaked the acetophenone into the polymer and also measured impurities of the added acetophenone. However, it is not clear if acetophenone created inside the polymer would have had the same effect [23]. This leads us to another subject of the influence of curing the by-products, which were, for example, studied by F. Aida et al. [24]. This study was done on an unknown polyethylene and discussed the influence of the by-products. It was demonstrated that during the time when these by-products evaporate the electrical losses and the electrical resistivity will change. This was only measured using the by-products created using dicumyl-peroxide.

T. Hayani et al. evaluated the influence of these by-products on the leakage current under DC stress [25]. He recommended removing them all in order to reduce the leakage current. Under practical conditions this will not be possible. One might reduce these products to an acceptable level determined by the manufacturer and the customer. At this point we also need to consider that in a real cable a semiconductive layer will also be present. In most of the commercial compounds supplied, at least for higher voltages, these compounds have a different peroxide as cross-linking agent, bis-(*tert*-butyldioxyisopropyl)benzene. This peroxide is a bis-peroxide and has, besides methane, completely different reactive by-products to dicumyl-peroxide, which are also far more difficult to remove. These by-products, present in a lower amount, will also migrate into the insulation during cross-linking and further processing of the cable, influencing the electrical parameters of the cable. In addition, the low molecular weight polar components of the semiconductive compounds will migrate into the insulation, changing the properties of the latter. This was confirmed by a study of J.A. Diego et al. on a medium voltage cable with a strippable screen [15]. The migration of low molecular weight of the outer and inner semiconductive layers into the insulation is dependent first on the chemical structure of these layers and their molecular weight distribution. The temperature in the cross-linking tube and the time will influence the migration from the outer layer. Additionally, the type of cross-linking tube such as VCV, CCV, or MDCV has an influence on the stress introduced into the cable and this factor will also influence the electrical properties of a cable [26].

Choo et al. studied the space charge accumulation under temperature and field gradient on a commercial medium voltage cable [27]. If we now take into account that we do not know anything about the construction of these cables, we understand how carefully we need to be interpreting results into a commercially produced cable for DC application, since the insulation and semiconductive materials are specially selected for this application and it is recommended to use a nonpolar screen with acetylene black for the inner and outer semiconductive screen, as described in the patent from former Alcatel Cable [28].

11.5.1.3 Cleanliness

As mentioned before the early XLPE compounds were quite dirty. This is one of the reasons why the so-called water tree tests have been developed. There are several tests used in the industry to evaluate compounds, but only two tests on cables are recognized around the world to evaluate the cable performance in wet conditions. It should be highlighted that these tests accelerate the growth of water trees and changing one of the parameters will give different results. In IEEE 1407 the different parameters are highlighted. The worldwide accepted tests are either according to Cenelec HD 605 protocol or according to ICEA S-97-682. Both tests will give different results. The Cenelec HD 605 protocol admits two different frequencies, 500 Hz for 3000 hours, and 50 Hz for two years. In order to simulate this, a two-year wet aging test has been carried out with a current XLPE compound, which does not fulfill the requirements in terms of cleanliness for medium voltage application. Three of the six cables failed during the test a 3 Uo (Figure 11.9).

Recently, new processes have been used where other peroxides are used. The decomposition products might be different and their influence needs to be investigated.

Additionally, tests have been performed using the same compound once before the material handling was upgraded to a class 1000 clean room and once afterwards. Also here the influence of cleanliness on the long-term performance has been demonstrated (Figure 11.10).

11.5.2 Insulation Screen

In a lot of constructions the insulation screen is of the same quality as the conductor screen. This means that the insulation screen is fully bonded to the insulation. However, in certain countries the adhesion of the insulation screen is

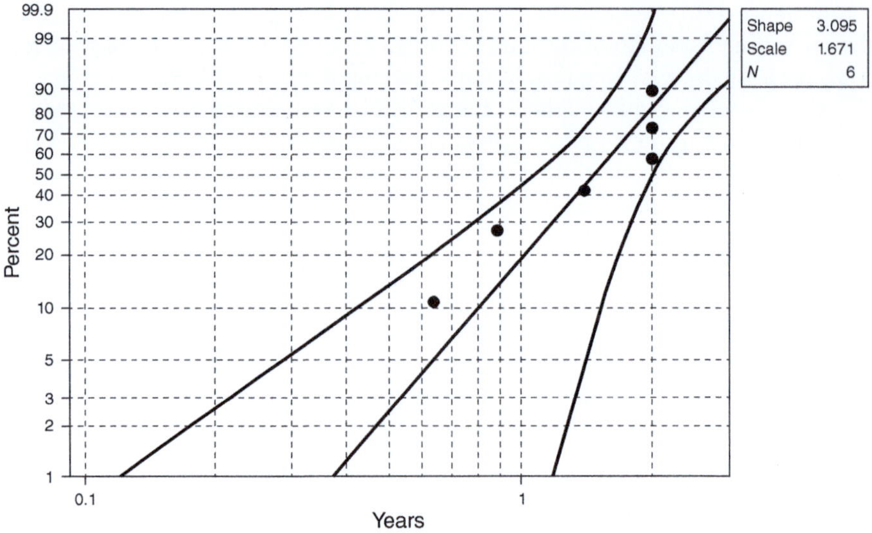

Figure 11.9 Influence of cleanliness on the results of a two-year test according to HD 605.

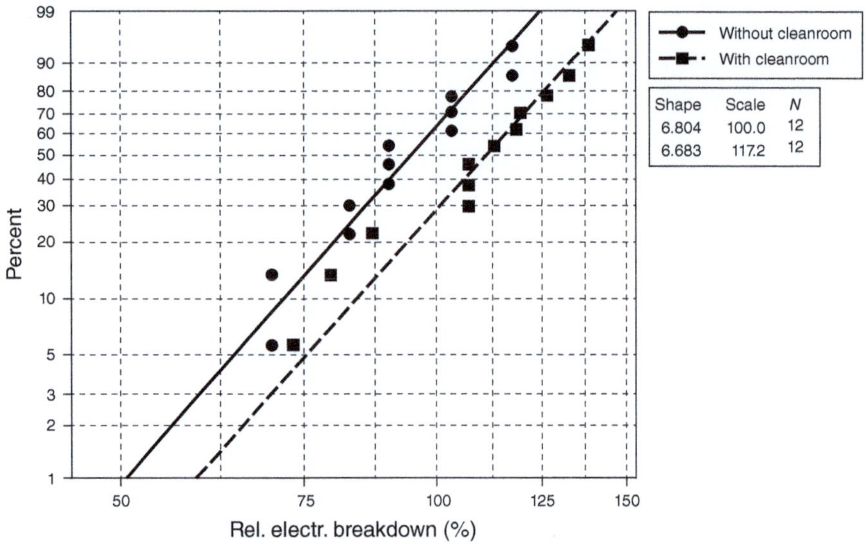

Figure 11.10 Influence of cleanliness in material handling on the relative results of a two-year test according to HD 605.

lower, which means that this screen is a so-called strippable screen and can be removed without using any tools. Strippable screens are only used in medium voltage cables.

In Figure 11.11 some disadvantages of a strippable screen are shown [29]. The partial discharge level of cables with a strippable screen is higher than that with bonded ones. It cannot be judged if this has an influence on the lifetime

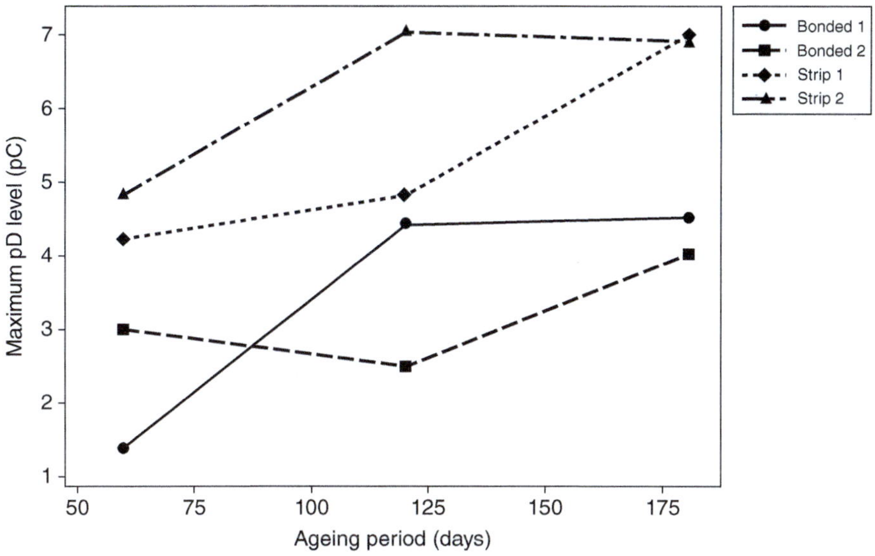

Figure 11.11 Partial discharge measurements on cables with bonded and strippable screens.

or the long-term performance of these cables. However, in Norway, which changed from strippable screens to bonded screens, this showed a lower SADIF value [30].

11.5.3 Metallic Screen

Much work has been done to determine the water tree retardancy of cable cores. There are also many studies evaluating this property of high voltage cables, which are laid into the sea bed. However, most of the tests have been done according to HD 620 with respect to HD 605 test protocol on cable cores, neglecting the fact that power cables are not laid without metallic and polymeric protection. This protection is not only necessary for preventing water penetration but also as a mechanical protection for the cable during laying and installation, and, later on, during the mechanical and thermomechanical movements throughout its entire life.

Cable cores produced with modern techniques and clean compounds still have more than 14 times the electrical design strength even after aging for two years in water. Studies carried out by H. Kober [31] highlighted that water in the screen is as dangerous as that in the conductor and therefore water should be blocked completely out of the cables.

Therefore, it is mandatory for high voltage cables to keep water out using a tight construction thereby avoiding the use of special water tree retardant compounds, since the losses in these compounds are higher than in high voltage insulation compounds.

This protection is done by metallic screens and a good polymeric jacketing. The polymeric jacket has the function of protecting the metal from corrosion and should also have a low water penetration.

Today, several different mechanical protections are applied to high and extra high voltage cables such as lead, corrugated copper and aluminum, aluminum or copper foil, and hybrid screens based on aluminum foil. The testing of the latter has been described in one of the latest Cigre brochures [32]. This paper will outline the advantages and disadvantages of the different screen materials, which normally have a double function: mechanical protection and carrying the short-circuit current in case of a short-circuit.

Naturally, one of the properties is also the weight of the cable, where the older constructions with lead have a significant disadvantage.

These screens also have a significant influence on the cost of the cable. However, one should consider that the cost of the cable is not the only criterion that should be taken into account when choosing the cable. It is more expensive to replace a cable after seven years of service than to pay more at the beginning.

Furthermore, the protection of the cable has to be adapted to the ground and the environment where the cable is laid.

11.5.3.1 Different Types of Metallic Protections

The metallic protection has two different tasks in the cable: first, protecting the cable against water penetration and mechanical protection; secondly, the electrical functions as equipotential screen, capacitive current collection, and draining and short-circuit draining. The main designs for high voltage cables for providing a moisture barrier function are extruded lead sheath or corrugated aluminum or copper sheaths, longitudinally welded corrugated sheath of aluminum or copper, longitudinal tape, and laminated foil of copper or aluminum. The latter are constructed with overlap or seam welded. There are also constructions that only provide short-circuit draining but not water-tightness like copper helically lapped tapes or concentric wires of copper and aluminum. These are not recommended for high and extra high voltage cables since water can penetrate into the cable.

11.5.3.1.1 Lead Sheath Cables The oldest metallic shield used is the lead shield, because originally all fluid filled cables had lead sheaths (Figure 11.12).

It is a well-known industrial process where the risks and problems during protection are well documented and known. Some types of lead alloy may fail by cracking after all extensions. This limits the choice of alloy, and the metallurgical

Figure 11.12 Typical design of an extruded lead sheath with copper wire.

structure associated with the type of extrusion press also has to be taken into account. For good application, manufacturers use an alloy with tin and cadmium or an alloy with tin and tellurium.

The cable should have a very good bending radius and be easily connected to accessories. The problems are in the vibration resistance, the weight, and the environmental friendliness of the cable. Additionally, the corrosion resistance against certain types of soil is not good and its resistance to DC stray current is limited and can result in cracking [33].

Stray voltage describes the occurrence of voltage between two objects that should not have any voltage difference between them. Small voltages are often measured between two grounded objects in distant locations, due to normal current flow in the power system. Large voltages can appear on the enclosures of electrical equipment due to a fault in the electrical power system, such as a failure of insulation.

Small stray voltages may never be noticed and may only be detected with a voltmeter. Larger voltages may have a range of effects, from barely perceptible to dangerous electric shocks.

11.5.3.2 Corrugated Screens

Another form of protecting cables is the use of corrugated metal screens. Here aluminum and copper have principally been used. There are also applications where smooth metal screens have been used. However, these cables are less flexible.

The corrugation process is performed either using a welding or pressing of the metal around the cable. It has already been used for oil filled cables and is therefore also quite well known in the industry.

The cables have the advantage of being lighter and more flexible than the commonly used lead sheath cables. It should be noted that the corrugation should be done in a linear form and not in the spiral way since this will influence the water barrier properties of the metallic screen. Only a linear corrugation will make the screen watertight (Table 11.2).

The test to measure watertightness of these screens has been performed many times and one of the test procedures is the ICEA T34-644. Cables with linear corrugation passed this test without any problem whereas the spiral ones failed.

11.5.3.2.1 Aluminum Corrugated Screens Aluminum is one of the lightest metals used in the cable industry. The advantages compared to lead are its significantly lower density, better thermal conductivity, and better electrical conductivity. Therefore, smaller screens can be used to drain the short-circuit current. It is also harder than lead and therefore provides better mechanical protection of the cable (Figure 11.13).

For shielding purposes it should be noted that aluminum is a paramagnetic material. This is a form of magnetism that occurs only in the presence of an externally applied magnetic field. Paramagnetic materials have a relative magnetic permeability greater or equal to unity, i.e. a positive magnetic susceptibility, and hence are attracted to magnetic fields. The magnetic moment induced by the

Figure 11.13 Typical design of a corrugated aluminum sheath.

applied field is linear in the field strength and rather weak. This gives it a disadvantage for magnetic shielding compared to lead and copper, which are both diamagnetic. This is the property of an object that causes it to create a magnetic field in opposition to an externally applied magnetic field, thus causing a repulsive effect. Specifically, an external magnetic field alters the orbital velocity of electrons around their nuclei, thus changing the magnetic dipole moment. This property has a strong influence when low magnetic fields are required.

Another disadvantage is that with a polyvinyl chloride (PVC) jacket corrosive protection is needed to avoid corrosion of the aluminum.

11.5.3.2.2 Copper Corrugated Screens Copper is the perfect waterproof metal barrier. It has the best resistance against corrosion and therefore needs no special protection against corrosion or oxidation. Copper is therefore used by preference in the naval industry. For example, the Statue of Liberty in New York is made out of copper for exactly that reason. The welding, specifically soldering to sleeve and connection pieces of flanges, is optimal. It has a better electrical conductivity than aluminum and therefore the corrugated copper sheath thickness can be made thinner.

It is also more flexible than aluminum given the cable has a better bending radius than any of the abovementioned constructions. Its resistance in relation

Table 11.2 Selected properties of selected metallic screens.

Property	Lead (Pb)	Aluminum (Al)	Copper (Cu)
Melting point (°C)	327	660	1084
Density (g/cm^3)	11.34	2.7	8.94
Electronegativity, according to Pauling	2.33	1.61	1.9
Magnetic property	Diamagnetic	Paramagnetic	Diamagnetic
Thermal conductivity (W/(m K))	35.3	237	401
Thermal expansion (μm/(m K))	28.9	23.1	16.5
Young modulus (MPa)	16	70	128
Mohs hardness	1.5	2.75	3

to vibrations is favorable. So it can be used close to highways and railway tracks. The disadvantage is its higher price and therefore in a lot of cases today aluminum is used.

11.5.3.2.3 Combined Aluminum Foil with Concentric Copper During the evolution of XLPE cables, a combined sheath using an aluminum foil with concentric copper wires as main short-circuit drain has been developed.

Here the aluminum foil is used as a metal barrier and is fully glued to the high-density polyethylene (HDPE) sheath. LDPE and PVC should not be used due to the softness of the plastic materials. Additionally, PVC should not be used since the water permeability is too high corroding the aluminum foil. The advantage of polyethylene sheath compared to the PVC has been presented in several publications, for example in [34]. In worst cases using PVC and aluminum foil, the latter had completely disappeared after 10 years in the ground leading to excessive cable failures. Even if the aluminum foil is not taken into consideration calculating the short-circuit capacity of the screen, a small amount of current is also drained throughout the foil. There are also constructions where the aluminum foil is replaced by a copper one.

11.5.4 Comparing Different Metallic Screens

The biggest advantage of copper and aluminum compared to lead is the weight of the cable. There is an additional advantage in the diameter of the cable. Both result in longer length on the reel, which will reduce transport and laying costs, since longer length can be transported to the construction side. Additionally, today the highest possibility of cable failures is still in the accessories of the cable system [35] (Figure 11.14).

The second difference is the bending radii of the cables. This results in a much easier use of the cables in the construction side. Especially in substations and other difficult environments, a shorter bending radius can be helpful during installation and operation of the cable system and may result in better reliability of the cable since it is less prone to mechanical stress.

In Figure 11.15 the relative transport ampacity of a 110 kV XLPE cable is shown. The difference between these four cable types is minimal and shows a

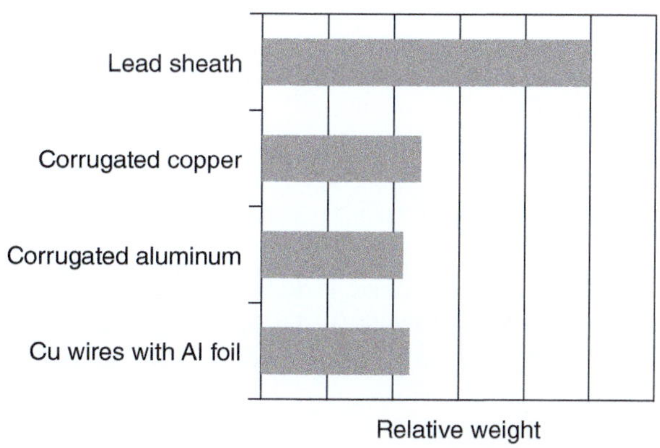

Figure 11.14 Relative weight of the same cable core with different screens.

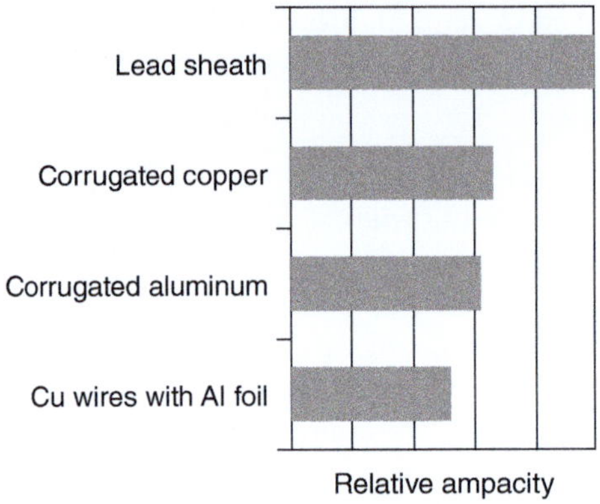

Figure 11.15 Relative ampacity of the same cable core with different screens.

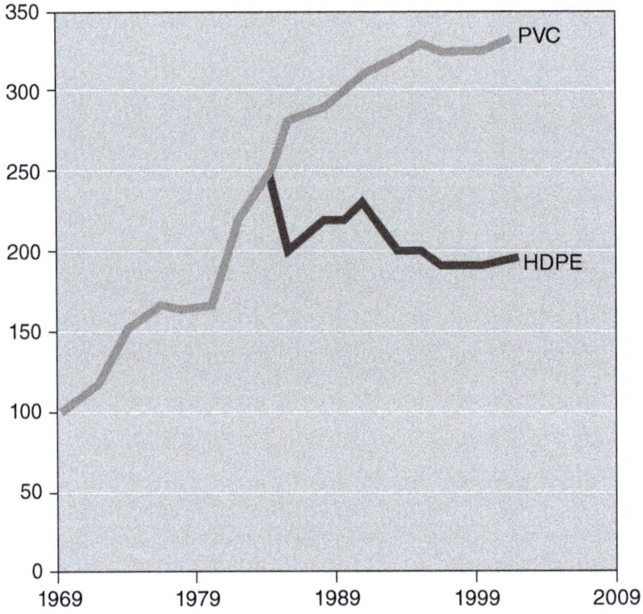

Figure 11.16 Laying costs of cables [36].

small advantage in the direction of the lead sheath. However, the advantage is not that significant that it would influence the choice of using a lead sheath.

11.6 Jacketing/Sheathing

Further on, a polymeric jacket is added via extrusion. These materials are PVC, linear low polyethylene (LLDPE), medium density polyethylene (MDPE), HDPE, polyamide (PA) or flame-retardant non-corrosive compounds (FRNC). PVC is generally seen as very soft and does not give much mechanical and moisture protection to a cable.

Figure 11.16 shows the decrease in laying costs in Germany when changing from soft PVC jacketing to harder HDPE jacket.

References

1 Langzeitprüfungen an VPE-Kabeln (1994). Einfluss der Prüfparameter auf die Alterung. In: *Elektrizitätswirtschaft Jg 93*, Heft 26 (ed. H. Kober).
2 Cigre Working Group B1-25. (2011). Advanced design of laminated metallic coverings. Report.
3 Tharby, R. The use of lead sheathed cables by British Telecom. In: .
4 Kirchner, M., Phillipcyk, C., Dammert, R., and Wald, D. (2003). The fifth generation of over sheath materials. *Paper, JiCable.*
5 Fothergill, J.C. and Nilsson, U. (2011). The measurement of very low conductivity and dielectric loss in XLPE cables. *IEEE Transactions on Dielectrics and Electrical Insulation* 18 (5).
6 Cigre Working Group B1-10. (2009). Update on service experience of HV underground and submarine cable systems. Cigre Report.
7 Wald, D. (2012). Polyethylen Mischung für HGÜ-Anwendungen, *RCC Conference*, Berlin, Germany.
8 Becker, P. (2001). Initiierungseffektivität von linearen und zyklischen Dialkylperoxiden in der Ethen-Hochdruckpolymerisation. Dissertation. Universität Gießen.
9 Smedberg, A., Hjertberg, T., and Gustavsson, B. (1997). Cross-linking reactions in an unsaturated low density polyethylene. *Polymer* 38 (16).
10 Akermark, G. (1999). *Vernetztes Polyethylene. VDI Kabelextrusion 199*. ISBN: 3-18-234217-7.
11 Smedberg, A. and Gustavsson, B. A cross-linking agent. EP 1944327A1
12 Thienel, P., et al. Einfluss der Verarbeitungsbedingungen auf das Eigenschaftsbild von Kunststoffen, Kautschuk + Gummi, 31. Jahrgang, Nr. 8/1978.
13 Wald, D. (2003). Mischungen für Wechsel- und Gleichspannungskabel. *VDI Kabelextrusion*. 3-18-234250-9
14 H. Orton, R. Hartlein, *Reliable XLPE Compounds, Borealis and Dow* 2005
15 Smedberg, A., Hjertberg, T., and Gustavsson, B. (2003). Effect of molecular structure and topology on network formation in peroxide cross-linked polyethylene. *Polymer* 44 (11).
16 Campus, A., Lafin, B., and Sterzynski, T. (1998). Structure – morphology modification of cable insulation polymers. *IEEE International Conference on Conduction and Breakdown in Solid Dielectrics*, Västeras, Sweden (22–25 June).
17 Cieloszyk, G.S. (1976). Polyethylene antioxidant fundamentals. BP Chemicals Brochure.
18 Fukunaga, K., Fukuma, Nagao, M., Hozumi, N., Kosaki, M., and Kohno, Y. (1991). Measurement of space charge distribution in cable insulation using the pulsed electro-acoustic method. In: *Third International Conference on Polymer Insulated Power Cables*, 520.

19 Campus, A. and Smedberg, A. (1998). Morphology of polyethylene for power cable insulation effects of antioxidant and cross-linking. *IEEE International Conference on Conduction and Breakdowns in Solid Dielectrics*, Västeras, Sweden (22–25 June).
20 Kjellqvist, J., et al. (2003). New XLPE insulation materials for HV and EHV cables. *Paper B2.2, JiCable.*
21 Andrews, T., Hampton, N., and Wald, D. (2006). The role of degassing in XLPE power cable manufacture. *IEEE Electrical Insulation Magazine* 22 (6).
22 D. Wald and W. Weisenberg (2005). Entgasen von VPE-Kabeln – ein notwendiges Übel? 11, 26–29.
23 Campus, A. and Matey, G. (1983). The influence of cross-linking temperature on the properties of cross-linked low density polyethylene. *PRI Conference on Polymers*, Manchester, UK (18–19 May).
24 Nakatsuka, T., Takahashi, T., Miyata, H., Yokoyama, A., Ishikawa, I., and Niwa, T. (1994). The effect on dielectric loss of polyethylene caused by acetophenone and cumylalcohol. *Conference Record of the 1994 International Symposium on Electrical Insulation*, Pittsburgh, PA, USA (5–8 June).
25 Aida, F., Tanimoto, G., Aihara, M., and Hoshokawa, E. (1990). Influence of curing by-products on dielectric loss of XLPE insulation. *CEIDP.*
26 Hayami, T., Yamanouchi, S., and Yatsuka, K. (1981). Leakage current characteristic of XLPE. *Conference on Electrical Insulation and Dielectric Phenomena.*
27 Diego, J.A., Belana, J., Orrit, J., and Canadas, J. (2011). Annealing effect on the conductivity of XLPE insulation in power cable. *IEEE Transactions on Dielectrics and Electrical Insulation* 18 (5).
28 Wald, D. and Zupan, S. (2011). Vpliv Tehnologije Izdelave na lastinosti kablov za viskoein yelo visko napetosti (Slovenian), (Influence of processing on the properties of high and extra high voltage cables, Cigre Slovenia).
29 Choo, W., Chen, G., and Swingler, S. (2011). Electric field in polymeric cable due to space charge accumulation under DC and temperature gradient. *IEEE Transactions on Dielectrics and Electrical Insulation* 18 (2).
30 Prigent, M. et al. (1995). Câble insultive structure. EP 0644588A1.

12

Coaxial Cables

Timothy Cooke

Product Management, Amphenol Broadband Solutions, 380 Tightsqueeze Industrial Road, Chatham, VA 24531, USA

12.1 History

One would think that coaxial cables are a relatively modern invention for the communications industry, but this is not the case. The concept for coaxial cables was first patented in Great Britain by Oliver Heaviside, a British Mathematician and Engineer, in 1880, while studying the skin effect of conductors. He was granted a British patent, No. 1407, on 6 April 1880. Four years later in 1884, Siemens and Halske would obtain a similar patent. These patents opened the door and opportunity for accelerated development of communications cables, and especially for coaxial cable design. The first patent in the United States for coaxial cables is 1 835 031. The assignee was American Telephone & Telegraph (AT&T). This patent, for Concentric Conducting Systems, was granted on 8 December 1931, with Lloyd Espenschied and Herman A. Affel listed as the inventors. This patent defined the coaxial cable design as well as the communication network that used it, with amplifiers, repeaters, and filters discussed. Coaxial cables are illustrated as the transmission line for telegraph, telephone, and TV applications, albeit the primary purpose was for transmitting television signals (Figure 12.1).

Coaxial theory actually dates back to the days of Maxwell in the nineteenth century and the exploration of electromagnetism. In the mid-1800s the earliest telegraph and first telephone cables spanning the Atlantic were actually a central conductor, with a gutta-percha (latex) insulator. Seawater was actually used as the outer conductor for this coaxial design. The bandwidth for these early cables was 1.0–1.5 Hz.

In 1937, initial field tests were conducted utilizing coaxial cables for television transmission between New York and Philadelphia. The result was the first live broadcast of the 1940 Republican National Convention. World War II halted the work on TV broadcasting, but after the war TV broadcasting research promptly resumed, with coaxial cable making it possible. Applications during this time ranged from communications to control cables, radar, and a myriad of others.

The Global Cable Industry: Materials, Markets, Products,
First Edition. Edited by Günter Beyer.

Figure 12.1 Oliver Heaviside. Source: Smithsonian Libraries, https://commons.wikimedia.org/wiki/File:Oheaviside.jpg.

12.2 Design: Components and Principles

Transmission of direct current (DC) requires two conductors to complete the circuit. One of the conductors is often referred to as the "go" wire, and the other as the "return" wire (Figure 12.2).

With radio frequency (RF) transmission, the coaxial cable becomes a transmission line where the "go" wire is the inner conductor, and the "return" wire is the outer conductor. These two conductors share the same longitudinal axis, hence the name coaxial cable. The outer conductor serves as an electrical shield for the cable and is typically covered by an outer jacket sheath.

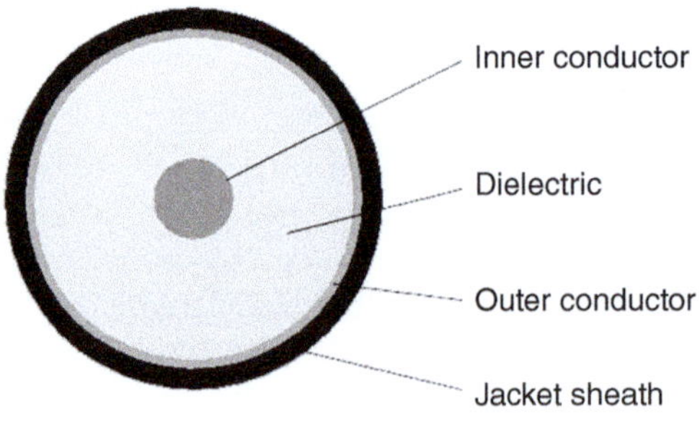

Figure 12.2 Coaxial cable components.

12.2.1 Components

The inner conductor is typically a copper or copper-clad material. Cost and weight may be deferred using cladded conductors, which take advantage of a concept known as the Skin Effect[1] and high-frequency signal transmission. Inner conductors can also be in the form of solid or stranded construction. Stranded conductors are utilized where a greater degree of flexibility is required for the application. Below are a few of the more common inner conductor materials used in basic coaxial cables:

1) Copper, solid
2) Copper, stranded
3) Copper-clad steel (CCS), solid
4) Copper-clad aluminum (CCA), solid
5) Silver-plated copper, solid
6) Others...

Separating the two conductors within a cable is done by an insulative material, better known as *dielectric insulator,* which maintains a uniform distance between the conductors. The uniform distance and structure are essential for a coaxial cable to exhibit the desired RF performance, i.e. characteristic impedance return loss and attenuation. The type of insulator used in the cable construction also provides mechanical strength and support for both the outer conductor and the overall cable design. The physical attributes of the dielectric medium have a significant influence upon the crush resistance stability, operating temperature, thermal stability, and flame-retardant properties of the coaxial cable. Both solid and foamed polyethylene (PE) dielectrics are flammable and are the primary fuel of this type of coaxial cable when exposed to flame. The type of dielectric insulator used in the coaxial design also has an equally important role to play with regard to the RF performance of the cable. Electrical properties for the materials that are identified are the dielectric constant and dissipation factor. The cable designer has to account for each of these features and select the component material best matched to the desired RF and mechanical performance of the coaxial cable and its application. The following are a few of the dielectric insulator options that are available today:

1) Solid low density polyethylene (LDPE)
2) Cross-linked polyethylene (XLPE)
3) Solid polytetrafluoroethylene (PTFE)
4) Solid fluorinated ethylene propylene (FEP)
5) Foam fluorinated ethylene propylene (FFEP)
6) Gas-injected foamed polyethylene (GIFPE)
7) Air/fused disk
8) Air/spiral wrap
9) Solid polypropylene (PP)
10) Foamed polypropylene (FPP)
11) Others...

The third component of a coaxial cable is the *outer conductor*. Just as with the inner conductor and dielectric insulator, there are numerous outer conductor

1 Skin effect is the tendency of an *alternating electric current* (AC) to become distributed within a *conductor* such that the *current density* is largest near the outer surface of the conductor and decreases with greater depths in the conductor – Wikipedia

options available, depending on the application and performance desired. The outer conductor serves to provide RF shielding for ingress and egress. The outer conductor also serves to complete and complement mechanical integrity of the coaxial cable design. The earliest coaxial cables only utilized one outer conductor whereas today engineers and designers use as many as four different layers of outer conductors to achieve either mechanical strength or enhanced shielding performance, or both. The four most common outer conductor layers used are as follows, and in order from dielectric, outward:

1) Aluminum, tape, laminate w/polyester/polypropylene strength member, longitudinally – bonded to dielectric
2) Aluminum, braid
3) Aluminum, tape, laminate w/polyester/polypropylene strength member, longitudinally – non-bonded
4) Aluminum, braid

Various types of outer conductor components are mixed or matched to achieve the desired performance. A sampling of the many available outer conductor components is given below:

1) Aluminum tube, smooth, solid or welded
2) Copper tube, corrugated, welded
3) Copper, braid
4) Copper, tape, solid, longitudinally or helically applied
5) Copper, tape, laminate w/polyester strength member, longitudinally or helically applied
6) Aluminum, braid
7) Aluminum, tape, solid, longitudinally or helically applied
8) Aluminum, tape, laminate w/polyester/polypropylene strength member, longitudinally or helically applied
9) Tinned copper, braid
10) Silver-plated copper, braid
11) Tin-soaked copper, braid
12) Others…

Together, the three components discussed thus far are required in the basic makeup of a coaxial cable. However, in industries where coaxial cables are used today, an additional outer jacket sheath for protection and/or flame retardancy is often part of the design. Coaxial cables may also include internal floodants for moisture or corrosion protection, armor for added physical protection, an integral wire for aerial installations, or even a fiber or twisted pair component attached to the cable sheath. Such additional components have greatly increased the use of coaxial cables and extended the longevity of the product based on performance and economics, when compared to other options available. The following are a few of the more common jacket sheath materials:

1) Polyethylene (PE)
2) Polyvinylchloride (PVC)

3) Polyvinylidene difluoride (PVDF)
4) Thermoplastic elastomers (SEBS, TPU, TPE, and others)
5) Fluorinated Ethylene Propylene (FEP)
6) Flame-retardant non-halogen (FRNH) compounds

From the myriad of materials now available for each component of the coaxial cable design, the configuration possibilities are seemingly without end. Add to this the fact that the RF performance and design can be configured to provide varying impedance and attenuation values, it then becomes easier to understand how coaxial cables continue to have significance in today's technology driven world (Figure 12.3).

Of the many various configurations of coaxial cables that are available, shown here are but a few (Figures 12.4 and 12.5).

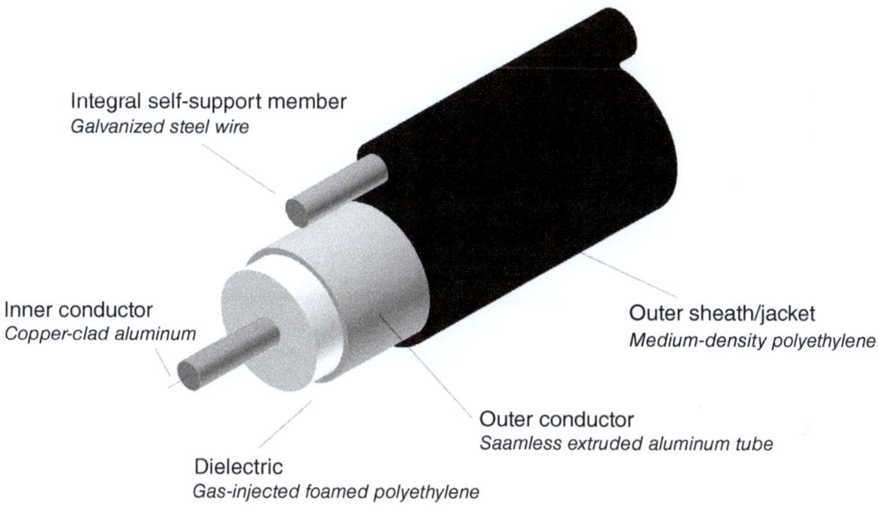

Figure 12.3 Typical hardline coax cable used in CATV installation.

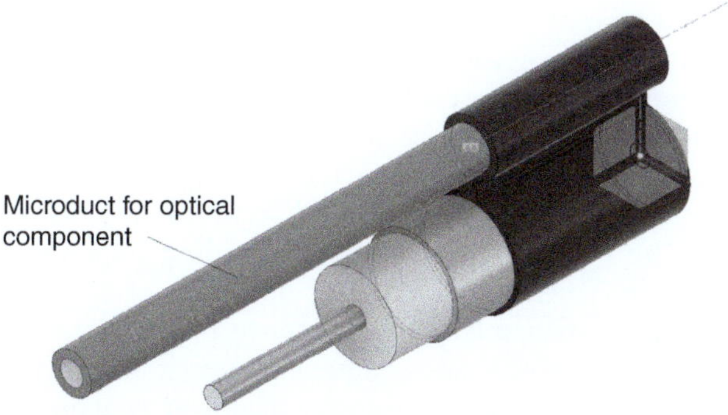

Figure 12.4 Hybrid coaxial cable.

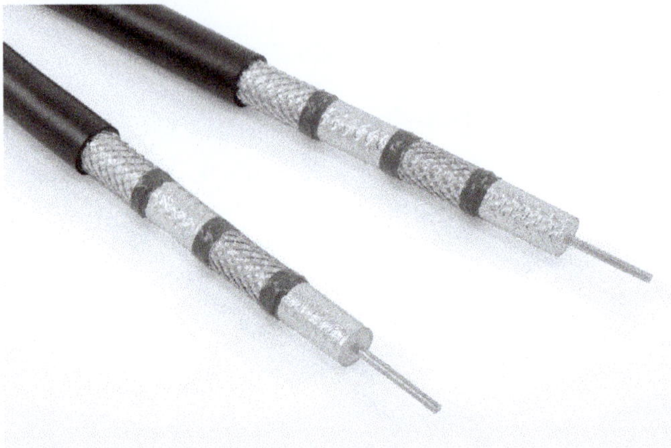

Figure 12.5 Drop coaxial cable, QuadShield. Source: Courtesy of Amphenol Broadband Solutions.

12.2.2 Principles

Now that the components and a few of the more basic materials are understood, attention will be focused on the basic principles of the design and function of a coaxial cable. Basically, coaxial lines or cables are a transmission system wherein electromagnetic waves are transmitted through a dielectric insulator bound by two conductive coaxial cylinders. This is accomplished via transverse electromagnetic (TEM) wave-field propagation within the coaxial cable. In this way, electrical energy, or signals, are transmitted from one location to another. The physics related to the function of coaxial cables is defined and explained in transmission line theory. For the purposes of this chapter, basic cable design formulary will be given and explained as related to general use and function.

12.3 Characteristic Impedance (Z_o)

Characteristic, or surge, impedance is the combined opposition to the flow of electrical energy within the coaxial line or cable. Impedance is a complex value, derived from the combined equivalent of the resistance, capacitance, and inductive and conductive values of the cable. This parameter is how the coaxial cable is generally defined, partly because it is derived from all of the other electrical parameters of the cable design. Characteristic impedance (Z_o) is expressed in ohms (Ω). The more common impedance values for today's coaxial cable are 50, 75, and 93 Ω. The type of coax and corresponding impedance will depend upon the application. To ensure the correct operation of a system using a signal source (e.g. a transmitter), a length of feeder (e.g. coaxial cable), and a load (e.g. an antenna), the coaxial cable impedance must match the source and the load. In this way the maximum transfer of power is achieved between the source and the coaxial cable, and then between the coaxial cable and the load.

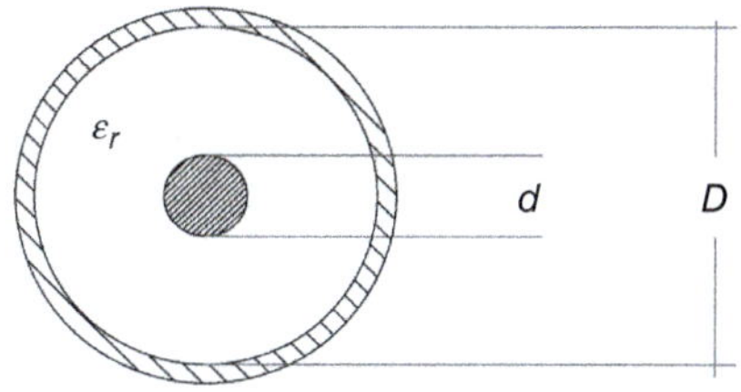

Figure 12.6 Coaxial cable impedance reference.

The best coaxial cable impedances in high-power, high-voltage, and low-attenuation applications were experimentally determined at Bell Laboratories in 1929. They were found to be 30, 60, and 77 Ω, respectively. For a coaxial cable with air dielectric and a shield of a given inner diameter, the attenuation is minimized by choosing the diameter of the inner conductor to give a characteristic impedance of 76.7 Ω [1]. When more common dielectrics are considered, the best-loss impedance drops to a value between 52 and 64 Ω. Maximum power handling is achieved at 30 Ω [2].

The approximate impedance required to match a center-fed dipole antenna in free space (i.e. a dipole without ground reflections) is 73 Ω, so 75 Ω coax was commonly used for connecting shortwave antennas to receivers. These typically involve such low levels of RF power that power-handling and high-voltage breakdown characteristics are unimportant when compared to attenuation. Likewise, with community antenna TV (CATV), although many broadcast TV installations and CATV headends use 300 Ω folded dipole antennas to receive off-the-air signals, 75 Ω coax makes a convenient 4 : 1 balun transformer[2] for these as well as possessing low attenuation. The arithmetic mean between 30 and 77 Ω is 53.5 Ω; the geometric mean is 48 Ω. The selection of 50 Ω as a compromise between power-handling capability and attenuation is in general cited as the reason for the number [1] (Figure 12.6).

Characteristic impedance calculations [3, 4]

Z_o	Characteristic impedance (Ω)
ε_r	Dielectric constant, dielectric
d	Diameter, inner conductor (in.)
D	Inner diameter, outer conductor (in.)
μ_{vac}	Magnetic permeability of vacuum (H/m)
ε_{vac}	Electric permittivity of vacuum (F/m)
V_p	Velocity of propagation (%)
η_{vac}	Impedance of vacuum

Constants

μ_{vac}	$4\pi \times 10^{-7}$
ε_{vac}	8.854×10^{-12}

2 https://www.markimicrowave.com/Assets/appnotes/balun_basics_primer.pdf.

Calculations

$$\eta_{\text{vac}} = \sqrt{\left(\frac{\mu_{\text{vac}}}{\varepsilon_{\text{vac}}}\right)}$$

$$Z_o = \left(\frac{\eta_{\text{vac}}}{2 \times \pi}\right) \times V_p \times \ln\left(\frac{D}{d}\right)$$

$$Z_o = 60 \times V_p \times \ln\left(\frac{D}{d}\right) \approx Z_o = \left(\frac{138.03}{\sqrt{\varepsilon_r}}\right) \times \log\left(\frac{D}{d}\right)$$

Example for a Series 6, Drop Coaxial Cable, derived Characteristic Impedance

$\varepsilon_r = 1.431\,54$; dielectric constant for foamed polyethylene [5]

$D = 0.180$ in.

$D = 0.0403$ in.

$$V_p = \left(\frac{1}{\sqrt{\varepsilon_r}}\right)$$

$$Z_o = \left(\frac{138.03}{\sqrt{\varepsilon_r}}\right) \times \log\left(\frac{D}{d}\right)$$

$$V_p = \left(\frac{1}{\sqrt{1.431\,54}}\right)$$

$$Z_o = \left(\frac{138.03}{1.1965}\right) \times \log\left(\frac{0.180}{0.0403}\right)$$

$$V_p = 0.83579$$

$$Z_o = 74.98$$

12.4 Velocity of Propagation

Velocity of Propagation or V_p is defined as the speed at which the RF signal travels through the coaxial cable with respect to the speed of light, which is 299 792 458 m/s in vacuum or free space. V_p is represented as a percentage of the speed of light in a vacuum. Using the example of a coaxial cable with a solid, LDPE dielectric, the V_p for a cable of this type is 0.66 or 66%. A similar cable, but which is constructed with a gas-injected, foamed polyethylene has a V_p of 0.836

Table 12.1 Dielectric Information.

Dielectric	V_p
Solid polyethylene	66%
Gas-injected foamed polyethylene (traditional)	82%
Gas-injected foamed polyethylene (new techniques)	90%+

or 84%. The higher the V_p of the coaxial cable, the faster the signals travel from the source to the load. In conjunction, the higher the velocity of propagation, the lower the attenuation losses of the coax design. These attenuation losses will be discussed later in this chapter, but the end result is that a smaller cable can achieve the same attenuation losses as a larger cable with a more dense dielectric or lower velocity of propagation. With technology and machinery evolutions, coaxial cable manufacturers have been able to produce dielectrics that yield 90% plus V_p for a GIFPE type. These achievements are part of the reason that coaxial cables continue to evolve and extend their relevancy in an ever-changing landscape (Table 12.1).

12.5 Capacitance

Capacitance of a coaxial cable is the ability of the cable to retain a charge. Therefore, the greater the capacitance value the more is the charge that can be stored and the longer is the time required for the signal to reach full amplitude. Capacitance of a coaxial cable is length dependent and expressed in pico farads/length (m). For design purposes, it is calculated utilizing the following formulary.

$$C = \frac{2 \times \pi \times \varepsilon_r \times 8.85}{\ln\left(\frac{D}{d}\right)} \approx C = \frac{2 \times \pi \times 1.431\,54 \times 8.85}{\ln\left(\frac{0.180}{0.0403}\right)}$$

$$C = 53.19\ \text{pF/m}$$

12.6 Attenuation

Attenuation is defined as the signal power loss within the coaxial cable, and is dependent upon the cable design, frequency, and length. The two main attributes of a coaxial cable that effect its attenuation are DC resistance and the dissipation factor of the dielectric insulator. After characteristic impedance, attenuation is the second most referenced attribute of a cable, as it defines the length of cable required to maintain the signal level within the closed network. The losses of a cable and accessories affect the need for amplification of the signal as well.

The units of measure for coaxial cable attenuation are neper per unit length, or dB per unit length. A neper is a unit of loss and gain based on the natural, or Napierian, logarithm. The decibel is a unit based on the common, or Briggsian, logarithm. The neper unit is 8.686 times larger than the decibel [6]. To obtain the total attenuation for a given length of cable, the conductive losses and dielectric losses are summed. The larger the cable the larger the conductive sources, thus reducing the attenuation losses.

Before we begin to explore the calculation of attenuation, there is one factor which is pertinent that should be explained. Conductive losses primarily reside along the inner conductor of the coaxial cable. This factor is referred to as skin effect or skin depth, which is defined as follows: "*Skin effect is the tendency of an alternating electric current (AC) to become distributed within a conductor such that the current density is largest near the outer surface of the conductor, and decreases with greater depths in the conductor.*"[3] As alternating current or RF signals increase in frequency, the energy travels generally along the surface of the conductor, penetrating the conductor only at lesser depths. This phenomenon has allowed manufacturers to utilize cladded conductors for applications where the DC resistance of the coaxial cables is not as critical as the losses at frequency. These materials can provide a conductor with less weight and greater tensile strength than a pure alloy conductor. Copper-clad aluminum and copper-clad steel are commonly used in the broadband industry to reduce weight on utility poles and infrastructure.

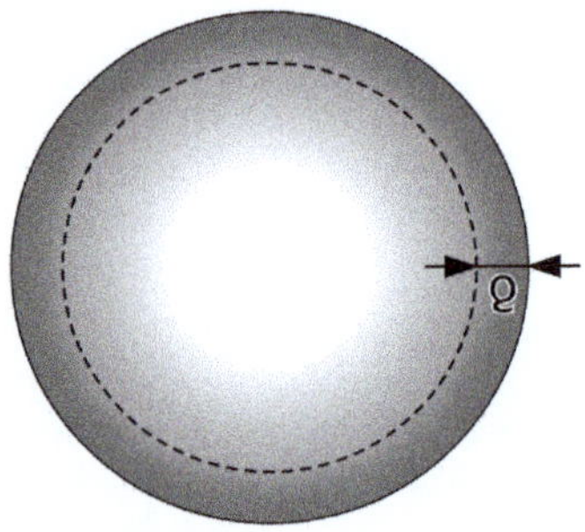

Copper-clad steel conductors are used to provide strength to drop coaxial cables, which allows installation for short spans without the need for a self-support member. The cladded materials are designed and manufactured to ensure that there is adequate copper content and depth to perform exactly as a solid conductor at frequencies of 5 MHz and greater. Copper tube conductors are used as the inner conductor of larger coaxial cables used in the cellular industry to reduce the weight and cost of the cable on the tower structure, and yet provide the desired RF performance.

Equations for skin effect are given below:

Constants for skin effect	Resistivity of conductor materials
ρ – resistivity of inner conductor (mW cm)	1.7241 μΩ cm (copper, soft)
f – frequency (MHz)	1.7930 μΩ cm (copper, hard)
μ – magnetic permeability of medium (H/m)	2.8282 μΩ cm (aluminum)
δ – skin effect or skin depth (in.)	1.6260 μΩ cm (silver)
	11.052 μΩ cm (tin)
	2.6525 μΩ cm (CCA)
	8.2100 μΩ cm (CCS – 21%)

3 Retrieved from Wikipedia website, 3 May 2019.

$$\delta = \left[\frac{\rho}{(\mu \times \pi \times f \times 10^6)}\right]$$

Example for copper conductor

$\rho = 1.7241$
$\mu = 4\pi \times 10^{-7}$
$f = 1000$

$$\delta = \left[\frac{\rho}{(\mu \times \pi \times f \times 10^6)}\right] \approx \delta = \left[\frac{1.7241}{(4\pi \times 10^{-7} \times \pi \times 1000 \times 10^6)}\right]$$

$$\delta = 2.09 \times 10^{-6} \text{ or } 2.09\ \mu m$$

Attenuation losses consist of two components: conductive losses and dielectric losses. The conductive losses make up the larger portion of the total attenuation, but dielectric losses still must be taken into consideration.

Equations for attenuation are given below:

Constants for attenuation	Values for example given
ρ_i – resistivity of inner conductor	1.7241 μΩ cm (copper, soft)
ρ_o – resistivity of outer conductor	2.8282 μΩ cm (aluminum)
f – frequency (MHz)	1000
σ – dissipation factor	7×10^{-5}
α – attenuation, total (dB/100 ft)	
K – attenuation of conductors	
L – attenuation of dielectric	
Z_o – characteristic impedance	75.0
V_p – velocity of propagation (%)	0.836 or 83.6%
I_c – inner conductor, diameter (in.)	0.0403
O_c – outer conductor, diameter (in.)	0.1800

$$K = \frac{3296}{Z} \times \left(\frac{\sqrt{\rho_i}}{I_c} + \frac{\sqrt{\rho_o}}{O_c}\right) \qquad L = \frac{0.884 \times \pi \times \sigma}{V_p}$$

$$K = \frac{3296}{75.0} \times \left(\frac{1.3130}{0.0403} + \frac{1.6817}{0.180}\right) \qquad L = \frac{0.884 \times 3.141\,59 \times (7 \times 10^{-5})}{0.836}$$

$$K = 0.184244 \qquad L = 0.00023$$

Total attenuation (dB/100 ft)

$$\alpha = (K \times \sqrt{f}) + (L \times f)$$

$$\alpha = (0.184\,244 \times \sqrt{1000}) + (0.000\,23 \times 1000)$$

$$\alpha = 6.06$$

12.7 Impedance Mismatch – Reflection Coefficients and Return Loss

Coaxial cable networks of today have what is referred to as a nominal impedance value, as it is essentially impossible for the cable and components to achieve exactly 75 Ω of impedance, especially across an ever-broadening frequency band, soon to be 5 MHz–3.00 GHz {*as a reference*}. Each component of a coaxial cable network has impedance mismatches, which are introduced by connectors, active/passive devices, and the coaxial cable itself. The intent is to minimize these mismatches and keep them from becoming too severe. There are multiple ways to characterize impedance mismatches, with the most common being return loss (R).

Return loss is defined in the IEEE Standard Dictionary of Electrical and Electronic Terms, Fourth Edition, as "*the ratio in decibels of the power incident upon the discontinuity to the power reflected from the discontinuity.*" When an impedance mismatch occurs, a portion of the incident power is reflected by the impedance mismatch toward the source. Reflected waves interact with the incident wave to produce "standing waves" in the transmission line [7].

Reflection coefficient and voltage standing wave ratio (VSWR) are a couple of terms used to characterize the severity of an impedance mismatch or return loss.

Reflection coefficient is defined as the ratio of reflected voltage, $E_{\text{reflected}}$, to the incident voltage, E_{incident}, and is typically represented by the Greek symbol Γ (gamma) or ρ (rho). The value of Γ can range from 0 for a reflectionless load to 1 for a completely reflecting load such as a short-circuit, open circuit, or pure reactance [8].

$$\Gamma = E_{\text{reflected}}/E_{\text{incident}}$$

Voltage standing wave ratio (σ or r) – "for a lossless transmission line at least $1/4\lambda$ long, the ratio of the maximum peak voltage anywhere on the line to the minimum value anywhere along the line is defined as the voltage standing wave ratio" [8].

The relationship between reflection coefficient and VSWR is as follows:

$$r = \frac{(1 + |\Gamma|)}{(1 - |\Gamma|)}$$

and

$$|\Gamma| = \frac{(r - 1)}{(r + 1)}$$

Return loss is a comparison of the power in the reflected wave to the power in the forward wave. It is the difference between these two power measurements, and the unit of measure is decibels (dB).

$$R_{\text{dB}} = 10\log_{10}\left(\frac{\text{incident power}}{\text{reflected power}}\right) \approx 20\log_{10}\left(\frac{1}{|\Gamma|}\right)$$

Structural return loss (SRL) – As coaxial cable is manufactured, a number of variables controlled by the extrusion process can cause the impedance to change.

Recall that the cable's impedance is a function of the cable's physical properties (conductor diameters, insulation dielectric constant), and if any of these properties changes, the impedance will change. For example, the dielectric material is extruded over the center conductor during the manufacturing process. As the dielectric is extruded, its diameter or dielectric constant can change and cause the impedance to change. Today's extrusion lines use automatic controls to continually monitor the line speed, capacitance, and diameter of the product. If one of these parameters varies from the specification values entered, the controller can adjust to correct. If it is unable to compensate, alarms sound to alert the operator that attention is required. This impedance change is extremely small and difficult to measure. If only one of these impedance changes occurs in the cable or if they occur at random intervals, the return loss will be good; but due to manufacturing processes, there may be many evenly spaced impedance changes, and return loss problems will arise. Reflections from these evenly spaced impedance changes add together at a frequency corresponding to a $1/2\lambda$ spacing. Although each impedance change may be very small, when they all add together, they cause a return loss "spike." These spikes can be narrower than 200 KHz. The return loss from these impedance changes is called the SRL because the impedance variations are due to structural nonuniformities in the cable [9]. Return loss is compared to the design impedance of the coaxial cable, whereas SRL is referenced to the measured average impedance.

12.8 Applications

Now that we have discussed the rudimentary concepts and designs for coaxial cables, we will turn our focus to the applications for coaxial cables, both past and present.

The coaxial lines developed by Heaviside were a copper wire inner conductor supported by an insulating disk inside a copper tube outer conductor. These disks maintained a consistent distance of the inner wire from the inside diameter of the copper tube. The dielectric was essentially air. The Heaviside coaxial transmission lines were used in the early transatlantic cables used to carry telegraph and later, telephone signals over vast distances.

In the period leading up to World War II, the British realized that a radio direction finding system later called RADAR, was needed and started to develop the *CHAIN Home Transmitter* system in 1935. The transmitter developed very high peak power and there was a need to transmit the signal to the array with minimal losses. Two hundred pound per mile copper cables were used in this construction as a balanced feeder arrangement. In the final installations, the telephone coaxial cables were used in lieu of the original 600 Ω balanced feeder wires. Again, coaxial cables at this stage of development were rigid or semi-rigid types (Figure 12.7).

Flexible coaxial cable or lines had not yet been developed. A Welsh Engineer, Edward Bowen, was working on the CHAIN Home Transmitter project. He requested a transfer in 1937 to work on the AI (Airborne Intercept), which

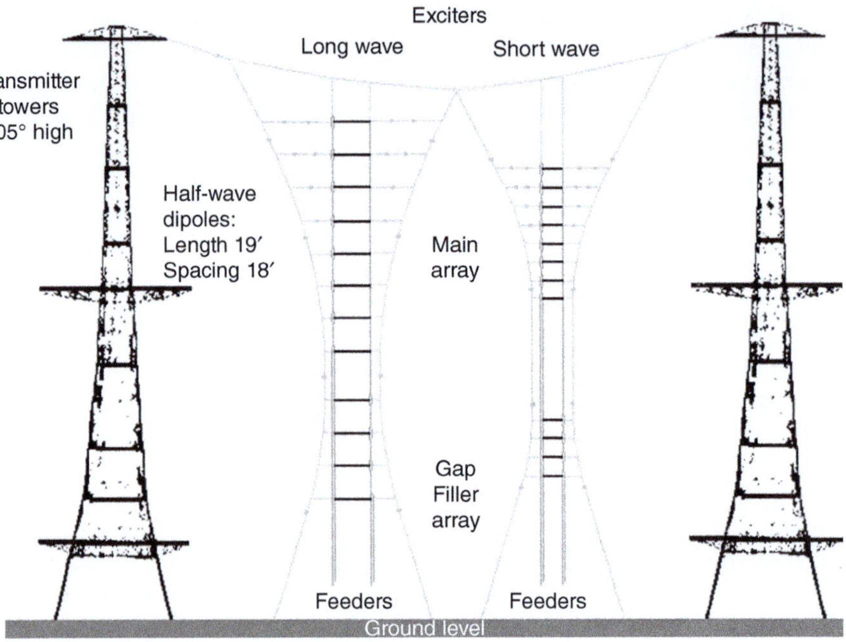

Figure 12.7 CHAIN home transmitter illustration.

resulted in airborne radar. Soon thereafter, it was realized that the current coaxial cables would not suffice for the application of being installed on aircraft flying at high altitudes and with tremendous vibrations. Imperial Chemical Industries had recently invented polyethylene, and Bowen approached them to determine if it could be applied as the insulator in a more robust coaxial cable line. He asked the ICI scientist to extrude a tubular layer of polyethylene over a stranded copper wire, with a constant diameter. The scientist agreed it could be done and soon provided Bowen with a short length of coaxial dielectric with inner conductor. This length was encased with a copper braided shield and wrapped tightly with electrical insulating tape. This was the first flexible coaxial cable. Development began to accelerate at this point; the cable now existed; however, no connector interface was yet available. Two companies developed connectors for the early flexible coax cables. They were Burndept, a British company, and Amphenol, an American company, which developed the PL-259 UHF[4] connector. The PL-259 connector is still manufactured and used today.

12.9 Video, TV, and Broadband Applications

Skip ahead to the end of World War II, and there is a huge inventory of coaxial cable that had been used for wartime communications, RADAR, and instrumentation applications. The government offered this inventory of coaxial cable to the

4 https://www.amphenolrf.com/connectors/uhf.html.

public at a fraction of its cost, providing the impetus for entrepreneurs to resume their efforts with TV distribution systems and broadcast. The advancements in flexible coaxial cables and connectors allowed these efforts to move at an accelerated pace. The first CATV systems were constructed in small towns around 1948–1949, in areas where off-air TV reception was poor or non-existent. As this fledgling industry began to flourish, the number of manufacturers who had invested heavily to provide hardware, connector interfaces and of course, coaxial cable also increased and grew. The cable was, and remained, a huge part of these technical advances. Early on, the CATV Broadband industry had standardized on 75 Ω impedance as the best option where power levels were lower than required for microwave transmissions.

The Society of Cable Telecommunications Engineers (SCTE) was formed in 1969 with the purpose of being dedicated to providing meaningful resources and programs for its members and the industry. One of the many contributions this organization has provided to the CATV industry and to coaxial cable manufacturers was the generation and adoption of standards, particularly for coaxial cable and interfaces. These specifications and associated test methods for the CATV industry facilitated interoperability and standard dimensions across the vendor pool. This interoperability insured that a coaxial cable from vendor A was compatible and able to be installed with a connector from vendors X, Y, or Z. Standardization so early within the industry allowed the systems to operate more efficiently and profitably, thereby promoting further investment into infrastructure and systems [10]. Other international standards organizations followed suit including such organizations as IEC [11] and CENELEC [12]. With the accelerated growth of the broadband communications sector, competition between the coaxial cable manufacturers increased, with each turning to their engineering resources to differentiate product while still maintaining interoperability. Enhancements such as extended frequency spectrum, low-loss performance, GIFPE dielectrics, multilayer shields/outer conductors, and unique jacket sheath compounds were developed in the atmosphere of competition. Coaxial cables of today for the broadband and broadcast industries have many of the attributes of earlier designs. Coaxial cable RF shielding performance today is achieved with only three of the four outer conductors detailed previously, where earlier coaxial cables with four outer conductor shields were necessary to meet specific RF shielding performance criterion. Today, the biggest revenue generation for traditional CATV operators is not video content, but rather hi-speed data utilizing the existing Hybrid Fiber Coax (HFC) infrastructure, including coaxial cables. Video, Telephony, Hi-Speed Internet, Cellular Back-Haul, etc... are all provided simultaneously over the network.

12.10 Automotive Coaxial Cable Applications

It is safe to say that the primary industries that make significant use of coaxial cable are communications and military applications. The automotive industry is challenging this notion with an ever-increasing need for coaxial cables. The

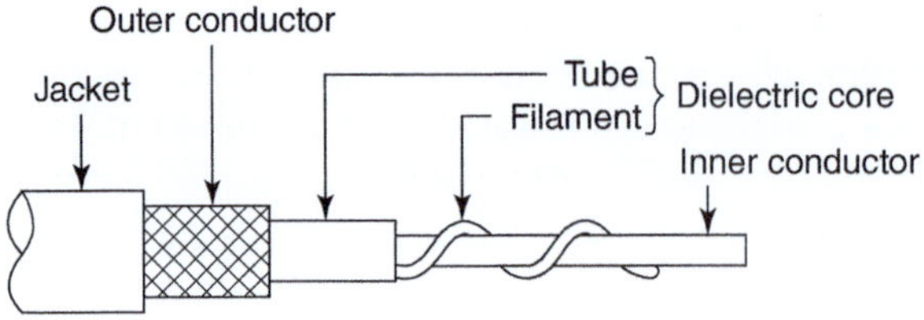

Figure 12.8 RG-62A/U, MIL-DTL-17/30D, 30 April 2010.

original use of coaxial cables in automobiles was to provide a connection cable between the AM/FM radio and the car's antenna. The primary coaxial cable used in this original application was RG-62A/U (see Figure 12.8), which is constructed with a monofilament dielectric. Today's automobiles, however, utilize more coax than the typical home. The list of automotive coaxial cable applications deployed today includes the following:

1) Advanced driver assistance systems (ADAS)
2) Global positioning system (GPS)
3) Satellite radio
4) WiFi/telematics
5) 5G/cellular communications
6) Vehicle to vehicle communications (C-V2X/DSRC)
7) Video entertainment
8) Camera/vision systems
9) AM/FM/digital radio

The automotive industry and the demanding applications have challenged and driven coaxial cable manufacturers to think outside of the box and become innovative both with product, and the technologies to manufacture these products. An example is the original application of radio antenna cable RG-62A/U as previously referenced. The MIL-STD for this RG (Radio Grade) cable details a polyethylene monofilament/air dielectric type. This particular coax performed well on clear sunny days without rain. But in rain storms or car washes water would collect at the base of the antenna and enter the coaxial cable dielectric. With the dielectric being a hollow polyethylene tube with the inner conductor suspended via the polyethylene monofilament, this was essentially a tube that allowed the moisture to collect and degrade the cable and performance. A major auto maker approached Times Wire & Cable, Inc., to develop a solution. Although solid polyethylene coaxial cables were developed and commercially available, there was a need for a foamed polyethylene dielectric with similar dimensions and RF characteristics. The challenge was to foam the polyethylene and maintain a closed-cell structure in the foam so as to impede the moisture flow in the dielectric. An added benefit to this new RG-62M (Modified) was the mechanical support the foamed dielectric provided over the tubed design. The RG-62M rapidly became the standard for the AM/FM radio antenna coaxial cable.

The automotive industry continually asks the coaxial cable manufacturers to test the limits of physics and manufacturing capabilities. As the number of applications for coaxial cable in automobiles has grown, so too have the requirements

Table 12.2 Dielectric and jacket materials [13].

Material	Operating temperature range	Use (dielectric or outer jacket)
Polytetrafluoroethylene (PTFE)	−75 to +250 °C	Dielectric or jacket
Polyethylene	−40 to +85 °C	Dielectric or jacket
Foamed polyethylene	−40 to +100 °C	Dielectric
Foamed or solid ethylene propylene jackets	−40 to +105 °C	Dielectric
Fluorinated ethylene propylene (FEP)	−70 to +200 °C	Dielectric or jacket
Polyvinylchloride (PVC)	−40 to +85 °C	Jacket
Ethylene chloro trifluoroethylene (ECTFE)	−65 to +150 °C	Jacket
Polyurethane	−100 to +125°C	Jacket
Perfluoroalkoxy (PFA)	−65 to +260 °C	Jacket
Nylon	−60 to +120 °C	Jacket
Ethylene propylene	−40 to +105 °C	Jacket
High molecular weight polyethylene	−55 to +85 °C	Jacket
Cross-linked polyolefin	−30 to +85 °C	Dielectric or jacket
Silicone rubber	−70 to +200 °C	Jacket
Silicone impregnated fiberglass	−70 to +250 °C	Jacket
High temperature nylon fiber	−100 to +250 °C	Jacket

for the cables used. With fuel economy always at the forefront for the consumer, the need to reduce the weight and profile of the cable is paramount. Coaxial cables used today range from 1.8 mm over the outer jacket to 6.0 mm. A 1.8 mm (0.071 in.) diameter cable, being quite small, is considered a micro-coax. Both mini and micro-coaxial cables are common in today's applications.

Another challenge put forth by the auto industry is the increased operating temperature of the coax due to the placement of these cables in engine compartments and along exhaust components. The original cables used in this industry had an operating temperature of 60 °C. As automobiles and the under the hood space become smaller and the space for routing these cables is at a premium, the temperature range in which they operate had to increase. The progression with respect to operating temperature is as follows: +75, +85, +90, +105, and +125 °C. This progression is achieved by using more thermally stable materials in both dielectrics and outer jacket sheaths. Some of the materials used to increase a cable's thermal properties are depicted in the Table 12.2.

Many of the advancements made with coaxial cables for the automotive industry are now coming full circle and being integrated into the coaxial cable used in the communications industry. The need for miniaturization and higher operating temperatures have been evidenced in newer DOCSIS 3.1 Optical Nodes, incorporating a modified version of an RG-179, with quad-shield outer conductor and +105 °C operating temperatures.

12.11 Conclusions

Although the applications for the various types of coaxial cables may be different, they share commonality in the properties of the coaxial cables used in them. The quantity and quality of signal processing required for military, communications, and automotive applications continue to push the frequency spectrum growth. Although digital compression technologies and software may pack more data within each bit of spectrum, the upper frequency band continues to expand as well. Cables that once specified an upper frequency limit of less than 1.0 GHz now are being specified to operate and perform to upper frequencies of 1.8, 3.0, 4.5, and even 6.0 GHz.

About 20 years ago it was thought that fiber was going to accelerate coax's end of life in short time. And, while advancements in optical fiber craftsmanship and technologies have fostered fiber growth, it has yet to replace coaxial cable in many of the original applications. Engineers and scientists continue to seek ways to capitalize on existing coaxial networks and creative ways to utilize the metallic conductors. Consider, for example, the fact that optical fiber does not have a means to provide the power for a network. This could be viewed as a limiting factor of a total fiber network. Coaxial cables, on the other hand, can serve industries with both signal transmission and power distribution (Figures 12.9 and 12.10).

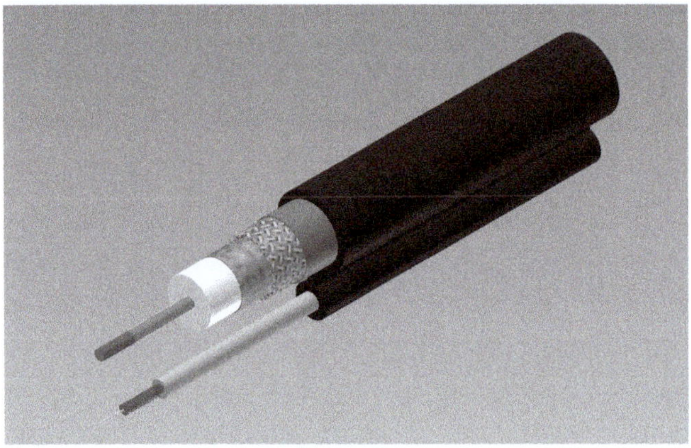

Figure 12.9 Coax-fiber hybrid design.

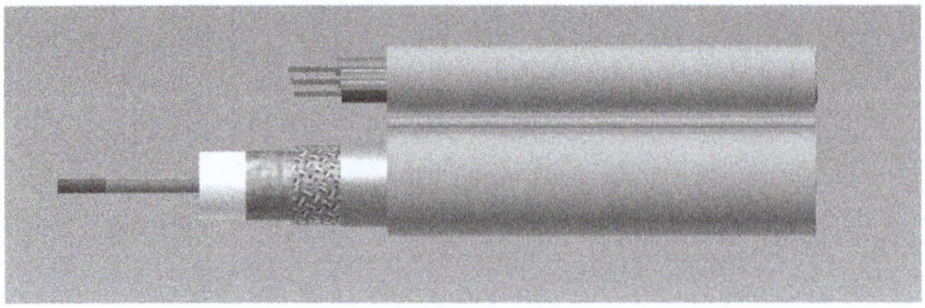

Figure 12.10 Coax-UTP hybrid design.

References

1 (2009). *Why 50 Ohms?* Microwaves 101. Retrieved from 25 January 2012.

2 (2008). *Coax Power Handling*. Microwaves 101. Retrieved from 25 January 2012.

3 Gray, D.A. (1968). *Handbook of Coaxial Microwave Measurements*. GenRad, Inc.

4 Schaum's Outline Series. *Theory and Problems of Transmission Lines*, Robert A. Chipman, 1968

5 Simpkins, R. (2000–2005). *Resultant Dielectric Constant from Historical Empirical Data and Analysis for Series 6 Coaxial Cables*. Times Fiber Communications, Inc.

6 Gray, D.A. (1968). *Handbook of Coaxial Microwave Measurements*, Section 1.4. GenRad, Inc.

7 Hranac, R. (2017). *Taken from Return Loss*. Broadband Library.

8 McElroy, G. (2019). *Handbook for Radio Communications*. American Radio Relay League, Inc.

9 Amato, A.J. (1999). *The Cable Book*, Technical Note 1069. Times Fiber Communications, Inc.

10 Society of Cable Telecommunications Engineers (SCTE). http://www.scte.org.

11 International Electrotechnical Commission. https://www.iec.ch.

12 European Committee for Electrotechnical Standardization. https://www.cenelec.eu.

13 *Catalog 16*, 307. Times Microwave Systems, Inc.

13

Optical Fiber Cables

Chen Baoping, Yao Di, and Qian Feng

Cable Business Unit, Fiberhome Telecommunication Technologies Co., Ltd., No.6, Gaoxinsilu, East Lake High-Tech Development Zone, Wuhan, Hubei Province 430205, China

Optical fiber communication uses light to transmit information, which is a revolution in the history of communication. Since the first optical fiber with attenuation of less than 20 dB/km was born in the United States in 1970, optical fiber communication has greatly developed around the world. Now, except for specific applications using satellite and microwave communications, the vast majority of the world's mainstream communications are done over fiber-optic networks.

Optical fiber is the carrier of information, but the tiny fiber can only be applied in engineering after it is made into an optical fiber cable. The cable material outside the fiber is mainly to protect the fiber from external damage. Unlike the electric cable, optical fiber cable transmits primarily signals rather than energy. The core component in the optical cable is an optical fiber, and the optical fiber is an extremely brittle non-conductive dielectric substance, which is mainly composed of glass and acrylic resin. Therefore, the physical characteristics and functions of the optical fiber determine the difference between the electric cable and the optical cable in terms of product design, main characteristics, test methods, and application methods. For this reason, when producing optical cable, we must ensure that the transmission reliability of the fiber is not affected by the cabling process.

Before we talk about the optical cable, let us take a look at the optical communication network system where the fiber is located.

13.1 Optical Communication System

In 1976, the transmission capacity of a single multimode fiber was 44.7 Mbit/s, and now its transmission capacity in a commercial fiber transmission system is 8 Tbit/s. In just 40 years, the transmission capacity of optical fiber transmission systems has increased by 170 000 times. Owing to the development over 40 years, the optical fiber transmission system has experienced a gradual evolution from Plesiochronous Digital Hierarchy (PDH), Synchronous Digital Hierarchy (SDH),

The Global Cable Industry: Materials, Markets, Products,
First Edition. Edited by Günter Beyer.

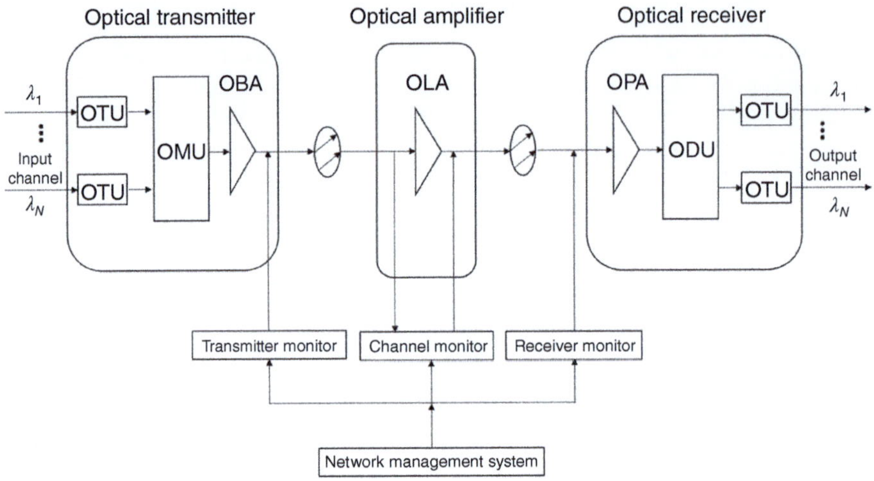

Figure 13.1 Schematic diagram of WDM communication system.

and Multi-Service Transport Platform (MSTP), Wave-length Division Multiplexing (WDM) to Optical Transport Network (OTN) [1].

Generally speaking, a complete optical communication network system includes three parts: an optical transmitting device, a transmitting medium, and a light receiving device. Meanwhile, the large transmission network can be divided into backbone network, metropolitan area network, access network, and other subsystems, and the subsystem can be further subdivided.

Figure 13.1 illustrates the working principle of an optical communication system by taking the WDM system of our common metropolitan area network as an example. Firstly, the optical transmitter transmits optical signals of different nominal wavelengths, and the optical signals are converted into standard wavelengths satisfying the WDM system through an Optical Transport Unit (OTU), and then are combined into a bundle of multiplexed optical signals by an Optical Multiplexer Unit (OMU). The final combined optical signals are amplified by an Optical Booster Amplifier (OBA) and then injected into the optical fiber for transmission. The optical signals of each channel of the optical fiber are amplified by an Optical Pre-amplifier (OPA) and decomposed by an Optical Demultiplexer Unit (ODU), and then input to the respective optical receivers.

In signal processing, the WDM system uses an optical amplifier to amplify the optical signal. Optical amplifiers are divided into OBA, Optical Line Amplifier (OLA), and OPA. The OBA realizes the amplification of the optical signal, the OLA relay amplifies the optical signal, and the OPA improves the sensitivity of the optical receiver.

The transmission of light in the optical fiber uses an independent wavelength as an optical monitoring channel, which transmits an optical monitoring signal and carries the network element management and monitors information of the WDM system.

Modern optical transmission networks are becoming more advanced and more automated. WDM's network management system is an integrated management platform that not only manages the performance of optical amplifiers, wavelength division multiplexers, wavelength converters, and optical monitoring channels, but also implements specific management for the performance, fault, configuration, and security aspects of the device. The information of the network management system is carried by the optical monitoring signals in the optical monitoring channel.

13.2 Fiber Main Features and Transmission Theory

The development of optical communication systems is inseparable from the role of optical fiber, and also puts higher requirements on the performance of optical fiber. From the integration of communication systems from PDH to OTN and future information and communication technology (ICT), and the development needs of 5G, the fiber has also experienced the development process from multi-mode fiber to ordinary single mode fiber (G.652), G657 bending fiber for FTTx, ultra-low loss fiber (ULL), and low-loss large-area fiber (G.654E) meet the performance requirements of fiber and bandwidth development. The development history is shown in Figure 13.2. To maximize the protection of optical fiber, the manufacturing requirements for optical cable are also increasing in order to meet the optical communication requirements of various applications.

This chapter begins with the principle of optical fiber communication and elaborates on the classification, manufacturing process, and development trend of the optical fiber.

13.2.1 Principle of Total Reflection and Optical Transmission Waveguide

It is well known that light travels as a straight line in the homogeneous medium like a ray. When light passes through a smooth interface from a medium with a refractive index of n_1 to another medium with a refractive index of n_2, a part of the light is reflected back to the n_1 medium by the interface, and the other part of the light is refracted into the n_2 medium through the interface. The reflection and refraction of the light on the heterointerface are shown in Figure 13.3:

The incident light and the refracted light are separated on both sides of the interface, and their angles with the interface normal N are the incident angle θ_1

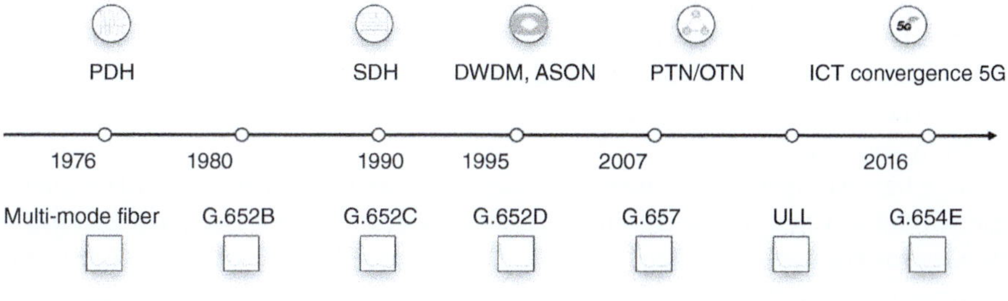

Figure 13.2 Development of fiber.

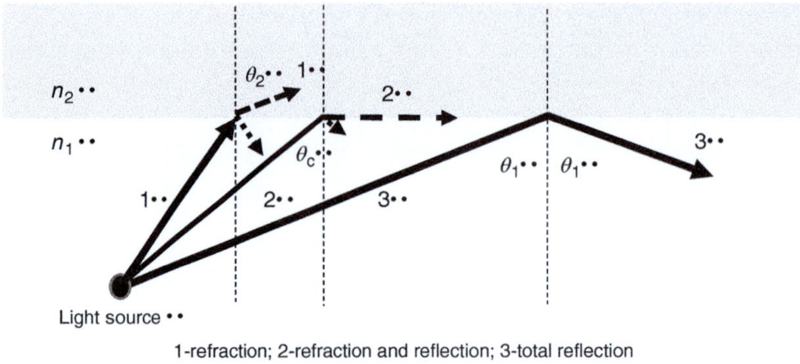

Figure 13.3 Schematic diagram of total reflection of light.

and the refraction angle θ_2, respectively, and the plane defined by the incident light and the interface normal N is called the incident surface.

Both the reflected light and the incident light are in the plane of incidence, and the angle of incidence is equal to the angle of reflection, and the relationship between the angle of incidence and the angle of refraction is expressed by the law of refraction proposed by Snell [2]:

$$\frac{\sin \theta_1}{\sin \theta_2} = \frac{n_2}{n_1} \tag{13.1}$$

where n_1 is the refractive index of the medium 1 and n_2 is the refractive index of the medium 2.

When light is emitted from a light-dense medium with a larger refractive index to a light-diffusing medium with a small refractive index (Figure 13.3, $n_1 > n_2$), the angle of refraction is larger than the angle of incidence. When the incident angle θ_1 is increasing, the refraction angle θ_2 also increases.

When the angle of refraction θ_2 is equal to 90°, the corresponding incident angle θ_1 is denoted by θ_c and is called the critical angle. Obviously, incident light with an incident angle greater than the critical angle will be completely reflected back to the optically dense medium from the dielectric interface without refraction. This phenomenon is known as total reflection.

The optical fiber for communication is a glass fiber designed according to the principle of total reflection, and has a higher core refractive index and lower refractive index of cladding. In order to protect the surface of the optical fiber and improve the tensile strength, a coating layer such as the polypropylene resin, thin silicone resin, or polyurethane is coated on the bare optical fiber. A typical fiber structure is shown in Figure 13.4, where a is the core radius and b is the cladding radius.

If the light is regarded as an electromagnetic wave, the optical fiber is a dielectric optical waveguide having a circular cross section, which encloses the light in its core for transmission. From the core to the cladding, the refractive index of the fiber should follow a certain decreasing law. Since light is an electromagnetic wave, its propagation law follows Maxwell's equations (Equations (13.2–13.7)).

Figure 13.4 Schematic diagram of optical fiber.

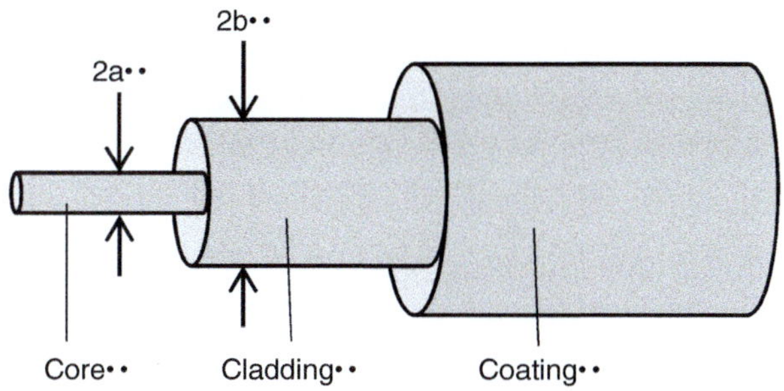

Considering that the fiber is a transparent medium (has no free charge, no conduction current, and no magnetism), Maxwell's equations are as follows:

$$\nabla \times \vec{E} = -\frac{\partial B}{\partial t} \tag{13.2}$$

$$\nabla \times \vec{H} = -\frac{\partial D}{\partial t} \tag{13.3}$$

$$\nabla \times \vec{D} = 0 \tag{13.4}$$

$$\nabla \times \vec{B} = 0 \tag{13.5}$$

where

$$\vec{\mathrm{D}} = \varepsilon \vec{E} \tag{13.6}$$

$$\vec{\mathrm{B}} = \mu \vec{H} \tag{13.7}$$

In the formula above, $\nabla$is the gradient operator, and $\vec{\mathrm{D}}$ and $\vec{\mathrm{B}}$ are the electric displacement vector and the magnetic induction intensity, respectively. ε and μ are the dielectric constant and magnetic permeability of the optical waveguide material, respectively [3].

With the development and continuous improvement of fiber design and manufacturing technology, we can research and develop many new types of optical fibers under different physical dimensions, refractive index distribution, and other physical variables by using Maxwell's equations as the theoretical basis, such as photonic crystal fiber (PCF), various types of doped fiber, and orbital angular momentum fiber.

13.2.2 The Classification and Main Features of Fiber

13.2.2.1 Classification of Fiber

Fibers can be classified according to their material composition, the refractive index profile, transmission wavelengths, fiber internal transmission modes, and fiber use.

In terms of fiber use, most of the fibers currently in use are communication fibers for signal transmission on communication lines. In addition to the application to communications, some special fibers are employed in special applications such as medical, electrical, aerospace, and energy.

According to the difference in the transmission mode of light in the optical fiber, the communication fiber can be further divided into single-mode fiber and multimode fiber. Single-mode fiber can only transmit one mode, and multimode fiber can transmit more than two modes.

On the basis of the classification of the International Electrotechnical Commission (IEC), single-mode fibers are classified into B1.1, B1.2, B1.3, B2, B4, B5, and B6 fibers, and multimode fibers are classified into A1, A2, A3, and A4 fibers. The International Telecommunication Union-Telecommunication Standardization Sector (ITU-T) uses another set of rules to name single-mode fiber and multimode fiber. The correspondence between IEC and ITU-T is shown in Table 13.1 [4–8]. The following sets out the development of single-mode fiber and multimode fiber according to the IEC fiber nomenclature.

13.2.2.1.1 Development of Single-Mode Fiber From the 1980s to the 1990s, the most dominant single-mode fiber was the conventional B1.1 (dispersion unshifted single-mode optical fiber) with two low-loss wavelength windows at 1310 and 1550 nm. However, the dispersion at 1550 nm is relatively large, and the system rate is limited by the dispersion. It is generally necessary to use an erbium-doped fiber amplifier (EDFA) to amplify the optical transmission distance.

B1.2 (cut off wavelength shifted single-mode optical fiber) has an improvement on B1.1 fiber. The zero-dispersion wavelength is still at 1310 nm, but the cut off wavelength is shifted to the long-wave direction relative to the B1.1 fiber, thus improving the attenuation at 1550 nm and the resistance ability to macro-bending and micro-bending of the fiber.

B1.3 (extended wavelength band dispersion unshifted single-mode optical fiber) is also an improvement on B1.1 fiber. It has eliminated the OH-absorption peak in the fiber, then expanding the effective wavelengths to all wavelengths in the range1280–1625 nm. Therefore, B1.3 is called a full-wave fiber, and it is the mainstream fiber of current optical communication network.

B2 (dispersion shifted fiber) is based on G.652 fiber. It moves the zero-dispersion point from the 1310 nm window to the 1550 nm window, and simultaneously achieves the lowest attenuation coefficient and dispersion coefficient at 1550 nm. However, due to the use of EDFA in dense wavelength division multiplexing (DWDM), the large core optical power density leads to nonlinear effects, especially four-wave mixing, which is disadvantageous for multi-wavelength operation.

B4 (non-zero dispersion shifted single-mode optical fiber) is appropriate for WDM at 1550 nm, which overcomes the nonlinear effects of four-wave mixing and cross-phase modulation of B2 fiber at 1550 nm. B4 can support long-distance transmission. It becomes more and more widely used in long-distance trunk lines and broadband access networks.

B5 (non-zero dispersion wideband transport single-mode optical fiber) is still a non-zero dispersion shifted fiber. It has a wider operating bandwidth than B4 fiber and a smaller dispersion slope, which can significantly reduce the dispersion compensation cost of the DWDM system.

Table 13.1 Classification of communication fiber and map from IEC nomenclature to ITU-T recommendations.

Fiber type	Fiber designations	IEC designations	ITU-T designations
Single-mode fiber	Dispersion unshifted single-mode optical fiber	B1.1	G.652A/B
	Cutoff wavelength shifted single-mode optical fiber	B1.2_b	G.654B
		B1.2_c	G.654C
		B1.2_d	G.654D
		B1.2_e	G.654E
	Extended wavelength band dispersion unshifted single-mode optical fiber	B1.3	G.652C/D
	Dispersion-shifted single-mode optical fiber	B2_a	G.653A
		B2_b	G.653B
	Nonzero dispersion shifted single-mode optical fiber	B4_c	G.655C
		B4_d	G.655D
		B4_e	G.655E
	Nonzero dispersion wideband transport single-mode optical fiber	B5	G.656
	Bending loss insensitive single-mode optical fiber	B6_a1	G.657A1
		B6_a2	G.657A2
		B6_b2	G.657B2
		B6_b3	G.657B3
Multimode fiber	62.5/125 μm multimode optical fiber	A1b	—
	50/125 μm multimode optical fiber	A1a.1	G.651
		A1a.2	
		A1a.3	
		A1a.4	
	100/140 μm multimode optical fiber	A1d	—
	A2 multimode optical fiber	A2	—
	A3 multimode optical fiber	A3	—
	A4 multimode optical fiber	A4	—

B6 (bending loss insensitive single-mode optical fiber) improves the bending characteristics of the fiber based on B1.3 fiber. It is appropriate for the complex and curved environment of the client network such as the access network and FTTx, and it can be completely compatible with B1.3 fiber [9].

With the continuous development of optical fiber technology, single-mode fiber is developing in the direction of ultra-low loss and ultra-large effective area

Table 13.2 The map from IEC nomenclature to ISO/IEC 60811 denominate for multimode fiber.

IEC nomenclature	ISO/IEC 60811 nomenclature
A1b	OM1
A1a.1	OM2
A1a.2	OM3
A1a.3	OM4
A1a.4	OM5

to meet the requirements of optical communication systems of 100G, 400G, and even larger capacity.

13.2.2.1.2 Development of Multimode Fiber Although single-mode fiber is the mainstream fiber used for communication currently, in the history of fiber development, the initial fiber that was developed by scientists with attenuation less than 20 dB/km is a multimode fiber. Compared with single-mode fiber, multimode fiber has larger loss and small bandwidth, but it has significant cost advantages in connection, light source, and energy consumption. Therefore, multimode fiber are widely used in short-distance connection, local area network (LAN), and data centers.

As shown in Table 13.2, multimode fiber is classified into A1, A2, A3, and A4 fiber, of which A1 fiber is further divided into A1a (50/125 μm), A1b (62.5/125 μm), and A1d (100/140 μm). According to the International Standardization Organization (ISO)/IEC 60811 standard, A1 fiber usually adopts nomenclature as shown in Table 13.2 [4, 10]. Both A1a and A1b could meet the need for OM1 and OM2 network transmission.

As traditional multimode fiber is no longer sufficient for the requirements of the continuous increase in transmission rate and 40G and 100G Ethernet technologies, OM3, OM4, and OM5 fiber have emerged.

The overfilled launch modal bandwidth of OM3 is higher than 1500 MHz km at 850 nm and 500 MHz km at 1300 nm. The transmission distance of OM3 can reach 300 m, which can not only meet the needs of 10G dual-fiber serial transmission, but also provide future-oriented parallel optical channel for 40G/100G migration.

OM4 improves the fiber bandwidth at 850 nm based on OM3. The overfilled launch modal bandwidth of OM4 is higher than 3500 MHz km, and the maximum transmission distance can reach 550 m.

OM5 broadens the bandwidth characteristics at 953 nm based on OM4, and it is fully compatible with OM4. OM5 supports at least four low-cost wavelengths in the 850–950 nm range, and optimizes the support for new short-wavelength division multiplexing (SWDM) applications, reducing fiber quantity and achieving higher speed.

With the gradual popularization of multimode fiber applications, multimode fiber with excellent transmission performance and bending insensitivity is a future development trend.

13.2.2.2 Main Features of Fiber

As mentioned in the Section 13.2.2.1, the communication fiber comprises single-mode fiber and multimode fiber. The single-mode fiber has smaller mode dispersion, but it still has material dispersion and waveguide dispersion. In order to meet the high requirement of spectral width and stability of light source, single-mode optical fiber needs to have narrow spectral width and better stability. The multimode fiber has a bigger core diameter and a large numerical aperture, which can couple more optical power from the light source, but the mode dispersion is large, which limits the frequency of transmitting signals, and the problem becomes more serious when the distance increases.

Based on the differences in transmission modes and characteristics, the characteristics requirements for single-mode fiber and multimode fiber are also different. IEC and ITU-T have classified the characteristics of single-mode fiber and multimode fiber into four categories: dimensional requirements, mechanical requirements, transmission requirements, and environmental requirements, which are shown in the Table 13.3 [4–8].

In practical applications, we need to select different types of optical fibers according to the specific need. For example, single-mode optical fiber is typically used in long-distance transmission network, and in the case of short transmission distance, frequent optical power splitting and large optical power, the multimode fiber is often selected.

13.2.3 Development of Optical Fiber Technology

With the advancement of optical communication technology, the types of optical fibers are becoming ever more abundant. The communication optical fiber technology has evolved toward ultra-low loss, ultra-large capacity, and ultra-long distance. In addition, the application of optical fiber is increasingly extensive, and the special optical fiber is still expanding. Here are a few new types of fibers that are currently needed or will be needed in the near future.

13.2.3.1 Large Effective Area and Ultralow Loss Fiber

With the rapid development of broadband, data center, and cloud computing, the transmission network is facing tremendous pressure from the explosive growth of business traffic. Ultrahigh speed, ultra-large capacity, and dynamic flexibility have become the future development trend of optical transmission technology. At present, telecom operators and equipment manufacturers are actively promoting the trial and deployment of 400G technology. The 400G WDM transmission technology is bound to become the development direction of the next generation high-speed optical transmission system. However, since the transmission capacity of optical fiber has approached the Shannon limit of the channel, increasing the spectral efficiency further would require a higher signal-to-noise ratio. The optical signal noise ratio (OSNR) limitation and nonlinearity caused by the 400G

Table 13.3 Main features for single-mode and multimode fiber.

Fiber features		**Single-mode fiber**	**Multimode fiber**
Dimensional requirements	Cladding diameter	√	√
	Cladding non-circularity	√	√
	Core-cladding concentricity error	√	√
	Primary coating diameter	√	√
	Primary coating non-circularity	√	√
	Primary coating-cladding concentricity error	√	√
	Fiber length	√	√
Mechanical requirements	Proof test	√	√
	Tensile strength	√	√
	Coating strippability	√	√
	Stress corrosion susceptibility	√	√
	Fiber curl	√	√
Transmission requirements	Attenuation coefficient	√	√
	Chromatic dispersion	√	√
	Cutoff wavelength	√	
	Mode field diameter	√	
	Macro-bending loss	√	√
	Polarization mode dispersion	√	
	Modal bandwidth		√
	Numerical aperture		√
	Differential mode delay		√
Environmental requirements	Damp heat	√	√
	Dry heat	√	√
	Temperature cycling	√	√
	Water immersion	√	√

transmission system will seriously affect the transmission distance, directly leading to a significant increase in the technical implementation and a rapid increase in cost of the 400G system.

The emergence of large effective area and ultralow loss fiber solves the problem of OSNR limitation and nonlinearity caused by 400G transmission system. The effective area of the large effective area fiber (G.654.E) is about 110–130 μm^2. By increasing the effective area, the nonlinear effect of the fiber is further reduced, and the optimum optical fiber power is increased, thereby prolonging the transmission distance. The ultralow loss fiber adopts pure silicon core technology, and the refractive index of the cladding is reduced by doping fluorine to reduce the refractive index of the cladding, thus achieving a loss of less than 0.170 dB/km at 1550 nm [11]. Fibers with lower attenuation could improve the OSNR of the system and increase the transmission distance.

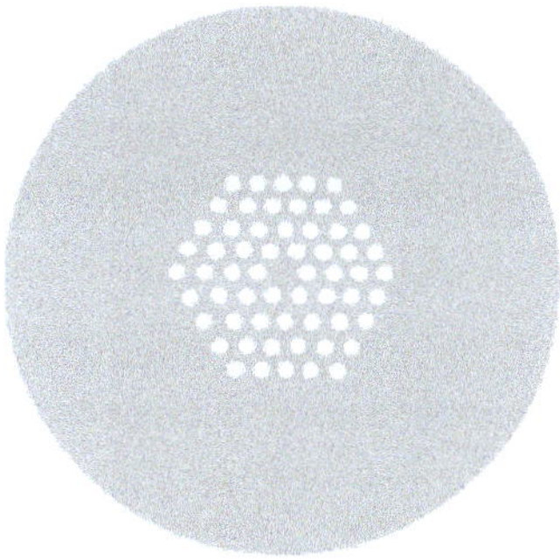

Figure 13.5 Photonic crystal fiber (typical).

13.2.3.2 Photonic Crystal Fiber

Photonic crystal fiber (PCF) is also known as micro-structured fiber (MSF). The photonic band gap PCF requires a strict periodicity of the cladding air hole structure, as shown in Figure 13.5. When this periodic structure is damaged by the core leading, a defective state or local state with a certain bandwidth is formed, and only light with a specific frequency can propagate in this defect area. This phenomenon is called photonic band gap effect. The core can be designed as a hollow structure under this light guiding mechanism. The PCF of this structure has extremely low nonlinear effects and transmission loss, and has great potential advantages in transmitting high-energy laser pulses and long-distance information transmission. Owing to these excellent characteristics, PCF will replace some existing conventional fiber, broadening the field of new fiber applications.

13.2.3.3 Next Generation of Communication Fiber

According to the current trend, the capacity of the optical network will be saturated in the short run. The data traffic generated by the Internet will exceed the information transmission capacity of the optical network. Space division multiplexing (SDM) technology will be the means to be adopted in the future. The SDM technology has three implementation forms: few-mode fiber, multi-core single-mode fiber, and multi-core and few-mode fiber.

13.2.3.3.1 Multi-core Fiber With the phenomenal increase in information demand, conventional single-core single-mode fibers have failed to meet the transmission capacity and transmission speed requirements. Some people have asked for new physical dimensions to fully increase transmission capacity and overcome the transmission capacity limitations of single-core single-mode fiber. However, the use of electrical time division multiplexing (ETDM) and WDM has reached its limit. Therefore, the latest optical fiber with ultrahigh capacity transmission based on SDM multi-core fiber has become the focus of research.

Conventional optical fiber is composed of a core and a cladding surrounding it. The so-called multi-core fiber has a plurality of cores in the same cladding layer, and the optical waveguide is transmitted in the core. A multi-core fiber is equivalent to a few traditional single-core fibers. Thus the large-density multi-core fiber increases the transmission capacity while reducing the additional installation space and cost of the cable.

In the long-distance transmission of multi-core fiber, the crosstalk between cores has little effect on the system transmission. The cross-section design of each core can be compatible with a large effective area and ultralow loss performance, and the transmission capacity is very easy to expand. The transmission experiments of seven-core fiber show that the transmission capacity can reach several hundred Tbps. The multi-core fiber solution realizes the signal transmission experiment of a single fiber with 1 Pbps, and breaks through the Shannon limit, i.e. the maximum rate at which information can be transmitted over a communications channel of a specified bandwidth in the presence of noise. Thus, the multi-core fiber is one of the ideal solutions for large-capacity transmission in the future.

Multi-core fiber is challenging with respect to the processes of manufacture, test, connect, and construction; it still has a large technical bottleneck for large-scale commercial use. Numerous factors need to be considered when making a multi-core fiber. The distance of the core spacing determines the degree of coupling. Too small a core spacing will cause serious crosstalk between the cores, resulting in loss of information energy and difficulty in distinguishing information. The thickness of cladding determines the density of the fiber and the ability to resist bending and torsion. In addition, the multidimensional multiplexing method based on multi-core fiber is developed. Therefore, optimizing core diameter, core spacing, and cladding diameter, and developing multi-core optical fiber with large effective area, high density, and large transmission capacity have always been the research direction and primary goal domestically and abroad.

13.2.3.3.2 Few-Mode Fiber As bandwidth requirements increase for transmission and access of fiber-optic communication systems, the nonlinear effects of single-mode fiber used in backbone transmissions make the system capacity gradually approach the Shannon limit. As a new type of optical multiplexing technology, the mode multiplexing technology of few-mode fiber has emerged, and it has become a research hotspot of optical communication due to its multichannel multiplexing, high spectral efficiency, and low nonlinear effect.

By using a limited stable mode in a few-mode fiber as the independent channel to transmit information, system capacity and spectral efficiency can be multiplied. Since the few-mode fiber is a fiber with a large core area and is sufficient to transmit parallel data streams by using several independent spatial modes, it can be used to simultaneously transmit information in multiple modes, further increasing the fiber transmission capacity. Ideally, the capacity of a few-mode fiber is proportional to the number of modes. Of course, a few-mode fiber amplifier is required in order to extend the transmission distance. The few-mode fiber amplifier has a controllable and mode-dependent gain to ensure that all space multiplexed channels are optimized.

The few-mode fiber is reused by increasing the mode of the propagating beam in the single-core fiber. The demand for energy consumption is not obvious under the premise of increasing the capacity. Meanwhile, the field use of the few-mode fiber is also compatible with the existing single-mode fiber. Compared with the current optical transmission system, the few-mode fiber is expected to become a "milestone product" that leads the leap-forward development of optical communication. The modular multiplexing system based on the few-mode fiber will become a research hot spot.

13.3 Cable Design and Manufacturing Technology

After the introduction above, we have understood that optical fiber is the core transmission medium of optical communication. However, in most scenarios, since the optical fiber is a tiny brittle material, it must be made into an optical cable for use.

Optical cable products are carefully manufactured through reasonable selection of materials, rigorous structural design, and advanced technical processes to ensure that each optical fiber in the cable meets the transmission medium requirements of the information transmission system. Based on the fundamental requirement of protecting optical fiber, three elements of cable structure performance are introduced: at working wavelength, providing sufficient small attenuation and sufficient bandwidth transmission channel for the optical signal of the optical fiber transmission system; ensuring the stability of transmission, machinery, environment, and other performance in the design working life under the given conditions; ensuring the overall economics of the fiber-optic cable in design, manufacturing, construction, operation, and maintenance.

The following is a description of the cable structure design theory and process technology.

13.3.1 Optical Cable Structural Design

13.3.1.1 Design Principle

Fiber crack theory is the basis of fiber-optic cable design. The possibility of natural fiber fracture is estimated by the crack theory, and then the lifetime of the fiber in the cable is predicted.

13.3.1.1.1 Optical Fiber Strength The basic composition of the fiber is quartz glass. The theoretical fracture strength of quartz glass is determined by the interatomic bond energy, which can be up to 1 mpsi (7 GPa). However, the cracks inside the glass greatly reduce its strength, which is only about 7 kpsi (50 MPa), while the strength of the optical fiber is between 100 and 800 kpsi (0.7–5.6 GPa) [9]. The reason for the relatively high strength of the fiber is due to the geometry of the fiber limiting the size and frequency of the crack. Improvements in modern fiber-optic process technology (material purity and process purification conditions) have greatly improved fiber strength.

To ensure that the drawn fiber has sufficient strength, the fiber must be screened for strength. According to the requirements of the standard, the strain applied to the fiber should be not less than 1% (approximately 100 kpsi screening strength, screening tension of 8.3 N), and the strain duration not less than one second.

There are many tiny cracks and defects on the fiber surface after drawing. Even if it is protected in time by two coatings, moisture, chemicals, cleanliness, and other factors in the coating, the environment will still affect the final strength of the fiber. Therefore, 1% strain screening is only used to preliminarily screen the weakest point in the longitudinal dimension, which cannot fully characterize the mechanical properties of the fiber, let alone estimate the service life of the fiber. To ensure the long-term reliability of the fiber in the application environment, several other strength tests also require fiber sampling in the form of a type test.

In fact, the screening process of the fiber itself is a dynamic fatigue process; during the screening process, the cracks of optical fiber will be increased under the effect of screening stress, thereby further reducing the strength of the fiber. The strength decrease of the optical fiber during dynamic fatigue can be described as follows [12]:

$$S_f^{n-2} - S_i^{n-2} = -\frac{1}{B}\int_0^t \sigma(t)^n dt \tag{13.8}$$

where S_i is the strength before fiber screening, S_f is the strength after fiber screening, σ is the stress applied during fiber screening, and n and B are crack growth parameters, respectively.

13.3.1.1.2 Optical Fiber Crack The distribution of cracks in optical fiber is random; usually the Weibull distribution is used to describe the fracture characteristics of optical fiber [9]:

$$F = 1 - \exp[(-L/L_0)(S/S_0)^m] \tag{13.9}$$

where F is the cumulative fracture probability for each sample, L is the sample length, L_0 is the length normalization factor, S is the fracture stress, and S_0 is the characteristic stress and represents the correlation value of the crack group.

Weibull curve, namely one kind of continuous probability distribution curve, describes the relationship between $\log\ln[1/((1-F))]$ and log S. The random orientation of many fracture stresses can form a straight line with the slope rate m. The Weibull curve of fiber fracture consists of several curves with different slopes rate m. Among them, the low-strength fracture is caused by the material and the manufacturing process, which is called the extrinsic crack; the densely distributed high-strength fracture is attributed to the binding force of silica, called intrinsic cracking. Owing to the non-uniform distribution of the surface crack of the fiber, the longer the sample fiber is, the more chances of larger cracks will exist, and the lower the fracture strength will be.

The crack tip of the fiber is formed in the stress concentration region, which is the most likely cause of fiber fracture. The stress intensity coefficient K_I is usually used to indicate the degree of stress concentration [9]:

$$K_I = Y\sigma\sqrt{a} \tag{13.10}$$

where Y is the constant, σ is the external stress, and a is the crack depth. At a certain crack point, with the increase of stress, the final K_I increases to its critical value K_{IC}, and the fiber fractures.

13.3.1.1.3 Stress Erosion Coefficient and Environmental Impact During the laying and service period, the surface crack of the optical fiber will continue to expand under the action of stress and moisture, thereby reducing the strength of optical fiber and finally leading to fracture of the optical fiber. This is the static fatigue process of the fiber, characterized by the velocity V of the crack propagation caused by stress erosion [9].

$$V = \frac{da}{dt} = AK_I^n \tag{13.11}$$

where a is the crack depth, A is the normalized material constant, K_I is the stress intensity factor that depends on the geometric size, depth, and applied stress of the crack, and n is the stress erosion coefficient or fatigue resistance parameter.

Both A and K are determined by the structure of the fiber, and can be regarded as a constant for a certain fiber structure. In addition to the fiber structure, the value of n is related to the environmental conditions of the fiber stress. Meanwhile, n is an environmental parameter that characterizes stress erosion and is an important factor that directly affects the service life of the fiber. The larger the value of n is, the stronger the fatigue resistance will be.

The lifetime of an optical fiber depends on the following three factors: crack, stress, and moisture, and value of n. Cracks in the optical fiber, especially on the surface, are the intrinsic factors of fiber fracture. The presence of stress and moisture promotes the propagation of the crack and eventually leads to the fracture of the fiber.

In order to prevent the fiber from being affected by stress and environmental erosion, in addition to optimizing the performance of the fiber during the manufacturing process, it is also necessary to avoid excessive tensile strain of the fiber in the manufacturing, laying, and operating environment during the cable design, thus avoiding contact between optical fiber and moisture to prevent the occurrence of stress erosion.

(1) To guarantee the mechanical strength of the cable
The cable design needs to consider the tensile strength and compressive strength to reduce the strain on the optical fiber during the period of manufacture, laying, and operation (including the strain caused by the bending), and also needs to optimize selection and design of the cable strengthening components, casing, sheath structure, and related materials to prevent stress erosion caused by fracture of the fiber.
(2) To prevent erosion by water and moisture
It is to prevent the fiber in the cable from being eroded by water and moisture, and avoid growth of the micro-crack, which may affect the mechanical strength and service life of the fiber. The cable structure should be designed to take effective measures against water and moisture intrusion, including the filling of all gaps in the cable, the moisture barrier, and the selection of relevant water-absorbing materials.

13.3.1.2 Design Calculation of Optical Cable Structure

In the design process of optical cable structure, whether outdoor or indoor cable, stranding loose tube type or not, one of the most important and primary design considerations is to ensure that the tensile characteristics and temperature characteristics of optical cables.

13.3.1.2.1 Classification of Tensile Elements and Stretch Calculation The tensile elements commonly used in optical cables can be classified into metal types and nonmetal types.

(1) Main metal tensile elements:

High carbon phosphating steel wire: A uniform, continuous, and strong phosphor layer is coated on the surface of the high carbon steel wire to effectively prevent oxidation and rust of the steel wire, and to reduce the hydrogen loss of the optical fiber, and ensure stable performance of the optical cable. Its high elastic modulus can greatly reduce cable strain and effectively protect the fiber.

Galvanized steel strand: In order to improve the carrying capacity of the optical cable, the cross-sectional area of the steel wire needs to be increased and flexibility of the optical cable improved. The steel strand can meet the above requirements. It can be placed in the middle of the core and can also be used as a messenger for self-supporting cables.

Medium and low carbon steel wire: The surface is coated with a uniform, continuous layer of zinc, usually used for cable armor.

Stainless steel wire and steel pipe: It is resistant to weak corrosive media such as air, steam, and water, and chemically etched media such as acid, alkali, and salt. Its mechanical properties are the same as those of phosphating steel wire, but it has high cost and less application.

(2) Nonmetal tensile elements:

Mainly includes fiber reinforce plastic (FRP), aramid fiber, glass fiber, polyester fiber, and its derivative materials.

GFRP/KFRP: GFRP (glass fiber reinforced plastic) is made of glass fiber and polyester resin or vinyl esterified resin. It is mainly used for the central reinforcement of outdoor non-metallic layer stranded optical cable, external reinforcement of the central tubular cable, and reinforcement of the indoor butterfly cable. KFRP (Kevlar fiber reinforced plastic) is made of Kevlar fiber and polyester resin or vinyl esterified resin. Its mechanical properties such as tensile strength and bending are better than those of FRP.

Glass yarn: A yarn made of glass fiber and polyester resin or vinyl greased resin, which can be used to strengthen the cable core or around the core, it also prevents mouse biting.

FRP tape: A composite tape made of glass fiber and polyester resin or vinyl esterified resin. It is longitudinally wrapped around the cable core and has moisture resistance, tensile strength, side pressure resistance, softness, good bending performance, and mouse bite prevention.

Aramid yarn (aromatic polyamide fiber): It has advantages of small elongation at break, high tensile strength, high modulus of elasticity, low linear expansion coefficient and negative value, good creep resistance, no melting, no combustion and resistance corrosion, etc. It is a very

Table 13.4 Comparison table of elastic tensile modulus of reinforcement materials used in optical cable.

Material	Young's modulus (GPa)
High modulus carbon single round wire	≥190
High modulus carbon steel strand	≥170
GFRP	≥50
KFRP	≥55
Glass fiber reinforced plastic tape	≥50
Glass fiber yarn	≥70
Aramid yarn	≥100

good nonmetallic reinforcing material, and is widely used in all-media self-supporting optical cables.

Table 13.4 lists the typical values of elastic tensile modulus of tensile components materials commonly used for optical cables [13].

(3) Stretch calculation

When the cable is subjected to tensile load, in accordance with the stress principle of parallel element model, the strain gauge of the cable should be equal to the strain of each component, so the amount of tension that each component of the cable is subjected to be distributed by the product value of the elasticity modulus and the cross-sectional area of the component. Suppose that there are n units of parallel elements in the cable; the tensile strength of the cable is calculated by the following formula:

$$F_c = \varepsilon \cdot \sum_{i=1}^{n} E_i \cdot A_i \tag{13.12}$$

where F_c is the cable tension resistance, A_i is the cross-sectional area of **i** element, E_i is the Young's modulus of **i** element, and ε is the allowable strain of the fiber-optic cable.

It can be known from the formula that according to the allowable strain of the optical cable, the material of the tensile element needs to be properly selected and matched, and the cross-sectional area of the tensile element is designed to meet the tensile strength of the optical cable required for laying and operating.

13.3.1.2.2 The Tensile Design for Central Bundle Tube Optical Cable The structural feature of the central bundle tube cable is that the cable and the bundle tube therein are simultaneously elongated during stretching. However, when the cable is laid or operated, it requires a certain tensile window (the maximum extension that an optical cable can withstand when the fiber is without stress) to ensure that the fiber does not generate additional attenuation. Therefore, the fiber should have a certain excess length in the bundle tube. The definition of the excess length in the tube is

$$\varepsilon = \frac{L_F - L_T}{L_T} \times 100\% \tag{13.13}$$

where L_F is the fiber length in the bundle tube and L_T is the bundle tube length [9].

When the tensile strength of the cable is constant, the tensile window (that is, the excess length of the fiber in the center-bundled tube cable) determines the maximum tensile force that the cable can withstand. Therefore, in order to improve the tensile properties of the cable, the length of the fiber in the bundle tube should be large. However, at a low temperature, the cable will shrink (the shrinkage of the fiber is negligible). If the excess length is too large, the radius of curvature of the fiber in the bundle tube will be too small, causing macro-bending loss. Therefore, in the design of the center bundle tube cable, by limiting the minimum curvature radius of the fiber in the beam tube at the extreme low temperature, the maximum allowable residual length of the fiber can be obtained.

When the excess length of the fiber in the bundle tube is not too large, it can be approximately assumed that the fiber has a sinusoidal distribution, and the minimum radius of curvature (at the room temperature condition) of the fiber is obtained at the sinusoidal peak [9].

$$R = \frac{s^2}{4\pi^2 r} \tag{13.14}$$

where s is the sinusoidal pitch and r is the degree of freedom of the fiber in the bundle tube [9]:

$$r = (d_{\text{in}} - d_f)/2 \tag{13.15}$$

where d_{in} is the fiber tube inner diameter of the bundle tube and d_f is the equivalent diameter of the fiber bundle [9]:

$$d_f = 1.16\sqrt{n} \cdot d_0 \tag{13.16}$$

where d_0 is the diameter of the single fiber and n is the number of the fiber in the bundle tube.

When calculating the excess length, the influence of the operating environment of the optical cable should be considered. In the low temperature environment, the shrinking of the center bundle tube cable (the fiber contraction is ignored) leads to the reduction of sinusoidal pitch **s**:

$$s_{\text{LT}} = [1 - (T_{\text{room}} - T_{\text{low}})\alpha_e] \cdot s \tag{13.17}$$

where s is the sinusoidal pitch in the room temperature condition, s_{LT} is the sinusoidal pitch in the lowest using temperature T_{low} condition, and α_e is the equivalent linear expansion coefficient of the optical cable [9].

Therefore, the minimum radius of curvature of the fiber in the center bundle tube cable is obtained at the lowest using temperature T_{low}:

$$R_{\text{min}} = \frac{[1 - (T_{\text{room}} - T_{\text{low}})\alpha_e]^2 s^2}{4\pi^2 r} \tag{13.18}$$

After determining the minimum radius of curvature, you can find the maximum allowable excess length fiber:

$$\varepsilon_{max} = \{(1 + K^2)^{1/2} \cdot [1 - \xi^2/4 - 3\xi^4/64 - 1]\} \cdot 10^3 \tag{13.19}$$

where $K = \left(\frac{r}{R_{min}}\right)^{1/2}$ and $\xi = \left(\frac{r}{R_{min}+r}\right)^{1/2}$ [9].

13.3.1.2.3 The Tensile Design for Layer-Stranding Optical Cable Stranding loose tubes optical cable is the most widely used type of fiber-optic cable. For this kind of cable, the bundle is stranded in a spiral on the central reinforcement element. In this structure, the stretching of the cable causes movement of the fiber in the bundle tube, and the fiber is not stressed, so the fiber has a natural stretching window. If the length of the fiber in the bundle tube is zero, in a static state, the fiber is located at the center of the bundle tube. When the cable is stretched (or under high temperature conditions), the fiber moves toward the inner wall of the bundle tube near the side of the reinforcement member. At this point, the stretch window of the fiber can be expressed as

$$\varepsilon_{tensile} = \frac{\sqrt{h^2 + 4\pi^2 b^2 - 4\pi^2 (b - r)^2}}{h} - 1 = -1 + \left[1 + \frac{4\pi^2 b^2}{h^2}\left(\frac{2r}{b} - \frac{r^2}{b^2}\right)\right]^{1/2} \tag{13.20}$$

where b is the stranding radius of the fiber tube, h is the stranding pitch, and r is the degree of freedom of the fiber in the bundle tube [9].

When the fiber has a positive excess length ε_0 in the bundle tube in the static position, the fiber is closer to the outer wall in the bundle tube, and has a pre-bias value **d** from the center line of the bundle tube [9]; **d** can be calculated by ε_0:

$$d = \frac{1}{2\pi}\sqrt{(4\pi^2 b^2 + h^2)(1 + \varepsilon_0{}^2) - h^2} - b \tag{13.21}$$

When the fiber has a positive excess length ε_0, the stretching window of the stranding loose tubes optical cable can be expressed as

$$\varepsilon_{tensile} = -1 + \left[1 + \frac{4\pi^2 b^2}{h^2}\left(\frac{2(r + d)}{b} - \frac{(r + d)^2}{b^2}\right)\right]^{1/2} \tag{13.22}$$

It can be seen from the above formula that the stretching window of the stranding loose tubes optical cable is mainly affected by the strand stranding pitch h, and by adjusting the value of h the stretching window of the fiber can be adjusted in a large range. On the other hand, the stretch window can also be increased by increasing the fiber excess length in the bundle tube before stranding. When using SZ stranding, there is an additional stretching window at the return point of the SZ stranding, which is about 1‰. Therefore, the stranding loose tubes optical cable adopts SZ stranding, which not only facilitates the bifurcation of the optical fiber in the optical cable but also has the existence of an additional window, thereby further improving the tensile performance of the optical cable.

For some optical cables with large tensile strain during operation, such as the all-media self-supporting optical cable (all dielectric self-supported [ADSS]), underwater optical cable, and optical fiber composite overhead ground wire (optical ground wire [OPGW]), stranding structure should be adopted to improve extended features.

13.3.1.2.4 Temperature Characteristic Design for Optical Cable Taking a stranding loose tube optical cable as an example, when the environment temperature in which the cable is placed changes, neglecting the radial thermal expansion coefficient of the reinforcement member and the change of the casing size, and the shrinkage after the sheath, in the case where only the length of the central reinforcement member changes in the axial direction, the amount of change in the stranding pitch of the cable is

$$\Delta h = \alpha \cdot (T - T_0) \cdot h \tag{13.23}$$

where α is the coefficient of linear expansion of the center reinforcement, $T - T_0$ is the temperature change value, and $T_0 = 25\,^\circ\mathrm{C}$ is the room temperature. The variation of stranded pitch is limited within the stretch/shrink window [14], according to the classic stretch/shrink window formula

$$\alpha \cdot (T - T_0) \leq \varepsilon \tag{13.24}$$

where ε is the stretch/shrink window of the stranding loose tubes optical cable.

When the temperature rises, that is, $T > T_0$, ε selects the stretching window:

$$\varepsilon_{\text{tensile}} = -1 + \left[1 + \frac{4\pi^2 b^2}{h^2}\left(\frac{2r}{b} - \frac{r^2}{b^2}\right)\right]^{\frac{1}{2}} \tag{13.25}$$

When the temperature decreases, that is, $T < T_0$, ε selects the shrink window:

$$\varepsilon_{\text{shrink}} = 1 - \left[1 - \frac{4\pi^2 b^2}{h^2}\left(\frac{2r}{b} + \frac{r^2}{b^2}\right)\right]^{1/2} \tag{13.26}$$

From the above formula, it can be seen that the allowable range of ambient temperature is proportional to the stretching/shrinking window of the stranding loose tube optical cable, and is inversely proportional to the linear expansion coefficient of the reinforcement. In the operation process of temperature cycling test, it is necessary to meet the extreme conditions of both the highest temperature and the lowest temperature at the same time. The temperature change can be satisfied by changing the stranded pitch to adjust the stretch/contraction window, and adjust the length of loose tube and other methods.

13.3.2 Manufacturing Process Route for Optical Cable

The manufacturing process from fiber to optical cable has varied types according to the structure of the cable. Usually it happens in accordance with the processing order from inside to outside. Take a typical stranding loose tube optical cable as an example; it includes the steps of fiber coloring/ribbon – loose tube or tight buffer – layer stranding – sheath, etc.

13.3.2.1 Technology of Fiber Coloring

Original fiber is of natural color, and one optical cable contains multiple fibers. In order to identify them in the processes of production, inspection, construction, operation, and maintenance, the method used is to form a chromatogram of 12 colors (blue, orange, green, brown, gray, white, red, black, yellow, purple, pink, and cyan) which is widely used in international standards.

Fiber coloring is a coating curing using a different color ink for a single coated fiber. Earlier, the thermal curing technology was adopted, that is, coating the surface of the fiber with a heat-curing ink of about 5 pm thick, and rapidly curing by a heating furnace or infrared light to form fibers of different colors. With the development of ultraviolet (UV)-curing materials and curing technology, UV-curing inks are applied to a coated optical fiber and rapidly cured by UV light irradiation, which is a commonly used coloring technology [15].

During the coloring process, the fiber with a natural color is released by the pay-off unit, and the optical fiber can always be placed under constant tension through the guide wheel and the take-up stretcher (stable tension of pay-off). The fiber is subjected to electrostatic precipitators to remove static electricity and dust before entering the coloring mold. The ink coating chamber is equipped with an optical fiber inlet and outlet mold, the coating chamber is filled with ink, and the passing optical fiber is colored. Then it enters the UV curing oven with N_2 protection and solidifies rapidly. Finally, it is wound on the take-up reel after the traction guide wheel and the pay-off tension dance wheel shown in Figure 13.6.

The color ring technology is usually used when the fiber capacity in the secondary coated casing (metal/non-metal) may exceed 12 fibers, such as the production of large core number of ordinary central tubular cable, optical fiber composite overhead ground (phase) wire, etc.; the fiber capacity will reach 24 cores, 48 cores, or higher, but the fiber chromatography is only 12 colors. The optical fiber coloring ring is a high-speed inkjet printer connected in series between the pay-off and the take-up of the coloring machine, and the color ring is periodically used on the optical fiber during the passing of the optical fiber.

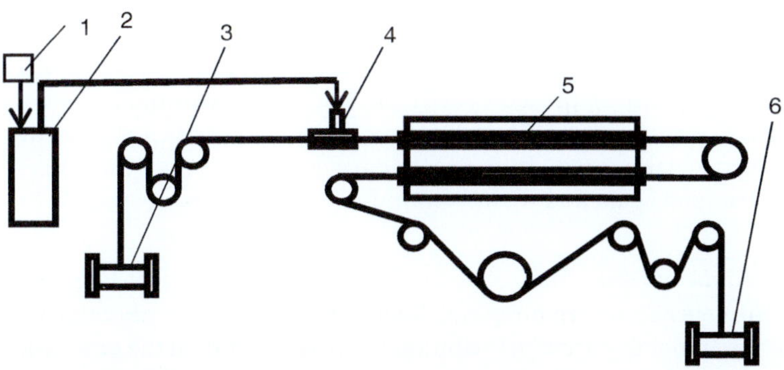

1-press air; 2-ink; 3-pay-off reel; 4-coating; 5-curing oven; 6-take-up

Figure 13.6 Operating principle fiber coloring.

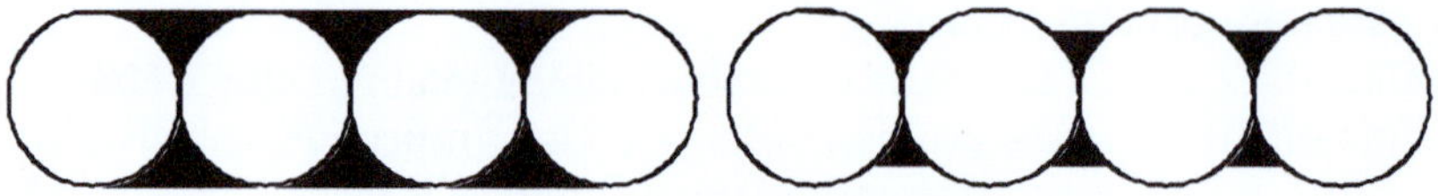

(a)

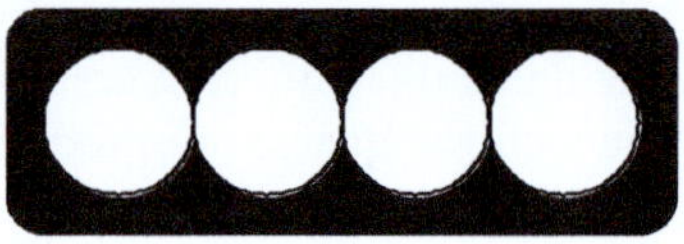

(b)

Figure 13.7 Structure of the fiber ribbon. (a) Typical edge-bonded ribbon. (b) Typical encapsulated ribbon.

13.3.2.2 Technology of Fiber Ribbon

When the core capacity of the optical cable is very large, the optical fiber ribbon is usually selected as the optical unit. The main purpose includes three aspects. Firstly, it is to increase the density of the optical fiber in the optical cable, reduce the outer diameter of the optical cable, save manufacturing costs, and save pipeline resources. Then the group connection technology saves construction welding time and improves construction efficiency. The last not the least, it is convenient to run and maintain the optical fiber with a group of vertical optical fibers through resin coating, pressure-coating mold, and UV light irradiation to form a "one" shape flat ribbon. According to the standards of the ribbon cable, common specifications of fiber ribbon are 4-core, 6-core, 8-core, 10-core, 12-core; even 24 cores are also available. There are two main types of fiber ribbon structures: encapsulated and edge-bonded, as shown in Figure 13.7. At present, the edge-bonded fiber ribbon is widely accepted in the market.

The operation principle of the fiber ribbon process is basically the same as the coloring process, but the fiber ribbon technology is more complicated than the fiber coloring, and the technical specifications have more demands and are more stringent.

13.3.2.3 Fiber Secondary Coating

Basically, the fibers used in the optical cable are either separate fibers or ribbon fibers. The secondary coating protects the fiber. The fiber is covered secondary with a tight buffer or a loose tube. The loose tube coating includes metal coating type and the plastic coating type. In the secondary coating process, the colored optical fiber is grouped again and then the number of the core increased.

13.3.2.3.1 Loose Tube of Fiber The secondary coating of the optical fiber is to package one (or several) discrete optical fiber (or optical fiber ribbons) in a bundle tube whose inner diameter of the bundle tube is larger than the equivalent outer diameter of the optical fiber bundle (or the optical fiber ribbon stack) to form a cable core element. The structure of the loose tube fiber is shown in (Figure 13.8):

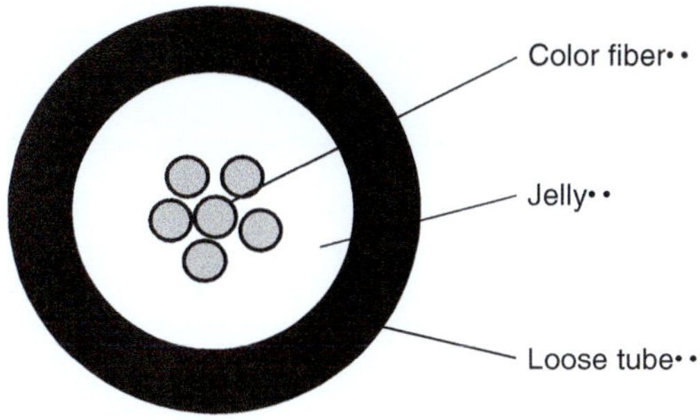

Figure 13.8 Structure of the loose tube fiber.

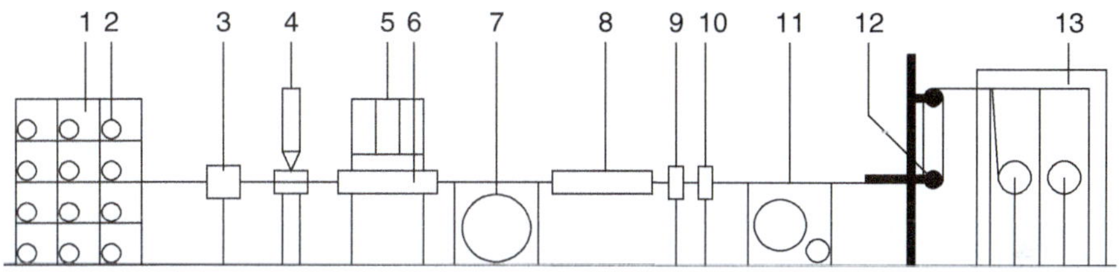

1-fiber pay-off device; 2-fiber guide wheel; 3-jelly filling device; 4-extruder; 5-control cabinet; 6-hot water tank; 7-extra long traction wheel; 8-cold water tank; 9-blow dryer; 10-caliper; 11-main traction device; 12-take-up dance wheel; 13-take-up device

Figure 13.9 Process schematic diagram of secondary coating production line of loose tube.

In the secondary coating of the optical fiber, the discrete optical fiber or the optical fiber ribbon first passes through the filling cavity of optical fiber paste, and the optical fiber paste is continuously transported by the metering pump under the action of the filling pump. When the optical fiber is carried with the optical fiber paste through the crosshead of the extruder, the extruder screw rotates and extrudes the molten secondary coating material to form a tubular shape through the mold, and then passes through the hot water tank, the excess long traction wheel, the cold water tank and so on, to form a loose tube fiber comprising a bundle of fiber (Figure 13.9).

The Technical requirements for the secondary coating of the optical fiber are that the outer diameter of the secondary coated bundle tube is uniform; the inner surface of the bundle tube is round and smooth; the bundle tube has a certain ability to resist lateral pressure; and the inner diameter of the fiber bundle tube, which effectively protects the fiber, is larger than the equivalent diameter of the fiber bundle (or the optical fiber ribbon stack), and has a certain gap. The length of the fiber bundle (or ribbon stack) is slightly longer than the bundle tube and has a positive excess length; the gap in the bundle tube is filled with a water blocking material, which may be dry or wet (oil-filled), to ensure that the entire end of the bundle tube is blocked. The tube can be fully chromatographic and used for identification in fiber-optic cables. Polybutylene terephthalate (PBT), polypropylene (PP), polycarbonate (PC), and other materials are often used as secondary coating materials.

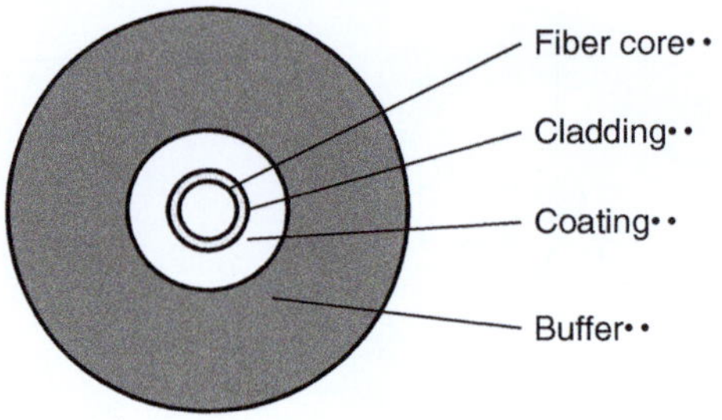

Figure 13.10 Structure of the tight buffer fiber.

13.3.2.3.2 Tight Buffer of Fiber The tight buffer fiber is affixed to the coated material on the coated fiber to moderately increase the diameter and mechanical strength of the fiber. Tight buffer fibers are commonly used in indoor fiber-optic cables. The structure of the tight buffer fiber is shown in (Figure 13.10):

The tight buffer fiber extrusion process is roughly the same as the loose tube process, except for the release of the single fiber. During the extrusion process, the optical fiber is preheated to remove surface moisture, and there is no filling gel between the primary coated optical fiber and the secondary coating layer, but in order to improve the peel ability, a special material may be applied between the two layers. And also, there is no excess length traction wheel in the production line. The outer diameter of the tight buffer fiber is mostly 0.9 mm. The most commonly used coating materials are polyvinyl chloride, nylon, low smoke zero halogen (LSZH) materials; and FEP can also be used in case of special requirements.

13.3.2.4 Stranding Cable Technology

The optical cable mainly consists of a cable core and a sheath. The cable core comprises the fiber unit, the filling unit, the tying yarn, the reinforcement member, the water blocking material, and if necessary, a tape layer, and other materials are integrated in a certain combination according to design requirements. Cabling is the main process in the optical cable production. By cabling, structural stability and good bending performance are achieved, and it forms a cable core with its excess length and fiber structure, which meets the design requirements. The stranding method is ST stranding and SZ stranding. SZ twisting has become the most used cable-forming technology in the optical cable manufacturing industry due to its fast cable-forming speed, high production efficiency, and low equipment cost. The cable forming process schematic with SZ stranding is shown in Figure 13.11.

13.3.2.5 Sheath Extrusion of Optical Cable

The sheath of the optical cable is similar to that of the electric cable. The outer sheath of the optical cable is the first protective layer of the cable to withstand various external influences. It must have excellent mechanical properties, environmental resistance, and chemical resistance, protect the core from external machinery, heat, and chemical and moisture effects, and ensure the transmission performance of the fiber.

There are many structural forms of cable sheath, which can be made of different sheath materials, such as polyethylene (PE), flame-retardant polyethylene

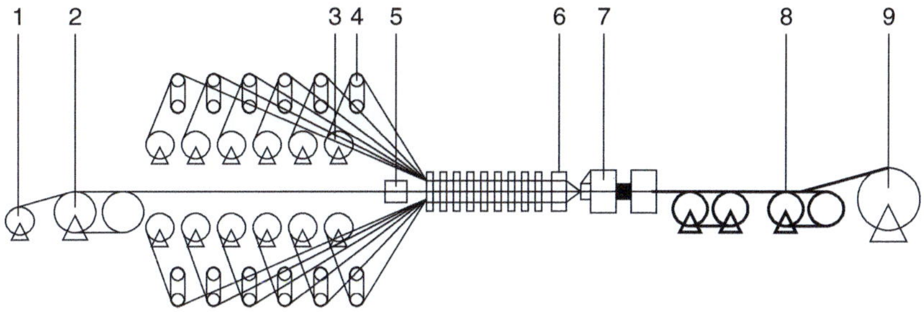

1-center reinforcement pay-off device; 2-tension dance wheel; 3-fiber unit; 4-fiber unit tension dance wheel; 5-pre-filling device; 6-SZ twister; 7-yarn head; 8-take-up tension dance wheel; 9-cable core take-up device

Figure 13.11 Cable forming process schematic with SZ stranding.

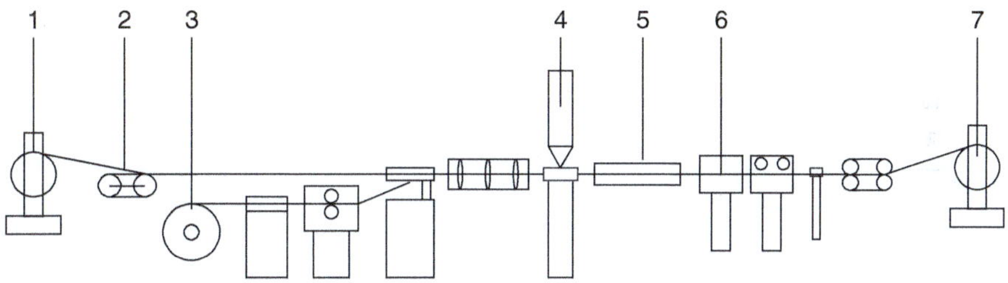

1-cable core pay-off device; 2-tension dance wheel; 3-tape pay-off device; 4-extruder; 5-water tank; 6-blow dryer; 7–cable take-up device

Figure 13.12 Schematic diagram of sheathing production line of optical cable.

compounds with halogenated flame retardants, polyvinyl chloride, and LSZH based polyolefin based cable compounds. Customers also require some special materials as sheathing materials for optical cables, such as anti-rodent sheathing material with capsaicin, nylon, and anti-tracking material. The sheathing production line should have the processing capabilities of a variety of thermoplastic materials. The schematic diagram of sheathing production line of optical cable is shown in Figure 13.12.

13.3.3 Optical Cable Standards and Key Test Items

The optical cable must have certain performances during operation. Generally speaking, in addition to optical performance requirements, optical cable performance should include mechanical properties, environmental performance, optical cable component performance, and electrical performance. IEC has detailed specifications for optical fiber cable performance testing. In addition to the IEC standards, ITU-T, ISO, Telecommunications Industry Association (TIA)/Electronic Industries Alliance (EIA), Insulated Cable Engineers Association (ICEA), Institute of Electrical and Electronic Engineers (IEEE), and other international standards organizations have also introduced the corresponding optical cable performance according to their respective standard systems.

Taking the IEC 60794-1 as an example, the mechanical properties of the optical cable include stretching, crushing, impact, bending, torsion, etc. The environmental performance of the optical cable includes temperature cycling, water penetration, dripping, aging, etc. The performance of the optical cable component includes tube kinking, component bending, twisting of the optical fiber ribbon, etc. Electrical properties of optical cable include short-circuit, lightning strikes, and electrical continuity of the metal components of the optical cable [16, 17]. In addition, the optical performance of the cable is a key technical point of cable performance. Before and after cabling, it is usually required to test the characteristics of fiber attenuation, dispersion, polarization mode dispersion (PMD), and cutoff wavelength and others. The following is a brief introduction to the most important test items in the cable manufacturing process – cable attenuation curve, PMD test principle, and cable tension curve.

13.3.3.1 Cabled Fiber Attenuation Test

A commonly used method of the fiber attenuation test is the backscatter method, which uses the optical time-domain reflectometer (OTDR) to test the attenuation coefficient of the fiber. Figure 13.13 shows a typical fiber backscatter signal curve when tested with OTDR, in which *a* is the Fresnel reflection zone (the blind zone) at the input end, *b* is the constant slope zone, *c* is the discontinuity caused by local defects, joints, or coupling, *d* is the reflection caused by the fiber defect, the secondary reflection residual wave, etc., and *e* is the Fresnel reflection at the output end. The OTDR curve of the standard fiber has no discontinuity at *c* point and a reflection peak at *d* point. The test curve is inclined and as the distance increases, the total loss increases. The average loss (dB/km) of the segment is obtained by dividing the total loss (dB) by the total distance (km).

13.3.3.2 Test Principle and Methods for Cabled Fiber PMD

With the increasing capacity and transmission rate of the communication system, PMD of optical fiber becomes a necessary index. The fundamental mode

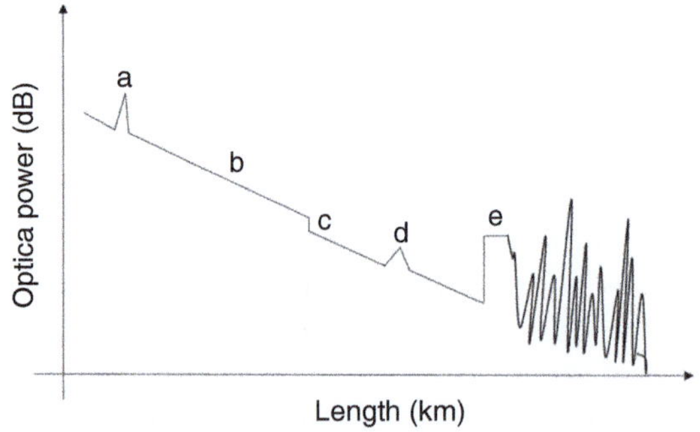

a-Fresnel reflection zone; b-constant slope zone; c-discontinuity caused by local defects, joints, or coupling; d-reflection caused by the fiber defect, the secondary reflection residual wave, etc.; e-Fresnel reflection at the output end

Figure 13.13 Typical OTDR characteristic curve.

in a single mode fiber has two mutually perpendicular polarization states, and the transmission speeds of the two states are exactly the same in an ideal fiber. However, due to the geometric asymmetry of the axial direction of the fiber, one fundamental mode becomes a fast mode, and the other becomes a slow mode, and the delay difference between the two modes is PMD. The existence of PMD causes pulse propagation, which limits the transmission rate and distance in the fiber transmission system.

The measurement methods for PMD mainly include stokes parameter evaluation (SPE), interferometry (INTY), and fixed analyzer (FA) method [18]. SPE is a benchmark test method for measuring PMD. The test principle is to measure the change of the output state of polarization (SOP) with wavelength at a certain wavelength interval in a certain wavelength range. The change can be characterized by the rotation of the SOP vector on the Jones Matrix Eigenanalysis (JME) or the Poincare Sphere (PS), and the PMD measurement results are obtained by analysis and calculation.

Interferometry is the second alternative test method for measuring PMD. In this method, when one end of the fiber is illuminated by a broadband source, the autocorrelation function or cross-correlation function of the electromagnetic field is measured at the output to determine the PMD.

Fixed analysis is the third alternative test method for measuring PMD. The test principle is that when the polarization direction of the input light remains fixed and the wavelength changes, the direction of the main principle state of polarization (PSP) of the output light field also changes. Through an FA, the change of polarization state (SOP) with wavelength is converted into the output power with peak-to-valley fluctuation as a function of wavelength, and the PMD can be determined according to the relationship between the output power spectrum and the group delay difference.

At present, many commercial instruments can measure PMD directly, and the PMD coefficient can be directly obtained through the calculation formula inside the instrument. Among the typical test instruments, Agilent's N7788BD optical device analyzer uses the SPE method to measure PMD. EXFO's FTB-5500B uses the INTY method to measure PMD, and JDSU's MTS-6000 uses the FA method to measure PMD [18].

13.3.3.3 Cable Tensile Curve

Optical cable tensile test is a necessary test item in the design of the optical cable. In the manufacturing process of optical cable, the verification of new products and batch quality inspection need to be confirmed by a tensile test.

Figure 13.14 is a typical optical cable tension curve. In the figure, the lower curve is the stress–strain curve of the fiber and the upper curve is the stress–strain curve of the optical cable. In the figure, T_0 is the stress corresponding to the initial strain of the fiber, and the cable strain corresponding to this force is the structural excess length of the fiber in the cable. It can be seen that the cable strain has a nonlinear section before entering the linear zone, which is because the cable is in a relaxed state at the beginning of the tensile test, and more nonlinearity means the cable is slacker at that period. In addition, because the tensile properties of the cable are primarily dependent on the tensile

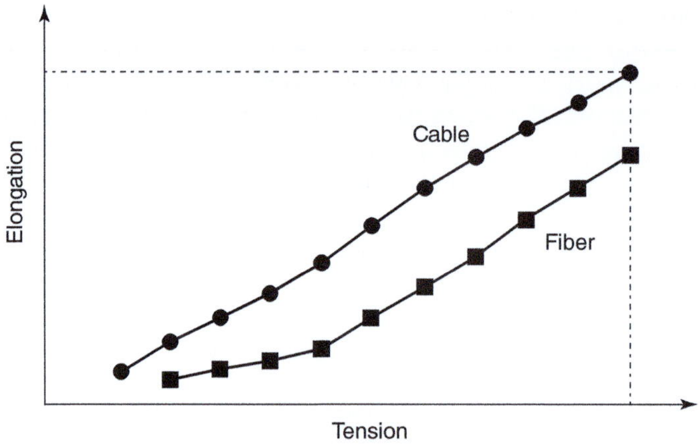

Figure 13.14 Diagram of cable tension curve.

properties of the central reinforcement in the cable, the central reinforcement may be a small redundancy in the cable relative to the plastic sheath of the cable. When the tensile force is applied to the optical cable, due to the above reasons, it is easy to cause the cable sheath to be stretched and elongated first. Thus, a large cable strain occurs after a small pulling force is applied. Thereby it forms a nonlinear region of the test curve of the initial test section.

Special optical cables used in specific situations, according to the actual use environment, require certain tests. For example, indoor optical cables need to have flame retardance and fire resistance test; the submarine cable needs to be tested for hydrostatic pressure resistance. The OPGW and ADSS used in the overhead power lines need to be tested for wind vibration, galloping, and over-pulling. The optical cable used in nuclear power plants needs to be tested for radiation resistance. All these test measurements can be queried from relevant standards.

13.4 Classification of Optical Cables and Typical Application

13.4.1 Classification of Optical Cables

According to the actual application scenarios of optical cables and the international standard such as IEC and others, optical cables are mainly divided into indoor optical cables, outdoor optical cables, aerial optical cables along electrical power lines, and microduct cabling for installation by blowing, etc. Table 13.5 lists some types of optical cables involved in IEC 60794 [19–21].

Owing to the variety of optical cables, optical cable classification of IEC standards does not list all the optical cable types. With the development of optical cable technology, the varieties and technical indicators of optical cable are still constantly updated. The following is a brief introduction to the main types of optical fiber cables in the table.

Table 13.5 Classification of optical fiber cable.

Indoor optical cable	Simplex and duplex cables
	Multi-fiber optical distribution cables
	Ribbon cables
	A4 fiber cables
	Simplex and duplex cables for use in terminated cable assemblies
Outdoor optical cable	Duct, directly buried and lashed aerial optical cables
	Self-supporting aerial cables
	Optical cables for lakes, river crossings, and coastal applications
	Sewer cables
	Gas pipe cables
	Drinking water pipe cables
	Outdoor optical fiber cables for rapid/multiple deployment
Aerial optical cables along electrical power lines	OPGWs (optical ground wires)
	ADSS (all dielectric self-supported) optical cables
Cabling for installation by blowing	Microduct cabling for installation by blowing
	Fiber units for installation by blowing

13.4.1.1 Structure Characteristics of Indoor Optical Cable

The indoor optical cable has the characteristics of a small number of fibers, good bending performance, convenient branching, etc. Therefore, the optical unit, the cable core structure, and the sheath have a large difference compared with the outdoor optical cable. Here are a few of the typical types of fiber-optic cables:

(1) Fiber unit
The fiber unit refers to a basic unit containing optical fiber. The common optical unit of the indoor optical cable may be a discrete optical fiber (simplex and duplex), a tight tube fiber, a ribbon fiber, a micro-cluster fiber bundle, a fiber secondary bundled tube, and enhanced performance fiber unit. These fiber units can be used independently as optical cables and are widely used in optical device applications.
(2) Simplex/duplex optical cable
Simplex optical cable is the most commonly used optical cable type in indoor communication cabinets, optical distribution frame (ODF) racks, and short-distance fiber-optic connection assemblies. It is usually applied after processing the connectors at both ends of a certain length of cable. The tight buffered fiber is first extruded in the manufacturing, and a layer of aramid yarn is coated around the tight buffered fiber as a reinforcement and buffer, and then the sheath is synchronously extruded outside the aramid layer. According to the type of sheath and armor, it is divided into simplex

jumper cable, simplex self-supporting cable, spiral steel with light armored cable, etc.

The duplex optical cable is also based on the tight buffered fiber, and adopts an 8-type structure, which can be conveniently separated into two simplex soft jumpers.

(3) Distribution cable and break-out cable

Tight buffered fiber of the distribution cables is made of fiber units. The fiber unit and the aramid yarn are combined into a cable core and then extruded into a sheath. Simplex cable of the break-out cables is made of fiber units, usually with FRP as the center reinforcement. It is stranded and sheathed finally. These two types of cables contain more fiber, and the cable strength is better, and is usually used in premises cabling.

13.4.1.2 Structure Characteristics of Outdoor Optical Cable

(1) Fiber unit

Outdoor optical cables have more types of fiber units than indoor optical cables, and the number of cores tends to be larger. According to the bundle mode of the optical fiber and with or without protection, the fiber unit of the outdoor optical cable can be divided into a discrete optical fiber, a discrete optical fiber bundle, a micro-cluster optical fiber, and a ribbon optical fiber. Discrete optical fibers are distinguished by coloring. When a larger core tube is produced, the discrete fiber can be grouped in units of 12, which is a discrete fiber bundle. Twelve fibers can be extruded to form a semi-tight structure for grouping, which is a micro-cluster fiber. The ribbon fiber uses a UV-curable resin to cure the discrete fiber to increase the fiber capacity. The number or other mark on the fiber ribbon can increase the number of cores to a greater extent.

Secondary coating protection is performed on the different types of fiber basic units described above, that is, various types of fiber units are obtained.

(2) Cable core structure

The common optical cable core structure mainly includes a central loose tube type (the central loose tube is the cable core that contains the basic fiber units), a stranded loose tube type (a central reinforcement is placed in the center of the cable core, and loose tubes are twisted around the central reinforcement), and a slotted core type (the cable core adopts the skeleton groove structure, the optical fiber or the fiber unit is placed in the groove, and the reinforcing member is placed on the central axis of the skeleton).

(3) Cable sheath

The cable sheath is divided into a non-metallic sheath, a steel/aluminum–plastic composite tape adhesive sheath, a steel–plastic adhesive sheath with entrained steel wires, an outer reinforcing sheath with a longitudinally wrapped glass or aramid fiber composite tape, and a self-support cable sheath, etc. The cable may have one sheath layer or multiple sheath layers.

(4) Cable outer sheath

When the outdoor optical cable is used in a harsh environment, it is necessary to arm the optical cable and increase the outer sheath to further improve the mechanical strength of the optical cable. According to the application environment and different laying conditions, the common armor and outer sheath structure include a steel–plastic composite tape armored outer sheath, steel tape wrapped armored or steel wire armored outer sheath, non-metal (glass, aramid) fiber armored outer sheath, and flat FRP armored outer sheath, etc.

13.4.1.3 Structure and Technical Characteristics of Electric Optical Cable

In the power system, the optical cable that is laid along the high-voltage aerial electrical power lines mainly has two forms. One is the OPGWs, which is usually used for the line above 220 kV, and the other is ADSS optical cables, which is used in relatively low voltage applications.

All-dielectric self-supported optical cable is a nonmetallic all-dielectric optical cable. It is mainly used in power communication. It does not rely on steel wire ropes; it is self-supporting and suspended on power poles with voltage levels below 220 kV. The cable structure is divided into a central loose tube type and a stranded loose tube type.

The central loose tube type ADSS has a simple structure, a small diameter, and relatively smaller number of fiber. Since the excess length of the fiber of this structure is completely dependent on the excess length in the tube, the no-strain window of the fiber should not be too large. It is mainly used in applications where the weather conditions are good and the span is small. The stranded loose tube type ADSS can accommodate a large amount of fiber, and can obtain a satisfactory tensile no-strain window by adjusting the stranding pitch. However, compared with the central loose tube type ADSS, the outer diameter of the stranded loose tube type ADSS is relatively larger. The stranded loose tube type ADSS is influenced by cable weight, wind loading, and ice loading. During the design and laying process, it is necessary to consider the maximum meteorological load, the diameter, and the self-weight of the cable, and find a better balance point in the tension–sag characteristics of the cable and the span.

The tensile force of ADSS cable is much higher than that of the conventional optical fiber cable. High strength and high modulus aramid yarn is generally used in ADSS cable. Aramid yarn has the advantages of light weight, high strength, high modulus, high temperature resistance, low creep, and negative temperature linear expansion coefficient, etc.

Long-term operation of ADSS cable in the field or high electric field environment requires that the outer sheath material can resist electrical trace corrosion and also must have good resistance to UV aging and environmental stress cracking. This is usually achieved by modifying the black polyethylene sheathing material by adding additive materials.

The OPGW is a composite body containing an optical fiber component in an aerial ground wire of a high-voltage transmission line, which satisfies the performance requirements of the conventional ground wire and has the function of a communication optical cable. According to the structure, OPGW cable is divided into a central tube type, a central slotted core type, and a stainless steel tube stranded type. Since the stranded type structure can provide a large structural excess length and easily realize a large tensile window, it is widely used in the market, and is, of course, more expensive.

In addition to stainless steel tubes with fiber unit, OPGW consists of two stranded materials, aluminum clad steel wire and aluminum alloy wire. The aluminum clad steel wire mainly provides mechanical strength and carries a part of the electric current. The aluminum alloy wire mainly carries current, and has a certain tensile strength, and is also beneficial in reducing the weight of the cable. The OPGW is essentially a "lightning line," which has a high probability of being struck by lightning. While the melting point of aluminum alloy is low, it is prone to fusing when struck by lightning, resulting in major failures in communication failure and power transmission. Therefore, the main use of the market is the stainless steel pipe with layer stranding and the all-aluminum clad steel strand (high melting point) structure.

13.4.2 Introduction to Typical Cable Structure and Application Scenarios

13.4.2.1 Typical Cable Structure and Application Scenarios

According to the different cable structures and classifications, the application scenarios are various. The following briefly describes several common cable structures.

13.4.2.1.1 Steel Tape Armored Directly Buried Optical Cable The characteristic of the directly buried optical cable is that an aluminum–plastic composite tape adhesive sheath is armored by a double-sided chrome-plated plastic-coated steel strip and then polyethylene is extruded as the outer sheath. The typical cable structure is shown in Figure 13.15.

This kind of cable has good waterproofing, and pressure and tensile resistance. The cable has high metal content and good high or low temperature performance, and is particularly suitable to be directly buried. It is a typical type of directly buried cable commonly used in the world.

13.4.2.1.2 Slotted Core Type Optical Cable Slotted core type optical cable is widely used in Japan and Southeast Asia. The structure of the optical cable is shown in Figure 13.16. The fiber ribbon is placed in a skeleton groove made of high-density polyethylene (HDPE), and the center of the skeleton is a single steel wire or a stranded steel wire. A layer of water blocking tape is wrapped around the skeleton, and a double-coated aluminum plate is longitudinally wrapped and then a

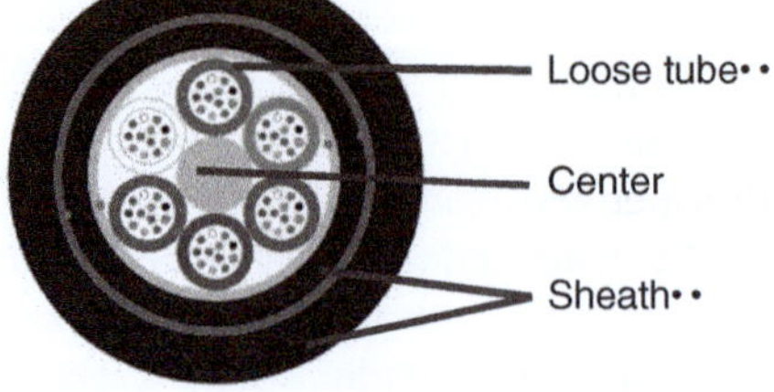

Figure 13.15 Typical steel tape armored directly buried optical cable.

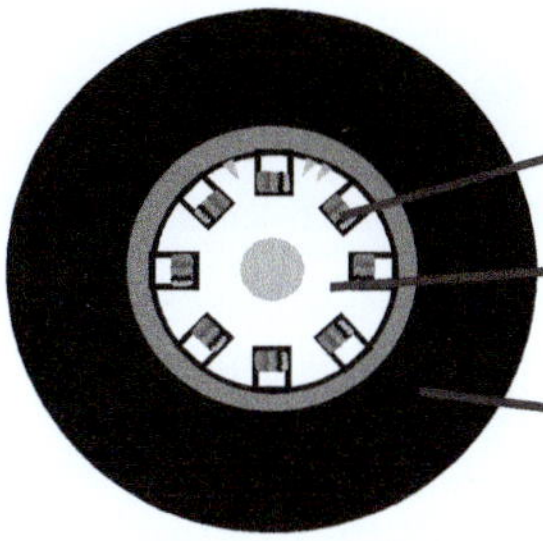

Figure 13.16 Typical slotted core type optical cable.

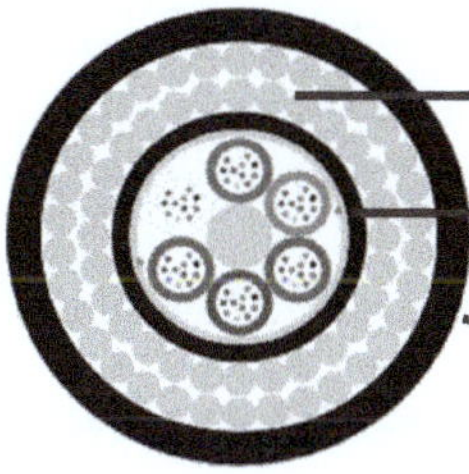

Figure 13.17 Typical structure of the underwater optical cable.

polyethylene (PE) sheath is extruded. A rip cord is placed between the aluminum plate and the water blocking tape to facilitate stripping.

This kind of optical cable is suitable for laying in the duct, and the ribbon in the cable is generally four cores, six cores, or eight cores. Core density of this kind of optical cable is high. The optical fiber cable can achieve 2000 core or even larger core number.

13.4.2.1.3 Optical Cable for Lakes, River Crossings, and Coastal Applications The underwater optical cable can be regarded as an ordinary duct optical cable with the double-layer thin round steel wire placed on the sheath, and then the polyethylene outer sheath is extruded. The typical structure of this cable is shown in Figure 13.17.

This kind of cable is mainly used for underwater laying. It has better tensile and pressure resistance. According to the value of the tensile strength, it is divided into products of less than 4 t or more than 4 t. The product with tensile strength of 4 t or less is mainly used for less stable riverbeds and beaches, flow rates greater than 3 m/s, or mainly navigable rivers, etc. It can also be used in mountainous climbing sections. The products with tensile strength of more than 4 t are mainly used for unstable riverbeds and beaches with serious erosion, and the rivers with width exceeding 500 m.

13.4.2.1.4 Gas Pipe Cable The requirements of the optical fiber cable with flame retardance and fire resistance are basically the same as those of the ordinary optical fiber cable; the difference is that the sheath of the ordinary optical fiber cable is a polyethylene material, and the sheath of the flame-retardant cable is a flame-retardant material. When there is a fire resistance demand, it is necessary to add fireproof reinforcing materials such as mica tape.

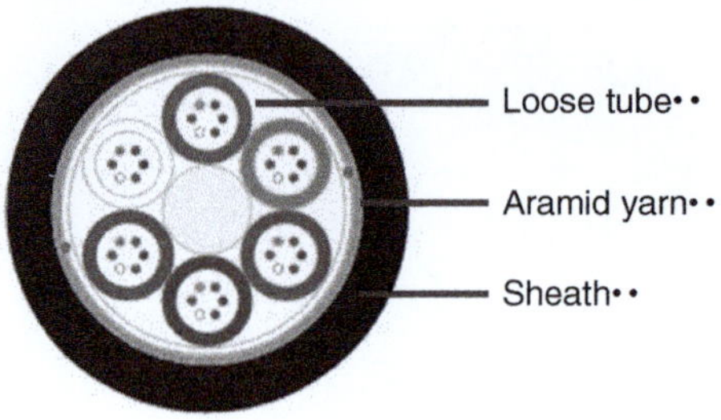

Figure 13.18 Single-layer ADSS optical cables along electrical power lines.

This type of optical cable is often used in subways, tunnels, and entry points of core nodes and important aggregation nodes. Generally, the length of each entry point optical cable should not be less than 500 m.

13.4.2.1.5 ADSS Optical Cable Along Electrical Power Line As mentioned above, the ADSS cable is a nonmetallic all-dielectric optical cable. It can be regarded that two layers of reinforcing aramid yarn are stranded on the outside of the sheath of a common nonmetal cable. Finally, a polyethylene (PE) jacket or an electric corrosion resistant (AT) jacket is cabled. The structure of the single-layer ADSS cable is shown in Figure 13.18.

The ADSS cable is suitable for self-supporting aerial laying along electrical power lines. Because of the strong tensile strength of the cable, it can also be used for large-span scenes such as rivers and mountains. In South America, Southeast Asia, and other places, aerial fiber-optic cables are usually laid on the same pole as other cables, and ADSS structures with self-supporting characteristics are also preferred.

The design of the ADSS cable needs to consider the maximum tensile tension (rated tensile strength, RTS) and maximum working tension of the cable according to the cable application span and weather conditions to ensure reliable application of the cable under wind and ice conditions.

13.4.2.1.6 Microduct Cabling for Installation by Blowing Owing to the factors of continuously increasing costs of artificial laying, tight duct resources, and expansion easily, it is increasingly common to install optical cable by blowing. There is no big difference in the structure between air blowing micro-cable and ordinary fiber cable. Air blowing micro-cables are thinner and lighter, and are usually constructed with nonmetallic reinforcements and without armor. The flexibility and surface friction coefficient of the cable can meet the need of air blowing. It can lay a micro-tube in the laid communication tube and then blow the micro-cables in the micro-tubes. The one-time distance of air blowing can reach 1 km or more.

The characteristic of small outer diameter of the air blowing micro-cables can effectively utilize the existing tube space. It is suitable for a scene with a long length and without decreasing the fiber number, such as a trunk line and a metro backbone layer. But it is not suitable for a broadband access project in the community.

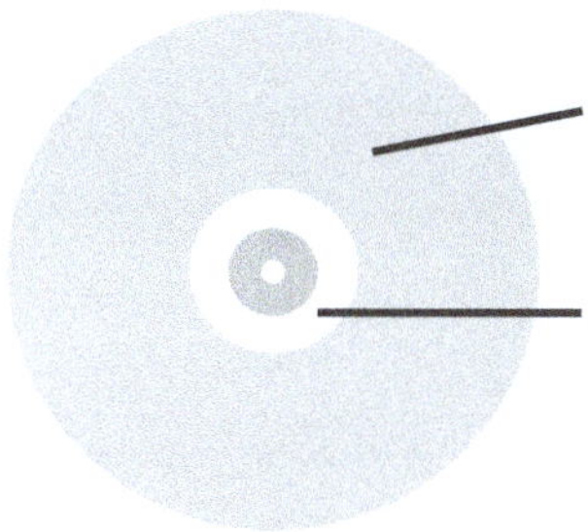

Figure 13.19 Invisible optical cable.

13.4.2.2 Introduction to New Special Optical Cable

In addition to the optical cables mentioned in the IEC standards, with the popularization of FTTx and the advancement of 5G technology, many new optical cables for special scenes, such as invisible bow-type optical cables, rodent resistance optical cables, and optical and electrical hybrid cables have appeared in recent years.

13.4.2.2.1 Invisible Bow-Type Optical Cable The invisible bow-type optical cable is suitable for indoor–outdoor use in FTTx applications. It is a simplex invisible cable unit in the middle of a traditional bow-type optical cable, similar to a tight buffered fiber. It is wrapped in a transparent jacket with a diameter of 0.9 mm on the outside of the uncolored G.657 fiber. The cable structure is shown in Figure 13.19.

Cable outer diameter is thin and completely transparent. When laying, the invisible optical fiber cable is generally glued and fixed along the perimeter of the wall and the door and window trim strips. The cabling is concealed and beautiful. When the drop optical fiber cable is a bow-type optical cable, the indoor part can tear and cut off the reinforcing member and the sheath, only leaving the inner invisible cable unit to be laid. It is suitable for users who have high requirements for aesthetics and concealment of indoor cabling.

13.4.2.2.2 Stainless Steel Tape Rodent Resistance Optical Cable The rodent resistance of outdoor optical cable is an important technical requirement. The rodent resistance effect of glass yarn, FRP armor, and rodent resistance sheath material is limited. In areas with severe rodent infestation, stainless steel tape rodent resistance cable structure is required.

The structure of the stainless steel tape rodent resistance cable is the same as that of the directly buried cable; the difference is that the corrugated steel tape of the armor layer for it is a thicker stainless steel tape. The stainless steel cable has high resistance to bite and corrosion. It is mainly used in wild farms, forests, and other places where voles, squirrels, and woodpeckers are more active. It can be laid as aerial type, in tube, or directly buried.

13.4.2.2.3 Optical and Electrical Hybrid Cable The optical fiber and the insulated copper wire are placed at the same time in the optical and electrical hybrid cables so that one cable can simultaneously solve the power and signal transmission. The structure of the optical and electrical hybrid cable is shown in Figure 13.20.

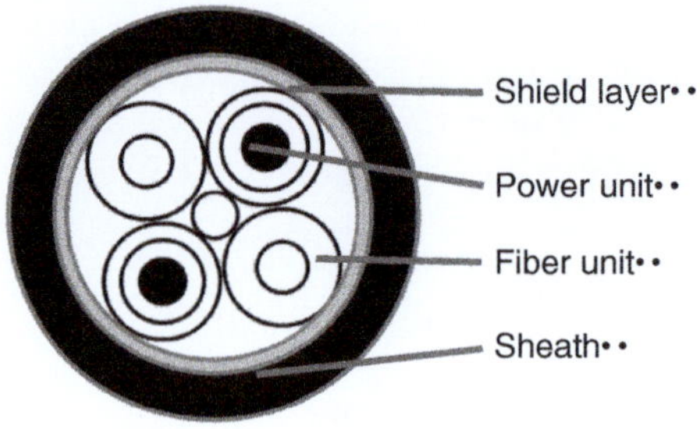

Figure 13.20 Optical and electrical hybrid cable in base station.

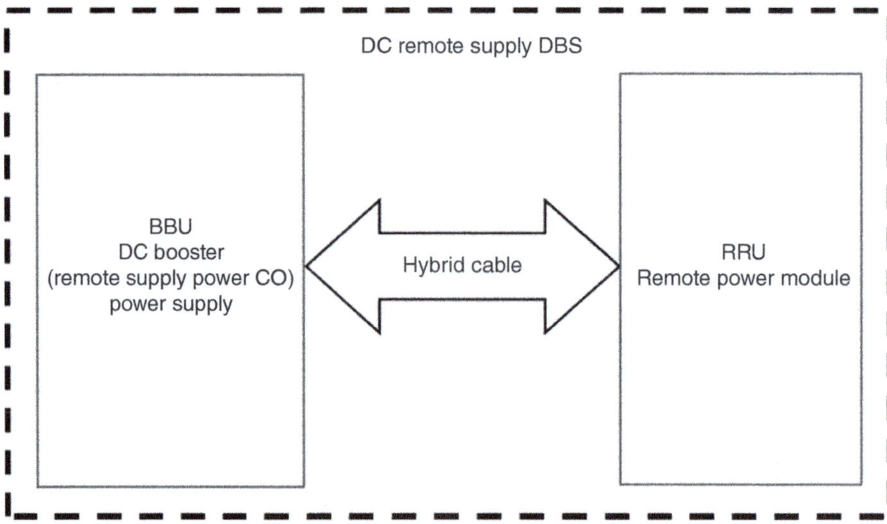

Figure 13.21 Optical and electrical hybrid cables in DC remote supply system.

The optical and electrical hybrid cable is mainly used to connect the building baseband unit (BBU) and the remote radio unit (RRU) in the distributed base station in DC remote supply system, as shown in Figure 13.21.The voltage of the DC remote supply system is usually 380 V DC. It can be laid as aerial type or in tube. From the perspective of safety, the optical and electrical hybrid cables line belongs to the strong electric line, so safety should be always paid attention to during construction and maintenance.

13.5 Scale and Market of Fiber and Cable Industry

After decades of development, the fiber and optical fiber cable industry has become more global, and more powerful multinational companies have emerged. As of now, the global cable manufacturers are estimated to be 400 or more, according to incomplete statistics. Among them, Asian optical fiber cable manufacturers account for 75% of the global optical fiber cable capacity, and China is the largest fiber consumer and largest manufacturing country. More than half of the world's optical fiber cables are produced in China. From the manufacturing level and production capacity of fiber preform, and in terms of fiber and optical cables, the top 10 leading manufacturers in the world are Corning (America), FiberHome (China), Prysmian (Italy), Furukawa (Japan), YOFC (China), Fujikura (Japan), Hengtong Optic-Electric (China), ZTT (China), Sterlite (India), and Futong (China). Most of these manufacturers have complete fiber production capacity and technical capabilities. Many manufacturers also have other industries such as electric cable, electromechanical equipment, communication systems, and glass products.

In 2018, global fiber preform capacity was about 650 million core kilometers, the fiber production capacity was about 600 million core kilometers, and the optical cable capacity was more than 550 million core kilometers. Among them, China is the most important contributor, accounting for more than half of the global fiber and optical cable consumption. In the past five years, strong demand for FTTx and 4G networks has supported the rapid development of the global optical fiber cable market. As of 2018, China and developed markets (USA, Europe, Japan, South Korea, etc.) have high FTTx and 4G coverage, especially in China, where more than 90% of residents have covered at least one 4G network.

In the next few years, the fiber and optical cable market in China and developed countries will develop at a lower speed. The demand in Southeast Asia, Africa, Latin America, and other regions will be the main driving force for the development of the global fiber and optical cable market. These developing markets are expected to carry out large-scale FTTx and 4G/5G network construction projects after 2020. And, to some degree, it compensates for the decline in demand for fiber and optical cables in China and other developed markets.

In addition to the construction needs of fiber networks in developing markets, the convergence of other ICTs, such as 5G, smart city/intelligent infrastructure, and data center construction, will become an increasingly important growth driver for the fiber and optical cable market. The fiber demand of 5G infrastructure is a key factor of fiber and optical cable demand in the future.

At present, 5G has entered the trial phase, and the United States, Japan, Europe, and China all have plans to deploy and build 5G networks on a large scale around 2020. 5G applications will require ultra-dense antenna networks that require large-core optical fiber cables and compact optical fiber cables in feeders and wiring segments. However, the deployment of 5G networks is a long-term process, and its scale, return on investment, and network security will affect the overall progress of 5G on a global scale.

References

1 Xianzhi, H. (2017). *Newest Researches on Communication Optical Fibers and Their System Applications*, 1e, 239–253. Wuhan, China: Wuhan University of Technology.
2 Azadeh, M. (2009). *Fiber Optics Engineering*, 32. USA: Source Photonics, Inc.
3 Liu, D. and Xiang, Q. (1995). *Fiber Optics*, 1e, 13–14. Beijing, China: National Defense Industry.

4 IEC 60793-2-10: 2019. Optical Fibers-Part 2-10: Product Specifications – Sectional Specification for Class A1 Multimode Fibers.
5 IEC 60793-2-20: 2015. Optical Fibers-Part 2-20: Product Specifications – Sectional Specification for Class A2 Multimode Fibers.
6 IEC 60793-2-30: 2015. Optical Fibers-Part 2-30: Product Specifications – Sectional Specification for Class A3 Multimode Fibers.
7 IEC 60793-2-40: 2015. Optical Fibers-Part 2-40: Product Specifications – Sectional Specification for Class A4 Multimode Fibers.
8 IEC 60793-2-50: 2015. Optical Fibers-Part 2-50: Product Specifications – Sectional Specification for Class B Single-mode Fibers.
9 Chen, B. (2016). *Design and Manufacture for Optical Fiber and Fiber Optical cable*, 3e, 61–69. Hangzhou, China: Zhejiang University.
10 GR-20-CORE (2013). *Generic Requirements for Optical Fiber and Optical Fiber Cable*. Telcordia Technologies, Inc.
11 Mu, C. (2017). *China Optical Fiber and Cable for 40 Years*, 1e, 88–94. Shanghai, China: Tongji University.
12 IEC TR 62048: 2014. Optical fibers- Reliablity- Power law theory.
13 Li, G., Ren, M., and Jiang, X. (2015). *Materials and Industry Development for Communication Optical Fiber and CableTongji University*, 1e, 191–207. Shanghai, China.
14 Zou, L. (2000). *Optical Fiber and Cable*, 1e, 49–51. Wuhan, China: Wuhan University of Technology.
15 Li, G., Ren, M., and Jiang, X. (2017). *Manufacturing Equipment and Product Development for Communication Optical Fiber and Cable*, 1e, 195–202. Shanghai, China: Tongji University.
16 IEC 60794-1-1: 2015. Optical Fiber Cables-Part 1-1: Generic Specification-General.
17 Zhang, Y. and Yue, L. (1995). *Optical Communication Instruments Basics and Applications*, 1e, 161–182. Beijing, China: Posts & Telecom Press.
18 IEC 60794-2: 2017. Optical Fiber Cables-Part 2: Indoor Cables- Sectional Specification.
19 IEC 60794-3: 2016. Optical Fiber Cables-Part 3: Outdoor Cables- Sectional Specification.
20 IEC 60794-4: 2018. Optical Fiber Cables-Part 4: Sectional Specification-Aerial Optical Cables Along Electrical Power Lines.
21 IEC 60794-5: 2014. Optical Fiber Cables-Part 5: Sectional Specification-Microduct Cabling for Installation by Blowing.

Index

a

accelerated aging time 277
accelerated wet aging 302
acetophenone (AP) 34, 295, 320
activation energy data 276, 277
ADSS cable 381, 384
aliphatic polyamide 41
aluminium corrugated screens 325
aluminium hydroxide (ATH) 171, 172, 175
angle of refraction 354
antimony trioxide (Sb_2O_3) 169–172
AP1000 nuclear power plant 258
aramids 41
aramid yarn 366, 367, 380, 381
armour layers 297
Arrhenius equation 27, 274, 275
ASEAN Smart City Network 12
attenuation, of coaxial cables 339
azodicarbonamide (ADCA) 28

b

balance of forces 90, 91, 95
band dispersion unshifted single-mode optical fibre
  extended wavelength 356
barrier zone 103
Beer–Lambert law 212
bending loss 357
Bernouilli law 211
bis-(*tert*-butyldioxyisopropyl)benzene 320
block copolymers 61, 62
blowing agents (BA) 36, 139–141
boiling water reactor nuclear power plant 256
bremsstrahlung 232–233
brominated phthalate plasticizers 170
burner calibration 197

c

cable
  aging parameters 276
  compounds 218
  conductor material 293, 294
  extrusion lines 85–87
  history of 311
  industry 1
  insulation
    extruded 295–296
    HVDC cables 296
    lapped 294–295
  sets 165
  sheath 117, 234, 380
  tensile curve 377–378
cable constructions
  in Asia and Australasia
    cable demand 10–11
    polymeric material demand 11–12
  demand, polymeric material 3–10
  electric 1
  in Europe
    cable demand 12–13
    polymeric materials demand 13–14

The Global Cable Industry: Materials, Markets, Products,
First Edition. Edited by Günter Beyer.

cable constructions (*contd.*)
investment 2
in Middle East and Africa
cable demand 14, 15
polymeric materials demand 14–16
in North America
cables demand 16–17
polymeric material demands 17–18
South and Central America
demand cables 18–19
polymeric materials demand 19–20
cabled fibre attenuation test 376
cabled fibre PMD 376, 377
cable lifetime aging performance evaluation
basic principles of accelerated aging of materials 272–277
capacitance of coaxial cable 339
capacity
functional capacity 241
measurements 127–128
CAP1400 nuclear power plant 260
Cargo inspection 227
catenary continuous vulcanization (CCV) 78
C–C bond 223
Cenelec HD 605 protocol 321
ceramifyable compounds 177
chain transfer agents (CTAs) 315–317
characteristic impedance, coaxial cables 336
chemical blowing agent (CBA) 139–141, 145
chemical foaming process 29
chemical stability 263, 267
chlorinated paraffins 70, 169, 171
chlorinated polyethylene (CPE) 36, 37, 224
chlorinated polymers
chlorinated polyethylene 36–37
polyvinylchloride 35–36
chordal length 235
circuit integrity 165, 166
class 1E equipment qualification
by analogical method 270
by analysis 269–270
by calculation 269
by combined method 270
by experience 270
sequence 277, 279
by testing 269, 270, 273
cleanliness in material handling 321, 322
coat hanger channels 117–119
coax-fiber hybrid design 348
coaxial cables 332
applications, history of 343–344
attenuation 339–341
automotive applications 345–347
capacitance 339
characteristic impedance 336–338
copper-clad steel conductors 340
dielectric insulator 333
direct current transmission 332
history of 331–332
impedance mismatch 341
inner conductor materials 332
jacket sheath materials 334
outer conductor 333, 334
principles 335
radio-frequency transmission 332
velocity of propagation 338–339
video, TV, broadband applications 344–345
Coax-UTP Hybrid Design 348
combined plot 136
communication cable 130, 166, 263
conductive losses 339–341
conductor design 293, 312
conductor screen 312–313
cone calorimeter 167, 183, 184
Construction Product Regulation (CPR) 178, 190
EN 13501-6 193
EN 50399 198–199
EN 60332-1-2 196–198
EN 60754-2 201–206
EN ISO 61 034-1 199–201
EN ISO 1716 196
continuous vulcanization (CV)
in saturated hot water stream 78–80

control cable 263
convective diffusion 108
cooling length
  analytical calculation
    for fine wires 123–124
    for fine wires 122–123
  finite difference calculation 123, 124
cooling trough 85, 86, 128, 143, 144, 149–151
copper corrugated screens 326
corrugated metal screens 325
Coulomb's law 89
CPR1000 nuclear power plant 260, 261
critical angle 354
cross-linked polyethylene (XLPE) 7, 280, 294–296, 314
  antioxidants influence 319–320
  cross-linking parameters and peroxide influence 320–321
  cross-linking process 317–318
  peroxides influence 318–319
cross-linking
  building construction 216
  electron-beam (E-beam) 217, 225
  fishing net 216
  high crosslink densities 216
  intermediate/medium crosslink densities 216
  irradiation 73
  low crosslink densities 216
  peroxide 217, 222–225
  peroxide system 71–72
  polymer chains 216, 217
  silanes 73, 217, 218
  sulfur system 71
cumylalcohol (CA) 320
curing 216
  of composite materials 227
cut-off wavelength shifted single-mode optical fibre 356

d

data transmission system (DTS) cable 264
diameter adaptation function
  DDR 120–121
  DRB 121
dielectric constant 28, 33, 37, 38, 134
dielectric losses 339–341
dispersion shifted fibre 356
dispersion unshifted single-mode optical fibre 356
dispersive mixing 108–110
distribution cable and break-out cable 380
distributive mixing 109–111
distributors
  coat hanger channels 117–119
  flattened distribution channel 119
  helical channel 119
dot notation 154
draw balance (DRB) 121
draw dawn ratio (DDR) 120, 121
drop coaxial cable 336, 340
Dulmage type mixing device 111
duplex optical cable 380
dynamic submarine cable testing 303, 308

e

elastomers 62
  stabilizers 70–71
electrical hybrid cable 385, 386
electrical time division multiplexing (ETDM) 361
electric cables
  constructions 1
  fire classes for 195
electric optical cable 381, 382
electromagnetic wave 336, 354
electron-beam (E-beam) 217
  advantages 244–245
  awareness 225
  capacity
    dose 241–242
    efficiency 241
    functional capacity 241
  compound design
    co-agents 243
    polymers 242–243
    raw materials 243
  control system 231–232
  costs indications 243–245
  disadvantages 245

electron-beam (E-beam) (*contd.*)
high voltage 228
parameters
amperage 237
radiation dose 237–241
voltage 235–237
principle of 227–228
safety
access doors 234
Bremsstrahlung/X-rays 232–233
IAEA 235
ozone 235
radiation and radioactivity 232
shielding and radioactivity 233–234
scan horn and beam window 228–229
temperature rise 242
UBHS 229–231
electron beam cross-linking 81–82
electron-beam welding (EBW) 227
electron lithography 227
electron microscopes 226
elongation at break (EAB) 274, 280
EN 13 501-6 193–196
EN 13501-6 192–193
EN 50399 193, 194, 198, 199
EN 60332-1-2 196, 197
EN 60754-2 201–206
EN 61034-2 194
EN ISO 61 034-1 199–201
EN ISO 1716 196
environmental stress crack resistance (ESCR) 23
environment resistance 265, 267
EPR nuclear power plant 259
equipment failure 267
Erbium Doped Fibre Amplifier (EDFA) 356
ethylene chlorotrifluoroethylene (ECTFE) 40
ethylene copolymer 22
ethylene-propylene-diene terpolymer (EPDM) 223
ethylene propylene rubber (EPR) 21, 251, 280, 294, 295
ethylene tetrafluoroethylene (ETFE) 38, 39
ethylene vinyl acetate (EVA) 8, 21, 224
European Construction Product Regulation (CPR) 9
extruded cables 78, 312
extruded insulation 294, 295, 297, 299
extruder 87
size 76, 78
extrusion defects 130
extrusion head
function 115–117
pressure *vs.* average temperature 121–122
extrusion screw
friction based feeding mechanism 89–90
improved feeding zone 95–97
simple modelling 90–95

f

factory acceptance (FAT) 302
Fast Fourier Transformation (FFT) 156
fast neutron breeder reactor 256, 257
fast reactor 256
feeding zone 88
channel depth and counter pressure 93–95
compression and friction 106–107
counter pressure on 95
extruder 91
improved 95–97
mass output calculation 93–95
volumetric output *vs.* screw flight angle 92–93
few-mode fibre 362, 363
fiber coloring 371
fiber crack theory 363
fiber length 368
fiber ribbon 372
fiber secondary coating
loose tube 372–374
tight buffer fibre 374
fiber unit 379, 380
fillers 66–68, 69
finite difference calculation 123, 124
fire growth rate (FIGRA) 178, 212, 213

Fire Performance Index 167
Fire Performance of Electric Cables (FIPEC) approach
  cable selection 185–186
  capability study 188–189
  correlation between real and large-scale test study 189–190
  full-scale fire test 186–188
  real scale fire tests 186
  real scale scenario 184–185
fire testing method 181
5G infrastructure 387
fixed analysis (FA) 377
flame propagation tests, wires and cables for 163–164
flame retardants (FR) 265, 278
  polyolefins 174–178
  PVC 169–174
  wires and cables for 161, 167–169
flame spread (FS) 162–164, 178, 189
flaming droplets 193, 200
flexible PVC compounds 35
flight angle 93
fluoro-ethylene propylene polymer (FEP) 37, 38
fluoropolymers 37
  ECTFE 40
  ethylene tetrafluoroethylene 38–40
  fluoro-ethylene propylene polymer 37–38
  perfluoroalkoxy polymer 39
  polytetrafluoroethylene polymer 38–40
  PVDF 40
foam extrusion
  blowing agents (BA) 139–141
  coaxial cables construction 133
  extrusion equipment
    cooling trough 149–151
    dosing station 145
    extruder and screw 143
    gas injection 145–147
    measurement devices 151
    melt transport 148–149
    physical foaming process 141, 143
  motivation 133
  permittivity 134
  polymer selection 136–139
  processing
    extrusion and die setup 152–153
    inline analysis, FFT 156–157
    maximum void and bubble size 154–156
    PBA handling 154
    process control modes 153–154
foam insulation 133, 144, 151, 153, 154
foam model factor 134
foam-outer skin (FS) 144
food processing 227
friction coefficients 89, 90
  channel depth 93–95
  constant screw 107
  on screw characteristics 105–106
FRP tape 366
full-scale fire test 186, 187
full-wave fibre 356
Furukawa Electric Co., Ltd 5

g

galvanized steel strand 366
gas injection 145
glass fibre reinforced plastic (GFRP) 366
glass yarn 366
global automotive investment growth 2, 3
green environmental protection products 287

h

halogen free flame retardant (HFFR) 8, 12, 13, 19, 21, 36, 277
heat release rate (HRR) 200
  flame spread classes 193
  by oxygen consumption technique 213
    FIGRA and SMOGRA 212–213
    mass flow rate 210–211
    methane burning 208–210
    smoke opacity 211–212
Heaviside, Oliver 331
heavy water reactors 256, 257
Hengtong Group 6
Henry's law 149

high carbon phosphating steel wire 366
high crosslink densities 216
high density polyethylene (HDPE) 7
 cable jackets 31, 32
high-speed development stage 250
homopolymers 22
horizontal continuous vulcanization (HCV) 78
hot capacitance device 151
HPR 1000 nuclear power plant 259
humidity resistance 267
HVDC cables 296
hybrid coaxial cable 335

i

IAEA 235, 254
impedance mismatch, in coaxial cables 341
indoor optical cable 378–380
in-line measurements 128–130
inner skin-foam (SF) 144
instrument cable 263
insulated cores 272, 295, 297
insulation defects measurements 126–127
insulation integrity 165
insulation screen 321–323
insulations, for cables 315
interferometry (INTY) 377
intermediate/medium crosslink densities 216
intrinsic cracking 364
invisible bow-type optical cable 385
irradiation process 81–82
irradiation resistance 280

j

jacketing/sheathing 328

k

Kevlar fibre reinforced plastic (KFRP) 366

l

lapped insulation 294, 295
latex 57
law of refraction 354
laying costs of cables 328
lead-based stabilizers 35
lead sheath cables 324
Leoni AG 6
light attenuation 199
light transmission 164, 352
light water reactors 256
linear low-density polyethylene (LLDPE) 7
 cable jackets 30
loose tube 372, 373
loss of cooling accident (LOCA)
 resistance 280
 test, 264 267
low crosslink densities 216
low-density polyethylene (LDPE) 7, 314
 production 315
  chain transfer agents, influence of 315–317
  cross-linked polyethylene 317–331
Low Smoke Zero Halogen (LS0H) 8
low voltage power cable 263

m

magnesium hydroxide (MDH) 171, 172, 175
magnitude, qualitative order of 96
Maillefer extrusion 89, 108, 111, 112, 118, 122
mass flow rate 210, 211
mass impregnated non-draining (MIND) 294
 cables 311
Maxwell's equations 354, 355
medium and low carbon steel wire 366
medium-density polyethylene (MDPE) 7
medium voltage power cable 263
melt filtration systems 113, 114
melt flow rate (MFR) 24, 99, 155
melt gear pumps 114, 115
melting zone 88

metallic protections
  corrugated metal screens 325
  designs for high voltage cables 324
  lead sheath cable 324–325
  tasks 324
metallic screens 323
  aluminium foil with concentric copper wires 327
  copper and aluminium *vs.* lead 327
  properties of 326
metallic sheaths 296, 297
metallocene technology 32
metal tensile 366
metering zone 88
  1D analysis 105
  representation of 104–105
methane burning 208, 210
microduct cabling 384
micromodule fiber optic cables
  polyester elastomer application 42
micro-structured fibre (MSF) 361
mixing zones 88
  aggregates type 108, 109
  devices selection criteria 109–111
  dispersive mixing 108–109
  distributive mixing 109
  Finathene HDPE 111–113
  homogenisation mechanism of material 108
  slow colour concentration 108
molecular diffusion 108
Monosil® process 219–220
Mooney viscosity 66, 68, 76
multi-core fibre 361, 362
multi-mode fibre, development of 358–359

n
nanocomposites 173, 176, 177
neper 339
Nexans 4–5, 14, 35
non-metal tensile elements 366
nuclear grade cables 262
nuclear power development stage
  high-speed 250
  rational 250–251
  stagnant 250
  verification and demonstration 249–250
nuclear power plants
  AP1000 258
  boiling water reactor 256
  cables characteristics
    combustion performance 265
    environment resistance 265–267
    low toxicity 267
    service life 265
    technical 258–260
  cable standards for 267
  CAP1400 260
  class 1E equipment qualification in 267–269, 273
  classification by safety level 263–264
  classification by technical types 264
  classification by use function 263
  classification by use occasion 264
  CPR1000 260, 261
  development trend
    in China 285–287
    in green environmental protection products 287–288
    in Japan 284
    in safety 287
  EPR 259
  excellent flame retardancy 288
  halogen-free 288
  heavy water reactor 257
  high transmittance 288
  history 261, 262
  HPR 1000 259
  low toxicity 288
  no corrosive gases 288
  pressurized water reactor (PWR) 256
  technical development of 257–258
  waterproof and ultraviolet proof 288
nuclear safety
  grade cables 262
  laws and regulations 268

o
1D analysis
  compression and friction 106–107

1D analysis (*contd.*)
friction coefficients 105–106
optical cable core structure 380
optical communication system 351, 352
optical fibre cables 351
for applications 383
characteristic design for 370
classification 355–358
crack 364
development 356, 358
gas pipe 383–384
industry 386–387
main features 359
optical transmission waveguide 353–355
principle of total reflection 353–355
sheath extrusion of 374–375
slotted core type 382–383
steel tape armored directly buried 382
strength 363–364
stress erosion coefficient 365
optical fibre technology
large effective area 359–361
next generation 361–363
photonic crystal fibre 361
ultra-low loss fibre 359–361
optical signal noise ratio (OSNR) 359, 360
optical transmission waveguide 353, 355
organoclay 8, 173, 176, 177
outdoor optical cable 380, 381
oxidation induction time (OIT) measurement 28
oxygen depletion factor 210

p

paper cables 311
paper insulated lead cables (PILC) 311
Paris Agreement (2016) 2
peak heat release rate (PHRR) 178
perfluoroalkoxy (PFA) polymer 39
periodicity of measurement analysis
in-line measurements 128–130
structural return loss 130
peroxide by-products 316
peroxide crosslinking process (CV-curing) 217
advantages 224
disadvantages 224–225
polymer selection 223–224
reaction 223
phosphate plasticizers 172
photonic band-gap effect 361
photonic crystal fibre 361
phthalate based plasticizers 169
plastic fibres 7
plasticizers 35, 36, 68–70
plastification 75, 97–104
Poiseuille flow 105, 114
polyamides (PA)
advantage 41
aliphatic 41
imide 282, 284
synthesis route 40
polyamide 12 41
polybutylphtherephtalate (PBT) 41
polycrystalline polymers 136
polyester elastomers 41
poly-ether-ether-ketone (PEEK) 281
polyetherimide (PEI) 281
polyethylene 312
for cable insulation 26–29
for cable jackets 29–32
classes 23
density effect 24
manufacturing process 25–26
melt index/flow rate effect 23
molecular weight distribution effect 24
repetitive units in 22
structure 22
polyimide (PI) 281
polymer blends 62
polymer-compositions 216
polymeric compounds 63
polymeric material demand
in Asia and Australasia 11
in cable insulation and jacketing 7, 8
in Europe 13
Furukawa Electric Co., Ltd 5
Hengtong Group 6

HFFR/LSOH 8
Leoni AG 6
LS Cable & System 6
in Middle East and Africa 14–16
Nexans 4–5
in North America 16–18
Prysmian 4
South and Central America 19
Southwire 5
Sumitomo electric industries 5
polyolefins 21
flame retardants (FR) 174, 175
polypropylene (PP) 8
cable industry applications 33
characteristic properties 33
manufacturing process 33
monomer units 32
polypropylene-based insulation 34
polysiloxanes 176, 177
polytetrafluoroethylene (PTFE) polymer 39
polyurethane 62
polyvinylchloride (PVC) 6, 7, 35–36, 168
polyvinyldifluoride (PVDF)
applications 40
emulsion polymerization process 40
vinylidene monomer building block for 40
power cable production 12
pressurized liquid continuous vulcanization (PLCV) 80
pressurized water reactor (PWR) nuclear power plant 256
primary plasticizers 68
productivity 241
protection of cable 324
Prysmian 4, 5, 10, 12, 14
Pythagoras Theorem 235

q

quality parameter
capacity measurements 127, 128
concentricity measurement 126
diameter and product circularity measurement 125, 126
insulation defects measurements 126, 127
periodicity of measurement analysis 128, 130
sheathing wall thickness measurements 128

r

radiation resistance 29, 262, 263, 265
random copolymer polypropylene 32
rational development stage 250
real scale fire tests 186
reflection coefficient 342–343
renewable energy 2
return loss 342
ring suspension cable 264
rodent resistance 385
rubber 8, 63
cables 62
coding 65
compounds 63, 69, 73, 75, 76
heat and oil resistance of 66
rubber elasticity 60, 61
rubber properties 66

s

Saxton type mixing device 111, 112
screw characteristics, friction coefficients on 105, 106
screw geometry 88, 89
sea trial test 302
secondary plasticizers 70
security cables, functional integrity for 165–166
self-contained fluid 294
self contained oil filled (SCOF) 294
self-skinning effect 152
Shannon limit 362
sheathing wall thickness measurement 128
shifted single-mode optical fibre
non-zero dispersion 356
short-wavelength division multiplexing (SWDM) 358
silane crosslinking 217
advantages 221
copolymers 220

silane crosslinking (*contd.*)
  disadvantages 221
  Monosil process 219
  reactivity of 220, 221
  Sioplas process 218
simplex optical cable 379
single screw extruder
  different functional screw zones 87–88
  screw geometry 88–89
Sioplas process 218
skin depth 339
skin-effect 332, 339
skin-foam-outer skin (SFS) 144
slotted core type optical cable 382, 383
SMOGRA 212, 213
smoke corrosivity 193
smoke density 278
  test 265
smoke opacity 211, 212
smoke production rate (SPR) 193, 200, 212
smoke suppressants 172, 173
solid extruded cables 312
Southwire 5
space division multiplexing (SDM) 361
stagnant development stage 250
stainless steel
  tape 385
  wire and steel pipe 366
sterilization 227
Stokes parameter measurement (SPE) 377
stranding cable technology 374, 375
stress erosion coefficient 365
strip-able screen 322
structural return loss (SRL) 130, 148, 342, 343
styrene-butadiene-styrene-triblock copolymers 62
submarine power cable
  accelerated wet aging 302
  accessories 299–300
  after installation or commissioning 302
  applications 291
  armour layers 297
  conductor material 293, 294
  design 292–293
  dynamic 303–304, 308
  factory acceptance (FAT) 302
  installation 306–308
  insulation 294–296
  manufacturing plants 298, 299
  metallic sheaths 296–297
  other cable elements 297–298
  power system 293
  pre qualification 302
  project specific and development/special 302
  protection and commissioning 306–308
  routine 301
  sample 301
  sea trial test 302
  sizing 305–306
  testing 301–303
  type 302
Sumitomo electric industries 5
Supersmooth screens 312
surface treating 227
system failure 267

t

telecommunication wire 127, 128
tensile design
  for central bundle tube optical cable 367–369
  for layer-stranding optical cable 369–370
  stretch calculation 366–367
tensile force 303–307
thermal resistance 280
thermoplastic compounds 63, 243
thermoplastic elastomers (TPE) 8, 60, 61
  block copolymers 62
  polymer blends 62
  structural models 62
thermoplastic polyester elastomers (TPEE) 41, 42
thermoplastic polyimide 281
thermoplastic polymers 21
  chlorinated polymers 35

polyolefins 21
thermoplastic polyurethane (TPU) 42, 168
thermoplastic vulcanizates (TPV) 62
13th Five-year Plan 10
thin film melting, barrier zone 103
thin film plastification
initiation of liquid film 98–99
melt flow rate 99
melting backside of flight 97–98, 100
melting length parameters 99–103
tight buffer fibre 374
time domain reflectometry measurement (TDR) 302
total heat relased (THRR) 178
total smoke production (TSP) 189, 212

u

ultra-high-capacity transmission 361
under-beam handling system (UBHS) 229
Unified National Fine thread (UNF) 143
United Nations Organisation 2
upside down procedure 75

v

velocity of propagation 338–339
verification and demonstration stage 249
vertical continuous vulcanization (VCV) 78
vinylchloride monomer (VCM) 35
voltage source converter (VSC) 295
voltage standing wave ratio (VSWR) 342
vortex-induced vibrations (VIV) 303
vulcanization technique 78
vulcanizing 216

w

wastewater treatment 227
water tree tests 321
wavelength division multiplexing (WDM) 361
WDM communication system 352
Weibull curve 364
wideband transport single-mode optical fibre
non-zero dispersion 356
wires and cables
flame propagation tests for 163–164
flame retardants (FR) 161–169
flammability of 167
polymers for flame-retardant 167–168
smoke, corrosivity and toxicity tests for 164

y

Young's modulus 367

z

Ziegler–Natta catalysts 26, 32
zinc borate 8, 172, 176, 177